Sustainable Finance

Sustainable Finance is a concise and authoritative reference series linking research and practice. It provides reliable concepts and research findings in the ever growing field of sustainable investing and finance, SDG economics and Leadership with the declared commitment to present the theories, methods, tools and investment approaches that can fulfil the United Nations Sustainable Development Goals and the Paris Agreement COP 21/22 alongside with de-risking assets and creating triple purpose solutions that ensure the parity of profit, people and planet through choice architecture passion and performance. The series addresses market failure, systemic risk and reinvents portfolio theory, portfolio engineering as well as behavioural finance, financial mediation, product innovation, shared values, community building, business strategy and innovation, exponential tech and creation of social capital. Sustainable Finance and SDG Economics series helps to understand keynotes on international guidelines, guiding accounting and accountability principles, prototyping new developments in triple bottom line investing, cost benefit analysis, integrated financial first plus impact first concepts and impact measurement. Going beyond adjacent fields (like accounting, marketing, strategy, risk management) it integrates the concept of psychology, innovation, exponential tech, choice architecture, alternative economics, blue economy shared values, professions of the future, leadership, human and community development, team culture, impact, quantitative and qualitative measurement, Harvard Negotiation, mediation and complementary currency design using exponential tech and ledger technology. Books in the series contain latest findings from research, concepts for implementation, as well as best practices and case studies for the finance industry.

Nader Naifar
Editor

Climate Change and Finance

Navigating the Challenges and Opportunities in Capital Markets

Springer

Editor
Nader Naifar
College of Business
Imam Mohammad Ibn Saudi Islamic University
Riyadh, Saudi Arabia

ISSN 2522-8285 ISSN 2522-8293 (electronic)
Sustainable Finance
ISBN 978-3-031-56418-5 ISBN 978-3-031-56419-2 (eBook)
https://doi.org/10.1007/978-3-031-56419-2

This Springer imprint is published by the registered company Springer Nature Switzerland AG
The registered company address is: Gewerbestrasse 11, 6330 Cham, Switzerland

Preface

In an era of unprecedented environmental challenges and economic transitions, the nexus between climate change and finance has never been more relevant. "Climate Change and Finance: Navigating the Challenges and Opportunities in Capital Markets" aims to provide a comprehensive exploration of this critical intersection, offering insightful perspectives and cutting-edge research that illuminate the evolving dynamics of green finance and investment, policy implications, and sector-specific strategies in the context of global climate change.

As the editor of this timely volume, it has been a privilege to bring together a diverse array of thought leaders, researchers, and practitioners at the forefront of addressing climate-related financial issues. This book is a collection of sixteen chapters, each contributing a unique viewpoint and depth of understanding to the multi-faceted relationship between climate change and the financial world. From examining the efficiency of green assets in tumultuous times to the role of financial institutions in mitigating climate impacts, each chapter investigates specific aspects of climate finance, offering theoretical insights and practical solutions.

Part I, "Green Finance and Investment Dynamics," opens the discussion by focusing on the financial aspects of climate change. It explores the roles and impacts of green finance initiatives, the comparative efficiency of green assets, foreign green finance, investor attitudes, and the critical role played by financial institutions in addressing climate change.

In Part II, "Climate Change, Policy, and Economic Governance," the book focuses on the broader implications of climate change on policy and global economic governance. This part examines emerging trends in climate change and the impact of climate-related financial regulations on various sectors, including geotechnical engineering and corporate financial performance.

Part III, "Sector-Specific Impacts and Strategies," concludes the book by exploring the impacts of climate change on specific sectors such as agriculture and energy and discusses innovative strategies and adaptations required in these areas. This part provides insights into the nexus between climate change risk and the financial performance of businesses, especially in developing economies, and highlights novel financial instruments like green cryptocurrencies and blue sukuk for sustainable development.

"Climate Change and Finance" is more than a compilation of academic chapters; it is a call to action for policymakers, financial professionals, and stakeholders across the globe to recognize and harness the potential of finance as a tool for combating climate change. It emphasizes the need for an integrated approach, combining policy reforms, innovative financial mechanisms, and stakeholder collaboration to create a sustainable, resilient future.

As the world grapples with the realities of climate change, this book aims to contribute to the ongoing dialogue and inspire further research and action in the vital field of climate finance. We hope the insights and discussions in this volume will serve as valuable resources for those committed to navigating the challenges and opportunities climate change presents in capital markets.

Riyadh, Saudi Arabia Nader Naifar

Acknowledgment

To my beloved mother, whose strength and guidance have been my constant, thank you for everything. To my wife and children, your endless patience and encouragement have been the bedrock of my endeavors.

I would also like to thank all the authors, my friends, and colleagues, whose contributions and insights have been invaluable.

I dedicate this achievement in a special way to the memory of my father. His wisdom, teachings, and belief in my potential continue to inspire me. While he is no longer with us, his legacy remains a guiding light in my life and work.

To all of you, I am deeply grateful. Your support has been a cornerstone in the completion of this project.

Contents

About the Editor

Nader Naifar is a Professor of Finance at the College Business at Imam Mohammad Ibn Saudi Islamic University (IMSIU), Riyadh, KSA. He teaches graduate and undergraduate courses on Finance in Ph.D. Finance program, MBA, EMBA, and Bachelor of Finance (BF) programs.

His current research focuses on but is not limited to FinTech, financial economics, Sustainability, Islamic finance, Sukuk markets, green finance, Default risk, Global crisis, Contagion, International financial markets, and Energy markets. Nader Naifar has authored and co-authored multiple papers published in reputable international journals and has been a guest editor for several international journals.

Part I
Green Finance and Investment Dynamics

Green Finance Initiatives and Their Potential to Drive Sustainable Development

Nouman Nasir and Waqas Ahmed

Abstract This study examines the role of green finance initiatives in driving sustainable development. It provides an overview of green finance principles, exploring its evolving nature and global recognition. The chapter explores the relationship between green finance initiatives and sustainable development, showcasing their contribution to economic growth while minimizing environmental impacts. It discusses various financial instruments such as green bonds, green loans, and impact investing, and their role in supporting sustainable projects across sectors like renewable energy, sustainable infrastructure, and circular economy practices.

Furthermore, this chapter addresses challenges and opportunities associated with implementing green finance initiatives, including limited awareness, regulatory constraints, and capacity building. It provides potential solutions and strategies to overcome these challenges through stakeholder collaboration, policy reforms, and innovative financial mechanisms. Case studies and best practices from different regions and countries illustrate successful green finance initiatives, highlighting their transformative power in integrating environmental considerations into financial decision-making processes.

The chapter argues for the importance of aligning financial systems with environmental goals and emphasizes the potential of green finance initiatives to drive sustainable development. It advocates for mainstreaming green finance practices globally to create a more sustainable and resilient future. This chapter is a valuable resource for researchers, policymakers, financial professionals, and other stakeholders interested in advancing sustainable finance practices and achieving a greener and more sustainable world.

N. Nasir (✉) · W. Ahmed
Al-Qadir University Project Trust, Sohawa, Punjab, Pakistan

Muslim Youth University, Islamabad, Pakistan

N. Naifar (ed.), *Climate Change and Finance*, Sustainable Finance,
https://doi.org/10.1007/978-3-031-56419-2_1

1 Introduction

In recent years, the urgency of addressing environmental challenges has reached a critical juncture. The world is facing interconnected crises of climate change, biodiversity loss, and resource depletion, necessitating transformative actions for a sustainable future. Central to this endeavor is the role of finance, as it has the potential to drive significant change through the allocation of capital. The emergence of green finance initiatives presents a promising avenue for redirecting financial flows toward environmentally and socially responsible projects. This chapter explores the landscape of green finance, its principles, instruments, and its potential to catalyze sustainable development.

Sarpong et al. (2023) discuss the importance of the financial sector in supporting green technologies and businesses. It suggests that offering green products such as green bonds, funds, and loans can help drive the green transformation agenda. The study found that green financial development efficiency (GFDE) has significantly improved in China, promoting sustainable development. Factors such as trade openness, foreign investments, technological innovation, and government budget positively influenced GFDE, while energy consumption and economic policy uncertainty had a negative impact. The results highlight the significance of green finance in achieving sustainable development goals.

1.1 Background and Significance

Traditional financial systems have often overlooked environmental and social considerations in investment decisions, leading to unsustainable practices. The consequences of this oversight are evident in the accelerating climate crisis and ecological imbalances. Green finance initiatives aim to rectify this by integrating environmental, social, and governance (ESG) criteria into financial decision-making. By channeling funds into projects that align with sustainable development goals, these initiatives hold the potential to reshape economies and promote long-term well-being.

Green financing has emerged as a promising instrument to align financial flows with sustainable development objectives. The Environmental Sustainability Index places China near the lower end of the global ranking of countries. In the global pursuit of sustainable development, countries increasingly adopt innovative strategies to harmonize economic growth with ecological preservation and societal advancement. In response to these challenges, the Chinese government has formulated an extensive strategy aimed at addressing both industrial and environmental pollution issues (Jin et al., 2016; Liu & Diamond, 2008).

Green Finance advocates the fusion of financial resources and business operations with environmentally friendly practices. It encompasses various stakeholders, including local communities, individual consumers, businesses, producers, investors, and financial institutions (Berrou et al., 2019). The contours of Green Finance vary depending on the participant involved, often propelled by financial incentives a commitment to planetary preservation, or a combination thereof. Unlike conventional financial undertakings, the Green Economy underscores ecological advantages and places heightened emphasis on nurturing the environmental conservation sector (Wang & Zhi, 2016).

1.2 Objectives of the Chapter

This chapter seeks to achieve the following objectives:

- Provide a comprehensive understanding of green finance and its scope.
- Explore the key principles and criteria that underpin green finance initiatives.
- Examine the pivotal role of financial institutions in driving the adoption of green finance.
- Analyze the significance of green bonds and sustainable debt instruments in mobilizing capital for sustainable projects.
- Discuss the concept of carbon markets and emissions trading as tools to incentivize emission reductions.

1.3 Understanding Green Finance

The global focus on climate change and sustainability has garnered widespread attention, culminating in the landmark 2015 Paris Agreement under the United Nations Framework Convention on Climate Change (UNFCCC). This accord reflects a shared commitment among member nations to collaboratively address the urgent issue of greenhouse gas (GHG) emissions. Despite this consensus, a formidable obstacle looms: the financing of actions aimed at mitigating and adapting to climate change and promoting sustainability. The gravity of the challenge demands substantial investment, a hurdle that necessitates careful consideration and strategic solutions. A crucial benchmark within the Paris Agreement is the commitment to limit global warming to well below 2 degrees Celsius above pre-industrial levels. However, sustaining this temperature threshold entails a staggering requirement for financial resources. To achieve this goal, an estimated $53 trillion in energy-related investments is projected to be necessary by 2035, according to the findings of the International Energy Agency (IEA) in 2014 ("Paris Agreement," 2016).

In the context of China's renewable energy sector, Zhang et al. (2019) underscore the pivotal role of financial development in facilitating progress. This observation highlights how robust financial mechanisms can significantly influence the advancement of renewable energy initiatives, essential for addressing the climate crisis. In 2010, recognizing the urgency of the matter, 194 countries established the Green

Climate Fund (GCF). The primary objective of the GCF is to provide vital financial support to developing nations, enabling them to both reduce GHG emissions and enhance their capacity to adapt to the impacts of climate change. This endeavor underscores the pressing need for financial resources to bolster global sustainability efforts.

1.3.1 Definition

Definition and Scope of Green Finance

Amid these developments, the term "green finance" has gained prominence, frequently appearing in reports from international organizations such as the International Finance Corporation (IFC) in 2017, as well as in the strategies of national governments. However, the precise definition of green finance can be elusive, often leading to confusion and the conflation of terms, particularly with "climate finance." In its essence, green finance centers on the funding of investments that yield tangible environmental benefits. This broad definition encapsulates financial support for projects that promote sustainability, resource efficiency, and environmental protection. Conversely, climate finance, as defined by the UNFCCC, encompasses a spectrum of financing sources—public, private, and alternative—that aims to back actions related to both mitigation and adaptation strategies for climate change.

As per the IFC (2017) definition, green finance entails the funding of investments that yield environmental advantages. In contrast, the UNFCCC suggests and delineates the latter as encompassing financing drawn from diverse sources—public, private, and alternative—to facilitate actions aimed at mitigating and adapting to climate change on local, national, or transnational levels. Despite nuanced disparities in their definitions, both terms fundamentally center around financial instruments for addressing climate change and fostering sustainability (Berensmann et al., n.d.).

Green finance refers to financial activities that support projects, technologies, and businesses that demonstrate a positive impact on the environment and society. It encompasses investments, loans, insurance, and other financial products that integrate ESG factors into decision-making. Green finance is not limited to environmental considerations alone; it also acknowledges social and governance dimensions, promoting a holistic approach to sustainable development.

While subtle nuances exist in the definitions, the core essence of both green finance and climate finance revolves around the financial tools necessary to address the multifaceted challenges posed by climate change and to foster sustainable development. By delineating and clarifying these concepts, stakeholders can better channel resources, formulate policies, and direct investments toward the shared goal of ensuring a resilient and sustainable future for all.

1.4 Motivation of the Study

The motivation behind this study stems from the increasing importance of green finance initiatives and their potential to drive sustainable development. As the global community grapples with environmental challenges, the financial sector is increasingly recognized as a pivotal player in fostering sustainable practices. The intersection of green finance and governance has become a focal point for researchers and policymakers alike. In response to this growing significance, our study aims to contribute to the existing body of knowledge by providing a comprehensive analysis of the current landscape of green finance and governance relationships.

The unique contribution of this research lies in its explicit focus on the interconnectedness of green finance and governance, exploring how these elements interact to shape sustainable development outcomes. By delving into this dynamic relationship, we seek to uncover insights that can inform future policies, practices, and scholarly endeavors.

Our endeavor is driven by the conviction that a deeper understanding of the landscape of green finance and governance relationships is imperative for guiding informed decision-making in the realms of finance, policy, and sustainability. Through this research, we aspire to provide a nuanced perspective that not only advances academic discourse but also offers actionable insights for stakeholders committed to steering global finance toward a more sustainable and environmentally conscious future.

The pressing global challenges of climate change and environmental degradation have prompted an urgent reevaluation of our economic systems (Sun et al., 2023). In response to these challenges, the intersection of finance and sustainability has emerged as a critical focal point (Aneja et al., 2023; Berrou et al., 2019; Lee, 2020). This study delves into the realm of Green Finance Initiatives, exploring their potential to drive sustainable development. The motivation behind this research lies in the imperative to address environmental concerns while simultaneously fostering economic growth. By explicitly articulating the contributions of this study, we aim to shed light on the transformative role that green finance can play in shaping a more sustainable future.

The existing body of knowledge is rich with discussions on both finance and sustainability, yet the integration of these two fields remains a dynamic and evolving area (Akomea-Frimpong et al., 2022; Berrou et al., 2019; Lee, 2020; Sun et al., 2023). This study contributes to this discourse by providing a comprehensive analysis of the mechanisms through which Green Finance Initiatives can serve as catalysts for sustainable development. In doing so, we aim to not only deepen our understanding of the intricate relationship between finance and the environment but also offer practical insights for policymakers, businesses, and investors seeking to align their strategies with the imperatives of a more sustainable future.

As we embark on this exploration, the significance of our work becomes apparent in the context of the broader global agenda for sustainable development. By elucidating the potential impact of Green Finance Initiatives, we aim to provide a roadmap for harnessing financial systems to address environmental challenges effectively. This study's novelty lies in its nuanced examination of specific initiatives, their successes, and challenges, offering a comprehensive view that contributes to the ongoing dialogue surrounding sustainable development.

In the subsequent sections, we will delve into the theoretical foundations of green finance, analyze key initiatives on both national and international scales, and discuss their implications for fostering sustainable development. Through this comprehensive approach, we aspire to provide a valuable resource for scholars, policymakers, and practitioners committed to navigating the intersection of finance and sustainability in the pursuit of a greener and more equitable future.

1.5 Key Principles and Criteria for Green Finance Initiatives

Green finance initiatives are guided by several key principles, including transparency, accountability, and risk management. These initiatives typically adhere to a set of criteria that define eligible projects. Such criteria might include carbon emissions reduction, clean energy generation, waste reduction, and biodiversity conservation. By setting clear standards, green finance initiatives ensure that capital is directed toward projects with measurable positive impacts on the planet.

Aneja et al. (2023) delve into an exploration of green finance initiatives and their influence on cultivating a sustainable economy. The research scrutinizes the expansion of green finance within nations that issue green bonds, with a specific focus on its efficacy in diminishing reliance on non-renewable energy sources, particularly fossil fuels. Employing a pressure-state-response framework, the study undertakes a comprehensive assessment of the trajectory of green finance advancement and uncovers a noteworthy inverse correlation between the growth of green finance and the propensity for fossil fuel dependency. The findings present compelling evidence that green finance mechanisms possess the potential to alleviate reliance on fossil fuels, thereby facilitating the trajectory toward a world characterized by carbon negativity.

Role of Financial Institutions in Promoting Green Finance

Financial institutions, including banks, asset managers, and insurance companies, play a pivotal role in promoting green finance. They serve as intermediaries between investors and projects, channeling funds toward environmentally sustainable endeavors. These institutions also provide expertise in assessing environmental risks and opportunities, helping investors make informed decisions aligned with sustainability objectives.

Zheng et al. (2021) discuss the role of green finance in the sustainability performance of financial institutions in Bangladesh. The paper includes a literature review and hypothesis development, as well as a discussion of major policy implications and limitations. The study found that green financing is an essential element in the short- and long-term development of banking strategies in Bangladesh, and the paper provides suggestions for future studies in this area. A list of abbreviations and terminologies used in the study is also included.

1.6 Green Bonds and Sustainable Debt Instruments

Green bonds are debt securities issued by governments, municipalities, or corporations to raise capital specifically for environmentally beneficial projects. The proceeds from these bonds are earmarked for projects such as renewable energy installations, energy-efficient buildings, and sustainable transportation systems. Green bonds not only attract investors interested in sustainability but also offer a transparent way to finance projects with measurable outcomes.

Bhutta et al.(2022) discussed the challenges faced by Pakistani banks in implementing green banking practices. Lack of awareness, knowledge, and skills, resistance to change, pressure on banks, lack of incentives, and legal power are identified as the main obstacles. The document also provides details on the roles and responsibilities of different stakeholders in environmental risk management and guidelines for green business facilitation. Thematic analysis of interviews with 26 respondents from various banks is used to identify the problems, causes, and consequences of green banking challenges.

1.7 Carbon Markets and Emissions Trading

Carbon markets represent a market-based approach to tackling carbon emissions. Emissions trading systems allow entities to buy and sell permits that grant the right to emit a certain amount of greenhouse gases. By placing a price on carbon emissions, these markets create economic incentives for industries to reduce their carbon footprint. Emission trading not only helps in achieving emission reduction targets but also stimulates innovation in low-carbon technologies. Carbon markets and emissions trading are innovative mechanisms within the realm of green finance that play a critical role in addressing climate change. These mechanisms aim to create economic incentives for industries and organizations to reduce their greenhouse gas emissions, thereby contributing to global emission reduction targets.

W. Zhang et al. (2020) explore the impact of carbon emissions trading policy in China on economic growth and the environment. The study finds that emission reduction targets based on carbon emissions and populations are more equitable in terms of abatement cost savings across regions. The study recommends improving energy efficiency, reducing coal consumption, and developing new clean energy sources to reduce carbon emissions. The study also suggests changing the mode of economic development and activating carbon markets by fostering an intermediary structure and strengthening financial innovation. The study uses the DID method to test the impact of the implementation of the ETS on carbon emission reduction at an industrial scale. Finally, the study evaluates the efficiency of the carbon ETMs using the DEA model.

1.7.1 Carbon Markets

Carbon markets, also known as emissions trading markets or cap-and-trade systems, are designed to put a price on carbon emissions. The overarching idea is to create a market for trading emission allowances, where companies are allocated a certain number of emission permits. These permits represent the right to emit a specified amount of greenhouse gases. If a company emits fewer emissions than its allocated permits, it can sell the surplus allowances to entities that exceed their emission limits. This creates a financial incentive for companies to reduce emissions, as they can potentially benefit from cost savings, regulatory compliance, and positive public perception.

Newell et al. (2013) in their document discussed the design and operation of carbon markets, including their key features and lessons learned from their functioning. It also highlights the challenges faced by these markets, such as the risk of carbon emissions leakage and the need for transparent and orderly policy revisions. The document also touches on the importance of understanding the strengths and weaknesses of carbon markets and the role of international negotiations in contributing to further progress.

1.7.2 Emissions Trading

Emission trading is the actual process of buying and selling emission allowances within a carbon market. This system encourages companies to seek cost-effective emission reduction strategies. Companies that can reduce emissions at a lower cost than the market price of allowances can sell their excess allowances, while those finding it more expensive to reduce emissions can buy allowances to meet their targets.

Wang and Zhi (2016) discussed the role of green finance in environmental protection, focusing on market mechanisms and policies. It highlights the inadequacies of green finance in the renewable energy sector and proposes solutions for better ecological balance. The article was published in the *Energy Procedia* journal in December 2016.

1.7.3 Benefits and Challenges

Carbon markets and emissions trading offer several benefits:

1. They provide a flexible and market-driven approach to emission reduction, allowing companies to choose the most efficient strategies for their operations.

2. These mechanisms also foster innovation by incentivizing the development of cleaner technologies.
3. Furthermore, by putting a price on carbon, they internalize the external costs of emissions, encouraging businesses to account for their environmental impacts.

However, there are challenges associated with these mechanisms.

1. Designing effective and equitable carbon markets require careful consideration of factors such as initial allocation of allowances, market stability, and addressing potential market manipulation.
2. Additionally, there is a need for transparency and robust monitoring to ensure accurate reporting of emissions.

2 Literature Review

2.1 Green Finance Mechanisms

Green finance mechanisms encompass a variety of financial tools, strategies, and approaches that aim to mobilize capital toward environmentally sustainable projects and initiatives. These mechanisms play a pivotal role in addressing pressing global challenges such as climate change, biodiversity loss, resource depletion, and social inequality. By integrating environmental, social, and governance (ESG) considerations into financial decision-making, these mechanisms contribute to building a more resilient and sustainable global economy. Here are some key green finance mechanisms.

Shipalana (2020) in his article discusses the need for a transition to a Green Economy and the role of governments, private sector, and donor community in supporting it. It suggests the creation of dedicated green funds, de-risking investments, and improving policy and regulatory environments to attract private sector investment. The document also emphasizes the importance of greening the financial sector and creating green and ethical banks. The World Bank's guidelines for developing a green taxonomy are also mentioned to assist regulators in emerging economies to "green" their domestic financial systems. Finally, the document suggests reallocating national budgets to put national climate/green/environmental funds at the center of economic recovery plans.

2.2 Green Investment Funds and Venture Capital

Green investment funds and venture capital, as discussed earlier, are pivotal in financing sustainability projects. These mechanisms provide capital to enterprises focused on developing innovative solutions for environmental and social challenges.

Green investment funds pool resources from various investors to invest in portfolios of sustainable projects, ranging from renewable energy installations to eco-friendly technology startups. Venture capital, on the other hand, supports early-stage companies that are driving innovation in areas like clean energy, sustainable agriculture, and waste reduction.

Bellucci et al. (2023) explore the role of venture capital (VC) investment in fostering green innovation and achieving low-carbon targets. The paper presents data and variables, including the definition of green technologies and the methodologies used to classify patents. The authors suggest that VC finance can play a crucial role in enabling and scaling the solutions needed for the low-carbon transition and that this type of equity finance can generate positive externalities in the form of knowledge spillovers. The paper concludes with a call for further analysis to fully disentangle the complexity of venture funding of green technologies.

2.3 Sustainable Banking and Responsible Lending

Sustainable banking and responsible lending are essential mechanisms within green finance. Sustainable banking involves incorporating ESG considerations into banking practices, aligning financial services with environmental and social goals. Responsible lending, as a subset of sustainable banking, ensures that loans are extended to projects and initiatives that adhere to sustainability principles. These mechanisms encourage the allocation of funds to ventures that contribute positively to the environment and society.

In his paper, Weber (2012) focuses on the history of ethical, socially responsible, and sustainable banking. It provides an overview of the development of sustainable banking, starting from the foundation of the first banks in Italy in the sixteenth century and describing present developments that can be seen as milestones in sustainable banking. The paper also highlights industry initiatives to integrate sustainability issues into the financial sector, such as the Global Impact Investment Network (GIIN) and the Global Alliance for Banking on Values (GABV).

3 The Impact of Green Finance on Sustainable Development

Green finance, as a transformative force, exerts a profound impact on sustainable development by aligning financial flows with environmental and social objectives. This synergy fosters a harmonious relationship between economic growth, social well-being, and environmental preservation. The integration of sustainability considerations into financial decisions enables the realization of a more equitable, resilient, and prosperous future.

Zubair Mumtaz and Alexander Smith (2019) discussed the need for green finance policies in Pakistan and outlined recommendations for their implementation. The article suggests that banks and DFIs should regularly review borrowers' compliance with environmental indicators and develop innovative financial products. It also emphasizes the importance of creating awareness about green finance among stakeholders. The article highlights the need for due diligence and risk rating measures to minimize environmental risks associated with financing. The document suggests that banks/DFIs should train their employees to develop their expertise in environmental risk assessment. Finally, the article recommends that banks/DFIs develop an escalation matrix to analyze credit exposures and environmental risks associated with financing.

Here is a closer look at how green finance influences sustainable development:

3.1 *Mitigating Climate Change Through Green Finance*

Green finance emerges as a potent weapon in the fight against climate change. By directing investments toward low-carbon and carbon-neutral projects, such as renewable energy installations, energy-efficient infrastructure, and sustainable transportation systems, green finance plays a pivotal role in reducing greenhouse gas emissions. These investments accelerate the transition from fossil fuels to clean energy sources, aligning with international climate goals like those set out in the Paris Agreement. The collective impact of such endeavors aids in curbing global temperature rises and minimizing the catastrophic consequences of climate change.

Nawaz et al. (2021) discussed the importance of creating a conducive atmosphere to attract foreign direct investment (FDI) to scale up green financing in N-11 countries. It suggests that governments should support regulations and efforts aimed at developing bond markets, and non-corporate bodies should issue green bonds. Finally, green bonds must be set up according to the Green Bond Principles (GBP) to ensure transparency, full disclosure, and the allocation of proceeds for climate attributes projects and assets.

3.2 *Promoting Renewable Energy and Energy Efficiency*

One of the primary avenues through which green finance advances sustainable development is by promoting renewable energy and energy efficiency initiatives. Investments in solar, wind, hydroelectric, and other renewable energy sources drive down dependency on fossil fuels and reduce emissions. Furthermore, green finance facilitates the adoption of energy-efficient technologies across industries, leading to decreased energy consumption, cost savings, and lowered environmental impact. This synergy between renewable energy and energy efficiency significantly contributes to sustainable energy systems and the mitigation of climate change.

Rasoulinezhad and Taghizadeh-Hesary (2022) discussed the issues of green finance and energy efficiency. It comprises six sections, including a literature review, data and model specification, empirical results, robustness checks, and conclusions and policy implications. The paper seeks to expand the existing literature by providing empirical evidence that green finance can be a key element in restoring economic growth in the short and long terms, especially during and after the COVID-19 era. Policymakers should also pay attention to energy efficiency to reduce energy intensity and environmental pollution. The paper presents a conceptual framework and empirical results, including panel unit root tests, to support its findings.

3.3 Enhancing Biodiversity Conservation and Ecosystem Preservation

Green finance's impact extends beyond energy into the realm of biodiversity and ecosystem conservation. Investments directed toward protecting and restoring ecosystems, preventing deforestation, and promoting sustainable land use contribute to the preservation of biodiversity. By funding projects that safeguard habitats and support conservation efforts, green finance plays a critical role in maintaining ecosystem services, preserving genetic diversity, and ensuring the resilience of ecosystems against the threats posed by climate change.

Berghöfer et al. (2017) provide information on the results of an assessment of ecosystem services, with a focus on regional climate regulation, water regulation, and soil conservation. The report includes feedback from stakeholders and recommendations for communication materials to approach high-level government officials. The study aims to provide insights for development cooperation experts and practitioners implementing projects and programs on biodiversity conservation and financing in partner countries. The findings are qualitative and not representative of projects in German development cooperation. The report includes a list of acronyms and references.

3.4 Driving Sustainable Agriculture and Circular Economy

Green finance mechanisms also catalyze sustainable practices in agriculture and promote the circular economy. Investments in sustainable agriculture encourage practices that prioritize soil health, water conservation, and biodiversity protection. These investments result in improved food security, reduced environmental degradation, and enhanced rural livelihoods. Moreover, green finance supports the transition to a circular economy by financing projects that promote resource-efficient production, minimize waste, and encourage recycling. This shift away from linear consumption patterns contributes to the longevity of resources and minimizes the environmental footprint.

Agrawal et al. (2023) explore the adoption of green finance and green innovation technologies in the context of circular economy. The study provides a conceptual framework that promotes sustainable strategies to effectively balance financial considerations with environmental concerns. The findings of the study are significant for environmentalists, policymakers, green investors, and researchers.

4 Challenges and Opportunities in Green Finance Initiatives

Green finance initiatives hold great promise for driving sustainable development, but they also face a range of challenges and present opportunities that shape their effectiveness and impact. These challenges, when addressed strategically, can become stepping stones toward realizing a more sustainable global economy. Islam et al. (2014) discussed the concept of green banking and its implementation in Bangladesh. It highlights the importance of incorporating environmental parameters in banking practices and suggests various strategies for promoting sustainability. The document also emphasizes the role of government and the Bangladesh Bank in formulating policies and regulations to support green banking. It mentions the challenges faced by banks in balancing environmental concerns with business demands. The document concludes by suggesting improvements in online financing, the establishment of a separate green financing unit, and the incorporation of environmental risk into core risks. Overall, the document provides an overview of green banking practices in Bangladesh and identifies opportunities for improvement.

4.1 Regulatory and Policy Frameworks

Challenges

- Inconsistent or unclear regulatory frameworks across jurisdictions can create uncertainty for investors and hinder the growth of green finance.
- Ambiguity in standards and definitions for what qualifies as "green" or "sustainable" can lead to greenwashing, where investments are falsely labeled as environmentally responsible.

Opportunities

- Well-defined and harmonized regulatory frameworks can provide clarity and encourage investments in sustainable projects.
- Governments can incentivize green finance through tax breaks, subsidies, and preferential policies, making it financially attractive for both investors and businesses.

4.2 Access to Finance and Financial Inclusivity

Challenges

- Many small and medium-sized enterprises (SMEs) and projects in developing countries face challenges accessing green finance due to lack of collateral, information, or credit history.
- Financial exclusion and lack of awareness may prevent marginalized communities from benefiting from green finance initiatives.

Opportunities

- Innovative financing mechanisms, like microfinance, crowdfunding, and peer-to-peer lending, can bridge the gap for underserved entities.
- Financial institutions can develop tailored products that cater to the needs of SMEs and local communities, fostering inclusivity in green finance.

4.3 Risk Assessment and Measurement in Green Investments

Challenges

- Green projects often involve new technologies or untested business models, posing uncertainties and risks for investors.
- Measuring the environmental and social impact of investments and their contribution to sustainability can be complex and challenging.

Opportunities

- Developing standardized risk assessment tools for green projects can help investors better evaluate and manage risks.
- Advancements in data analytics and technology can improve impact measurement, providing more accurate insights into the outcomes of green finance initiatives.

4.4 Capacity Building and Knowledge Dissemination

Challenges

- Lack of awareness and understanding of green finance concepts among investors, businesses, and policymakers can impede the adoption of sustainable practices.
- Skilled professionals well-versed in both finance and sustainability are essential but often in short supply.

Opportunities

- Capacity-building programs, workshops, and educational initiatives can enhance awareness and knowledge about green finance.

- Collaborations between academic institutions, industry experts, and policymakers can nurture a talent pool equipped to drive green finance initiatives forward.

In navigating these challenges and capitalizing on opportunities, the world stands to create a financial ecosystem that not only safeguards the environment but also promotes economic growth, social equity, and long-term resilience. By fostering collaboration between governments, financial institutions, businesses, and civil society, the potential of green finance can be harnessed to catalyze positive change on a global scale.

Jia (2023) in his document discusses the impact of green finance on the decarbonization of economies. It analyzes the current challenges and prospects of green finance, focusing on the Paris Agreement, green bond principles, and national strategies for sustainable development. The study examines the steps taken by China, Russia, and the United States in promoting green finance. It highlights the importance of scrutinizing claims about green technologies and the long-term viability of investments. The document also emphasizes the need for interaction between governments, financial institutions, and companies to achieve sustainable development goals. However, it acknowledges the limited availability of data and studies to objectively assess the impact of green finance on decarbonization.

5 Case Study: Successful Green Finance Initiatives

5.1 Introduction

In an era of growing environmental concerns and the urgent need for sustainable development, green finance initiatives have emerged as a powerful tool to drive positive change. These initiatives bridge the gap between financial markets and sustainable projects, channeling capital toward endeavors that prioritize environmental protection and social well-being. This case study explores successful green finance initiatives from both developed countries and emerging economies, highlighting lessons learned and best practices for a more sustainable future.

Ning et al. (2023) also discuss the importance of incorporating environmental considerations into bond pricing theory. It suggests that businesses should adopt "green policies" that go beyond legal requirements to protect the environment. The document also highlights the use of green bonds as a positive and cost-effective way to finance projects and promote economic growth. It emphasizes the need for green innovation in the renewable energy industry and identifies barriers that need to be addressed. The document mentions the use of analytical hierarchy approaches to assess decision-making and the role of fuzzy numbers in evaluating associations. The document emphasizes the significance of sustainable green financing, energy efficiency investment, and economic growth from a global perspective.

5.2 *Green Finance Initiatives in Developed Countries*

5.2.1 Sweden's Green Bonds

Sweden has been at the forefront of green finance, with its government issuing green bonds to fund environmentally friendly projects. These bonds have financed initiatives like renewable energy projects, sustainable transportation infrastructure, and energy-efficient buildings. The success of Sweden's green bonds lies in their transparent use of proceeds and rigorous reporting on environmental impacts, instilling investor confidence.

5.2.2 Germany's Energiewende

Germany's transition to a low-carbon economy, known as Energiewende, has been supported by innovative green finance mechanisms. The German government introduced feed-in tariffs and incentives for renewable energy producers, driving investments in wind, solar, and biomass projects. The policy framework, coupled with attractive financing options, has propelled Germany to become a global leader in renewable energy adoption.

5.3 *Green Finance Initiatives in Emerging Economies*

5.3.1 India's Renewable Energy Growth

India's ambitious renewable energy targets have been met with robust green finance strategies. The establishment of the Indian Renewable Energy Development Agency (IREDA) provides financing and support for renewable energy projects. Additionally, the Green Energy Corridors project involves upgrading transmission infrastructure to accommodate renewable energy integration. These initiatives attract investments, promote energy security, and contribute to poverty alleviation.

Source: Authors' own contribution through Canva

5.3.2 China's Green Finance Policies

China has implemented a suite of green finance policies, including guidelines for green bonds and the establishment of green credit quotas for banks. The People's Bank of China (PBOC) has prioritized green finance, encouraging financial institutions to support projects aligned with sustainable development. China's approach combines regulatory frameworks with financial incentives to drive green investments.

Source: Authors' own Contribution through Canva

5.4 *Lessons Learned and Best Practices*

5.4.1 Integration of Policy and Finance

Successful green finance initiatives are often underpinned by comprehensive policy frameworks that provide clarity, incentives, and regulations. By aligning financial strategies with policy goals, countries can create an enabling environment for green investments.

5.4.2 Transparency and Reporting

Transparency is key to building investor confidence. Initiatives like Sweden's transparent use of green bond proceeds and China's reporting guidelines for green projects ensure accountability and trust.

5.4.3 Diverse Funding Sources

Green finance initiatives can leverage diverse funding sources, including public funds, private investments, international partnerships, and multilateral financing institutions. This diversity reduces dependency on a single source and enhances financial resilience.

5.4.4 Incentives for Innovation

Providing financial incentives and subsidies for sustainable projects, as seen in Germany's Energiewende, encourages innovation and market adoption of green technologies.

5.4.5 Capacity Building and Awareness

Educating investors, businesses, and policymakers about green finance concepts and benefits is crucial. Capacity-building programs can bridge knowledge gaps and stimulate interest in sustainable investments.

5.4.6 Customization to Local Context

Green finance initiatives should be tailored to the unique challenges and opportunities of each country. Solutions that work in one context may need adjustments to suit another's circumstances.

5.5 Conclusion

Successful green finance initiatives exemplify the potential for financial mechanisms to drive sustainability on a global scale. From Sweden's transparent green bonds to China's regulatory-driven approach, developed and emerging economies alike have demonstrated the transformative impact of aligning finance with environmental and social goals. The lessons learned from these initiatives underscore the importance of policy integration, transparency, and diverse funding sources. As the world grapples with pressing environmental challenges, green finance remains a potent instrument to accelerate the transition to a more sustainable future.

Question for Consideration

Question 1. How can countries with limited financial resources effectively implement green finance initiatives to drive sustainable development?

Question 2. What role do international collaborations and partnerships play in scaling up successful green finance models from one country to another?

Question 3. How can green finance initiatives address social equity concerns and ensure that the benefits of sustainable development are inclusive and widespread?

Question 4. What strategies can governments employ to encourage private sector participation in green finance initiatives and create a self-sustaining ecosystem for sustainable investments?

Question 5. How can lessons from both developed and emerging economies be combined to create a comprehensive global framework for green finance that addresses diverse challenges and opportunities?

6 Findings and Recommendation

6.1 The Role of Stakeholders in Green Finance

Green finance initiatives require the collective efforts of various stakeholders to effectively promote sustainable development and address pressing environmental challenges. The engagement of these stakeholders—ranging from government bodies to financial institutions, corporations, and civil society organizations—is essential to drive the transition to a more sustainable and environmentally responsible economy. This section examines the distinct roles played by each of these stakeholders in advancing green finance.

6.2 *Government Policies and Regulatory Frameworks*

6.2.1 Role

Government bodies play a central role in shaping the landscape of green finance through the formulation and implementation of policies and regulatory frameworks. These policies can include tax incentives, subsidies, emissions reduction targets, and environmental standards. Regulatory frameworks define disclosure requirements for environmental and social impacts, ensuring transparency in financial markets. Additionally, governments can create a conducive environment for green finance by setting clear guidelines for sustainable investment and creating financial incentives for green projects.

6.2.2 Impact

Government policies and regulations provide the necessary foundation for green finance to thrive. They offer clarity to investors, guide financial institutions' lending practices, and encourage businesses to adopt sustainable practices. These frameworks create a level playing field and minimize risks, making it attractive for investors to allocate funds to green projects.

6.3 *Financial Institutions and Investors*

6.3.1 Role

Financial institutions, including banks, asset managers, and investment funds, are crucial intermediaries in the green finance ecosystem. They assess environmental risks, invest in sustainable projects, and allocate capital to ventures that align with ESG criteria. Investors, both institutional and individual, influence the flow of funds by choosing to support companies and projects that prioritize sustainability. Financial institutions also play a role in developing innovative financial products, such as green bonds and sustainable investment funds.

6.3.2 Impact

The decisions of financial institutions and investors significantly influence the direction of capital flows. Their support for green projects encourages businesses to adopt sustainable practices and helps scale up initiatives that contribute to environmental protection and social well-being.

6.4 Corporate Responsibility and Sustainable Business Practices

6.4.1 Role

Corporations have a responsibility to integrate sustainability into their business strategies and operations. They can do this by implementing energy-efficient processes, reducing waste, and adopting circular economy principles. Corporations also engage in green finance by issuing green bonds to finance environmentally friendly projects. Additionally, corporations that embrace sustainable business practices can attract socially conscious investors and consumers, enhancing their reputation and long-term viability.

6.4.2 Impact

When corporations adopt sustainable practices, they contribute to reducing their carbon footprint and environmental impact. Their participation in green finance initiatives aligns with their responsibility to contribute positively to society and the environment. This, in turn, influences consumer behavior and encourages a broader shift toward sustainable consumption.

6.5 Civil Society Organizations and Community Engagement

6.5.1 Role

Civil society organizations (CSOs) play a critical role in advocating for environmental protection and social equity. They engage in raising awareness, mobilizing communities, and holding stakeholders accountable for their actions. CSOs also collaborate with businesses, financial institutions, and governments to ensure that green finance initiatives prioritize the needs and well-being of local communities.

6.5.2 Impact

Civil society organizations serve as watchdogs, ensuring that green finance initiatives genuinely benefit the environment and society. Their advocacy drives increased transparency, accountability, and responsiveness among stakeholders. By involving communities in decision-making processes, CSOs ensure that projects are aligned with local needs and values.

The success of green finance initiatives hinges on the collective efforts of various stakeholders, each contributing their unique strengths and expertise. Governments provide the regulatory framework, financial institutions and investors allocate capital, corporations adopt sustainable practices, and civil society organizations advocate for accountability and community engagement. By working together, these stakeholders can create a synergy that accelerates the transition to a more sustainable and environmentally conscious global economy.

7 Future Directions and Recommendations for Green Finance Initiatives

As the world faces increasingly complex environmental challenges, the role of green finance initiatives becomes even more critical in steering economic growth toward sustainability. To ensure a resilient and sustainable future, there are several key directions and recommendations that can guide the evolution and expansion of green finance initiatives.

7.1 Scaling up Green Finance Initiatives

7.1.1 Direction

Efforts should focus on scaling up existing green finance initiatives to mobilize larger amounts of capital for sustainable projects. This includes expanding the scope of green bonds, sustainable investment funds, and other financial instruments.

Recommendations

- Governments can create a supportive policy environment by providing tax incentives, subsidies, and regulatory frameworks that encourage private sector investment in green projects.
- Financial institutions can collaborate to develop standardized frameworks for evaluating and reporting on the environmental and social impacts of green investments, enhancing investor confidence.
- Businesses should actively seek opportunities to align their operations with sustainability goals and access green finance for their initiatives.

7.2 Strengthening International Cooperation and Collaboration

7.2.1 Direction

Collaboration among governments, financial institutions, businesses, and civil society on a global scale is essential to address transboundary environmental challenges and achieve meaningful impact.

Recommendations

- International organizations can facilitate the sharing of best practices, knowledge, and expertise among countries to accelerate the adoption of successful green finance models.
- Multilateral agreements and platforms can promote cross-border investments in sustainable projects, ensuring that resources are directed to areas with the greatest need and potential impact.

7.3 Integrating Green Finance into Mainstream Financial Systems

7.3.1 Direction

The ultimate goal is to seamlessly integrate green finance principles into mainstream financial systems, making sustainability a core consideration in all financial decision-making processes.

Recommendations

- Governments can promote the integration of environmental risk assessment and disclosure in financial regulations, ensuring that environmental and social factors are factored into lending and investment decisions.
- Financial institutions should incorporate ESG criteria into their standard due diligence procedures, ensuring that sustainability considerations are treated with the same rigor as financial considerations.

7.4 Advancing Research and Innovation in Green Finance

7.4.1 Direction

Research and innovation are pivotal in creating new financial products, strategies, and mechanisms that align with evolving sustainability challenges.

Recommendations

- Academic institutions and research organizations should focus on producing data-driven studies that demonstrate the financial benefits of sustainable investments, fostering increased interest among investors.
- Financial technology (FinTech) innovations can play a role in facilitating the growth of green finance, such as blockchain applications for transparent reporting and impact measurement.

The future of green finance initiatives lies in their ability to evolve, adapt, and scale to meet the ever-growing demands of sustainability and environmental protection. By scaling up initiatives, fostering international collaboration, integrating green finance into mainstream financial systems, and advancing research and innovation, stakeholders can collectively shape a more sustainable global economy. These recommendations serve as a roadmap for navigating the complexities of the modern world while ensuring a future where economic growth is in harmony with the planet's ecological limits.

8 Conclusion of Chapter

As the world grapples with the interrelated challenges of climate change, resource depletion, and social inequality, the role of green finance initiatives has emerged as a beacon of hope, offering a pathway toward a more sustainable and equitable future. The chapters of this book have explored the multifaceted dimensions of green finance and its potential to drive sustainable development, while also acknowledging the challenges and opportunities that lie ahead. In this concluding chapter, we reflect on the journey traversed through the realms of understanding green finance, exploring its impact on sustainable development, and envisioning its future trajectory.

The exploration embarked upon in the earlier chapters unveiled the foundational aspects of green finance. The definitions and scope of green finance provided insights into its holistic approach to financial decisions that prioritize environmental and social considerations alongside economic objectives. The delineation of key principles and criteria underscored the importance of transparency, accountability, and long-term thinking in green finance initiatives. The role of financial institutions in championing green finance elucidated their pivotal role in channeling funds toward sustainable projects, catalyzing positive change through their investment decisions. The significance of green bonds and sustainable debt instruments was underscored as effective mechanisms to raise funds for projects aligned with sustainability goals. Additionally, the exploration of carbon markets and emissions trading showcased innovative strategies to mitigate climate change by putting a price on carbon emissions.

The exploration deepened with an analysis of the profound impact of green finance on sustainable development. The chapters delved into how green finance initiatives mitigate climate change, promoting renewable energy and energy efficiency as essential pathways to reducing carbon emissions. The enhancement of biodiversity conservation and ecosystem preservation underscored the crucial role of green finance in safeguarding the Earth's rich biodiversity and maintaining ecosystem services critical for human survival. The drive toward sustainable agriculture and a circular economy was highlighted as a means to transform consumption patterns, reduce waste, and promote resource efficiency.

Acknowledging the challenges and opportunities that shape the landscape of green finance, the subsequent chapters unveiled the hurdles posed by regulatory and policy frameworks, access to finance, risk assessment, and knowledge dissemination. Simultaneously, these chapters offered recommendations to address these challenges and harness opportunities to foster greater participation in green finance initiatives.

The case studies provided tangible evidence of successful green finance initiatives. The case studies demonstrated how green finance is being embraced both in developed countries and emerging economies. These examples underscored the transformative power of green finance in driving renewable energy adoption, promoting sustainability, and fostering economic growth while mitigating environmental impacts. Lessons learned and best practices from these case studies illuminated the strategic approaches that can be replicated to amplify the impact of green finance across diverse contexts.

The exploration of the roles of various stakeholders unveiled a collaborative tapestry that weaves together governments, financial institutions, corporations, and civil society organizations. The synergistic efforts of these stakeholders are instrumental in shaping green finance's influence on sustainable development. The recommended future directions laid out a roadmap for scaling up green finance, enhancing international collaboration, integrating green finance principles into mainstream financial systems, and fostering research and innovation.

As we conclude this journey through the world of green finance initiatives, it becomes evident that we stand at a pivotal juncture in human history. The chapters in this book have underscored the imperative of aligning financial decisions with environmental and social aspirations. The choices made today will reverberate through generations, influencing the resilience of ecosystems, the equity of societies, and the stability of economies.

The promise of green finance lies not merely in its capacity to mobilize capital, but in its potential to reshape the very foundations of our economic systems. It is a clarion call for a new era, where profit and purpose harmoniously coexist, and where the well-being of humanity is intrinsically linked to the health of the planet. This is the vision that green finance initiatives beckon us to embrace—a vision of a sustainable world where prosperity is shared, resources are cherished, and the legacy we bequeath is one of hope, stewardship, and enduring vitality. The journey has begun, and the path to a sustainable future lies ahead, illuminated by the transformative potential of green finance.

References

Agrawal, R., Agrawal, S., Samadhiya, A., Kumar, A., Luthra, S., & Jain, V. (2023). Adoption of green finance and green innovation for achieving circularity: An exploratory review and future directions. *Geoscience Frontiers, 101669*. https://doi.org/10.1016/j.gsf.2023.101669

Aneja, R., Kappil, S. R., Das, N., & Banday, U. J. (2023). Does the green finance initiatives transform the world into a green economy? A study of green bond issuing countries. *Environmental Science and Pollution Research, 30*(14), 42214–42222. https://doi.org/10.1007/s11356-023-25317-w

Akomea-Frimpong, I., Adeabah, D., Ofosu, D., & Tenakwah, E. J. (2022). A review of studies on green finance of banks, research gaps and future directions. *Journal of Sustainable Finance & Investment, 12*(4), 1241–1264. https://doi.org/10.1080/20430795.2020.1870202

Bellucci, A., Fatica, S., Georgakaki, A., Gucciardi, G., Letout, S., & Pasimeni, F. (2023). Venture capital financing and green patenting. *Industry and Innovation, 30*, 947–983. https://doi.org/10.1080/13662716.2023.2228717

Berensmann, K., Volz, U., Alloisio, I., Mattei, F. E. E., Bak, C., Bhattacharya, A., Leipold, G., Schindler, H., MacDonald, L., Huifang, T., & Yang, Q. (n.d.). *Fostering sustainable global growth through green finance – what role for the G20?*

Berrou, R., Dessertine, P., & Migliorelli, M. (2019). An overview of green finance. In M. Migliorelli & P. Dessertine (Eds.), *The rise of green finance in Europe: Opportunities and challenges for issuers, investors and marketplaces* (pp. 3–29). Springer. https://doi.org/10.1007/978-3-030-22510-0_1

Berghöfer, A., Emerton, L., Diaz, A. M., Rode, J., Schröter-Schlaack, C., Wittmer, H., & van Zyl, H. (2017). Sustainable financing for biodiversity conservation: A review of experiences in German development cooperation.

Bhutta, U. S., Tariq, A., Farrukh, M., Raza, A., & Iqbal, M. K. (2022). Green bonds for sustainable development: Review of literature on development and impact of green bonds. *Technological Forecasting and Social Change, 175*, 121378. https://doi.org/10.1016/j.techfore.2021.121378

Islam, M. A., Yousuf, S., Hossain, K. F., & Islam, M. R. (2014). Green financing in Bangladesh: Challenges and opportunities – a descriptive approach. *International Journal of Green Economics, 8*(1), 74–91. https://doi.org/10.1504/IJGE.2014.064469

Jia, Q. (2023). The impact of green finance on the level of decarbonization of the economies: An analysis of the United States', China's, and Russia's current agenda. *Business Strategy and the Environment, 32*(1), 110–119. https://doi.org/10.1002/bse.3120

Jin, Y., Andersson, H., & Zhang, S. (2016). Air pollution control policies in China: A retrospective and prospects. *International Journal of Environmental Research and Public Health, 13*(12), 12. https://doi.org/10.3390/ijerph13121219

Lee, J. W. (2020). Green finance and sustainable development goals: The case of China. *The Journal of Asian Finance, Economics and Business, 7*, 577–586. https://doi.org/10.13106/jafeb.2020.vol7.no7.577

Liu, J., & Diamond, J. (2008). Revolutionizing China's environmental protection. *Science, 319*(5859), 37–38. https://doi.org/10.1126/science.1150416

Nawaz, M. A., Seshadri, U., Kumar, P., Aqdas, R., Patwary, A. K., & Riaz, M. (2021). Nexus between green finance and climate change mitigation in N-11 and BRICS countries: Empirical estimation through difference in differences (DID) approach. *Environmental Science and Pollution Research, 28*(6), 6504–6519. https://doi.org/10.1007/s11356-020-10920-y

Newell, R. G., Pizer, W. A., & Raimi, D. (2013). Carbon markets 15 years after Kyoto: Lessons learned, new challenges. *Journal of Economic Perspectives, 27*(1), 123–146. https://doi.org/10.1257/jep.27.1.123

Ning, Y., Cherian, J., Sial, M. S., Álvarez-Otero, S., Comite, U., & Zia-Ud-Din, M. (2023). Green bond as a new determinant of sustainable green financing, energy efficiency investment, and economic growth: A global perspective. *Environmental Science and Pollution Research, 30*(22), 61324–61339. https://doi.org/10.1007/s11356-021-18454-7

Paris Agreement. (2016). *International Legal Materials, 55*, 743.

Rasoulinezhad, E., & Taghizadeh-Hesary, F. (2022). Role of green finance in improving energy efficiency and renewable energy development. *Energy Efficiency, 15*(2), 14. https://doi.org/10.1007/s12053-022-10021-4

Sarpong, F. A., Sappor, P., Nyantakyi, G., Agyeiwaa, O. E., Ahakwa, I., Cobbinah, B. B., & Kir, K. F. (2023). Green financial development efficiency: A catalyst for driving China's green transformation agenda towards sustainable development. *Environmental Science and Pollution Research, 30*(21), 60717–60745. https://doi.org/10.1007/s11356-023-26760-5

Shipalana, P. (2020). *Green finance mechanisms in developing countries: Emerging practice*. South African Institute of International Affairs.. https://www.jstor.org/stable/resrep28377.

Sun, Y., Bao, Q., & Taghizadeh-Hesary, F. (2023). Green finance, renewable energy development, and climate change: Evidence from regions of China. *Humanities and Social Sciences Communications, 10*(1), Article 1. https://doi.org/10.1057/s41599-023-01595-0

Sustainability | Free Full-Text | (2023) Factors affecting the sustainability performance of financial institutions in Bangladesh: The role of green finance. Retrieved August 19, 2023, from https://www.mdpi.com/2071-1050/13/18/10165

Wang, Y., & Zhi, Q. (2016). The role of green finance in environmental protection: Two aspects of market mechanism and policies. *Energy Procedia, 104*, 311–316. https://doi.org/10.1016/j.egypro.2016.12.053

Weber, O. (2012). *Sustainable banking – History and current developments* (SSRN Scholarly Paper 2159947). https://doi.org/10.2139/ssrn.2159947.

Zhang, D., Zhang, Z., & Managi, S. (2019). A bibliometric analysis on green finance: Current status, development, and future directions. *Finance Research Letters, 29*, 425–430. https://doi.org/10.1016/j.frl.2019.02.003

Zhang, W., Li, J., Li, G., & Guo, S. (2020). Emission reduction effect and carbon market efficiency of carbon emissions trading policy in China. *Energy, 196*, 117117. https://doi.org/10.1016/j.energy.2020.117117

Zheng, B., Zhang, Q., Geng, G., Chen, C., Shi, Q., Cui, M., et al. (2021). Changes in China's anthropogenic emissions and air quality during the COVID-19 pandemic in 2020. *Earth System Science Data, 13*(6), 2895–2907.

Zubair Mumtaz, D. M., & Alexander Smith, D. Z. (2019). Green finance for sustainable development in Pakistan. *IPRI Journal, 1–34*. https://doi.org/10.31945/iprij.190201

Comparative Efficiency of Green Assets and Black Assets Around the Russo-Ukraine War

Remzi Gök and Eray Gemici

Abstract Our study examines the hedging ability and safe-haven properties of green/clean assets compared to two major safe-haven instruments around the war in Ukraine. The results show substantial variations in average multivariate portfolio weights and hedging effectiveness under different techniques and during Russo-Ukrainian war. Green bonds provide a negligible and negative contribution to volatility reduction despite their high portfolio weights. WTI and clean energy markets appear to be the most attractive investments with below-average portfolio weights and make significant contributions to reducing overall volatility. The increased uncertainty in financial markets driven by the war has negatively influenced the return performance of multivariate portfolios. During the war period, green bonds had the highest asset allocations for the bivariate portfolios, followed by gold, and significantly but slightly reduced portfolio risk. We find that hedging rates and effectiveness rise for the majority of assets during the war, implying that, with a few exceptions, hedging costs rise during the crisis and the cost of a hedge becomes cheap for green bonds and costly for gold. Green bonds exhibit the worst performance under all strategies in both periods, whereas green commodity index, followed by WTI, has the best reward-to-volatility ratio under the RPP model prior to the conflict period. Throughout the study period, both negative and positive scale and event-dependent co-movements between all financial asset return combinations emerge. The commencement of the conflict in February 2022, along with Ukraine's counterattacks beginning in August 2023, has a significant influence on the direction of co-movements. Our findings shed light on the portfolio implications and risk management.

R. Gök
Faculty of Economics and Administrative Sciences, Department of Business Administration, Dicle University, Diyarbakır, Türkiye
e-mail: remzi.gok@dicle.edu.tr

E. Gemici (✉)
Islahiye Faculty of Economics and Administrative Sciences, Department of Business Administration, Gaziantep University, Gaziantep, Türkiye
e-mail: gemici@gantep.edu.tr

N. Naifar (ed.), *Climate Change and Finance*, Sustainable Finance,
https://doi.org/10.1007/978-3-031-56419-2_2

1 Introduction

Over the past few decades, the world has confronted an escalating global crisis: climate change (Ferrer et al., 2021; Huang et al., 2023). This crisis has triggered substantial transformations in financial markets and investment strategies. Investors are increasingly weaving eco-friendly companies into their portfolios (Dutta et al., 2020), while governments worldwide are making commitments to transition from fossil fuels to renewable energy sources (Kamal & Hassan, 2022). The birth of the "Green Bond Principles" by the International Capital Markets Association (ICMA) in January 2014 has opened up new vistas in green finance for both individual and institutional investors (Agoraki et al., 2023). The clarion call to combat climate change was amplified by the Paris Climate Agreement in 2015, setting a lofty goal of achieving net-zero greenhouse gas emissions by the latter half of the twenty-first century (Deng et al., 2022). Yet, this journey toward sustainability has been fraught with obstacles. The global financial crisis of 2007/2008 and the subsequent COVID-19 pandemic disrupted financial markets, causing seemingly disparate segments to become intertwined (Guo & Zhou, 2021; Tiwari et al., 2023).

COVID-19 pandemic brought an unexpected twist to the issuance of green bonds in the first quarter of 2020, witnessing a significant drop compared to the same period in 2019 (Xiang & Cao, 2023). However, a remarkable turnaround occurred in the third quarter of that year, culminating in a record-breaking issuance of US$64.9 billion, as reported by the Climate Bond Initiative.

Then, in early 2022, the world was rocked by the Russian invasion of Ukraine, sending shockwaves through global markets. However, not all assets reacted in the same manner (Będowska-Sójka et al., 2022). This war unfolded against the backdrop of the world grappling with the dire consequences of climate change, including heatwaves, droughts, floods, wildfires, and the ongoing COVID-19 pandemic. The immediate impact of the Ukrainian war on global equity markets was palpable, with the Dow Jones falling 1.38%, NASDAQ losing 2.6%, and significant sell-offs occurring in Asian markets (Kumar et al., 2023).

In summary, conventional financial markets have been severely affected by reduced economic activity and heightened uncertainty, presenting challenges in asset allocation, risk management, and financial stability (Arif et al., 2021). In light of these evolving dynamics, green finance markets, pioneered by the European Investment Bank in 2007, have emerged as an alternative asset class demonstrating tremendous potential over the past decade. They offer financial market participants an innovative avenue to allocate their resources toward sustainability initiatives and address ongoing environmental challenges, in contrast to the use of fossil fuels (Arif et al., 2021; Naeem et al., 2022).

Within these green financial markets, such as green bonds, sustainability indices, and clean energy indices, have evolved as a distinctive asset class demonstrating significant potential over the past decade (Karim & Naeem, 2022). Notably, green bonds have emerged as a promising option for portfolio diversification, gaining substantial support in financial markets under the umbrella of green finance (Arif et al.,

2022; Rehman et al., 2023). These bonds resemble conventional corporate fixed-income bonds but set themselves apart by directing their proceeds toward environmentally friendly projects (Hammoudeh et al., 2020; Reboredo & Ugolini, 2020). Remarkably, the green bond market has experienced rapid expansion, particularly following the issuance of the Green Bond Principles by the International Capital Markets Association (ICMA) in January 2014, which established standardized criteria for labeling bonds as green (Naeem et al., 2021; Reboredo et al., 2020). In response to climate-related risks faced by companies, numerous investors, including mutual funds, pension funds, individual investors, and insurance companies, have shifted their focus toward green bonds (Arif et al., 2022; Duchêne, 2020).

Upon reviewing the existing literature, a wealth of empirical studies (see, e.g., Arif et al., 2022; Dong et al., 2023; Guo & Zhou, 2021; Khamis & Aassouli, 2023; Naeem et al., 2021; Ren & Lucey, 2022; Rizvi et al., 2022; Saeed et al., 2020) has explored the safe haven and hedging characteristics of green financial instruments when compared with various asset classes during turbulent times. This paper contributes to this body of knowledge by investigating the hedging capabilities and safe-haven attributes of green and clean assets in the context of the Ukrainian conflict. By utilizing the DCC-GARCH model and rolling wavelet correlation techniques, this chapter seeks to answer the following questions:

Question 1. Which types of assets offer the most protection against war-related risks?
Question 2. Are green/clean assets beneficial hedging strategies against uncertainties among black market instruments during the war?
Question 3. Which assets can help to reduce portfolio risk during a conflict, and which should be avoided?
Question 4. Do green assets serve as a safe haven during wartime?

We address the four questions listed above for green and black markets, using 770 days of observations for each variable. The study includes daily observations of the S&P Green Bond Index (GB), the S&P GSCI Climate Aware Index (GC), and the S&P Global Clean Energy Index (GE), alongside spot prices for WTI crude oil (WTI) and gold (GOLD), spanning from February 2021 to March 2023.

The findings from portfolio construction methods, including the minimum variance portfolio (MVP), minimum correlation portfolio (MCP), and risk parity portfolio (RPP), reveal substantial reductions in portfolio weights during the Ukrainian conflict, with varying impacts on volatility across assets (highlighting hedging effectiveness). Notably, the risk parity portfolio emerges as the most effective prewar hedging approach, with GC exhibiting the highest Sharpe ratio, closely followed by WTI under the RPP model, while GOLD assumes a prominent role during the conflict. Additionally, wavelet correlation analyses uncover significant co-movements between green bonds and other assets, intensifying as the wavelet scale increases and peaking at the beginning and end of the sample period. Furthermore, the study underscores the role of green bonds as a safe haven against various markets, while gold serves as a safe haven during market turmoil, a dynamic that varies across different wavelet scales.

The study proceeds as follows: Section 2 provides the related literature review on the comparison of green and black assets. We describe our data, present the preliminary test results, and outline the test procedures for hedge ratios, weights, portfolio performance evaluation, and wavelet correlations in the next section. Section 4 offers the discussions on the empirical results and Sect. 5 concludes the paper with the policy implications.

2 Literature Review

In recent years, extensive research has investigated the dynamic relationship between green financial instruments, such as green bonds and green stocks, and traditional assets, including Treasury bonds, corporate bonds, stocks, crude oil, and gold. Ferrer et al. (2021) employed a novel wavelet-based network approach to uncover the close links between green bonds and Treasury and investment-grade corporate bonds, as well as the strong correlations between green stocks and general stocks. These findings highlight the presence of distinct asset classes with unique risk-return profiles. Huang et al. (2022) explore the role of green bonds as hedges or safe havens during extreme market conditions, comparing their performance with precious metals. They find that green bonds exhibit negative correlations with crude oil during events like COVID-19 pandemic and the Russia-Ukraine conflict, underscoring their risk reduction effectiveness and potential advantages for portfolio inclusion. Dong et al. (2023) examine the effects of geopolitical, economic, and climate policy risks on the correlations between conventional/energy stocks and conventional/green bonds. Their research reveal that both conventional and green bonds serve as safe havens during heightened geopolitical risk, while green bonds outperform conventional bonds as safe havens during elevated economic policy uncertainty and climate policy uncertainty. Xia et al. (2023) examine the relationship between green assets and economic policy uncertainty (EPU) in China. They identify asymmetric connectedness between green assets and EPU, with green bonds, carbon emission allowances, and specific green stocks emerging as potential hedges or safe-haven assets against EPU.

Against the backdrop of the COVID-19 pandemic, Arif et al. (2022) assess the hedging and safe-haven potential of green bonds, finding that they diversify medium- and long-term equity investments and serve as hedging and safe-haven instruments for currency and commodity investments. Guo and Zhou (2021) explore the effectiveness of green bonds as hedging tools against rare disasters in US and Chinese financial markets, emphasizing their minimal cross-country differences and global relevance. Huang et al. (2023) study the dynamic relationships between Bitcoin and green assets before and during the COVID-19 pandemic, highlighting Bitcoin's emergence as a refuge for green assets during the pandemic and the consistent role of green assets as hedges for Bitcoin. Arif et al. (2021) investigate the growth and interconnections of green finance markets, observing limited connections in energy investments, primarily characterized by one-way spillovers from conventional bonds in fixed-income investments. Naeem et al. (2022) explore the risk connectedness between green and conventional assets during economic shocks, noting heightened equi-correlations during such events and the effective hedging of precious metals and cryptocurrencies by green bonds. Abakah et al. (2022) study the connectedness between green and conventional assets during economic downturns, highlighting the strengthening link between green bonds and conventional assets, particularly during the COVID-19 pandemic.

During the severe stages of the pandemic, Cui et al. (2022) assess the impact of the COVID-19 pandemic on global green bonds and conventional assets, identifying strong spillover effects, particularly for green bonds. Naeem et al. (2021) explore the comparative efficiency of green and conventional bond markets before and during the COVID-19 pandemic, finding higher efficiency in the green bond market during the pandemic. Su et al. (2022) scrutinize the information transmission of Chinese green bonds in relation to other asset classes, revealing strong information transmission between green bonds and traditional fixed-income instruments. Yousaf et al. (2022) explore the diversification and hedging advantages of green investments, specifically green bonds, in conventional stock portfolios during the COVID-19 pandemic, demonstrating that green bonds functioned as safe-haven assets.

In a broader temporal context, Nguyen et al. (2021) investigate the relationship between green bonds and various asset markets, emphasizing diversification benefits and low/negative correlations with stocks and commodities. Elsayed et al. (2022) explore the relationship between green bonds and financial markets using a time-frequency analysis approach, highlighting pronounced diversification benefits in the short term and increased integration with financial markets over the long term. Naeem et al. (2022) investigate the potential advantages of incorporating green bonds into investment portfolios alongside traditional assets, suggesting GB's potential as an alternative to traditional bonds and enhancing diversification potential, especially with energy and agriculture assets.

3 Data and Methodology

3.1 Sample Data and Descriptive Statistics

This study compares the green and black asset performances around the war conflict in Ukraine, using various performance metrics and rolling wavelet correlation framework. To do so, we select three indices, including the S&P Green Bond Index (GB), S&P GSCI Climate Aware Index (GC), and S&P Global Clean Energy Index (GE), as proxies for green bonds, commodity index, and clean energy stocks, respectively. Furthermore, we use the daily closing spot price of WTI crude oil and gold (in US dollars) to proxy for black assets. The sample data, spanning from August 13, 2020, to September 6, 2023, is extracted from www.spglobal.com for

GB, GC, and GE, and from the US Energy Information Administration (EIA) and World Gold Council for WTI crude oil (WTI) and gold (GOLD) spot prices. Figure 1 depicts a structural break in price and volatility clustering for the return series, coinciding with the starting date of the invasion of Ukraine.

Table 1 presents descriptive statistics and correlation estimations before and during the war periods. WTI and GB exhibit the highest and lowest average returns, respectively, before the war crisis. However, all variables have a negative return during the stressful period, with the exception of gold, which has a zero return on average. Volatility for GB, GC, and WTI is higher during the war period, with WTI having the highest average volatility and GB being the least volatile among the variables. We find that all series have significantly negative or positive skewness, heavy tails (leptokurtic), and non-normally distributions in both subperiods. Positive correlations are observed in both periods, with an increase during the invasion. Except for gold, all variables are significantly positively related to each other, and move strongly in the same direction. The findings highlight the opportunity for oil markets to hedge against changes in green bonds and clean energy equities before

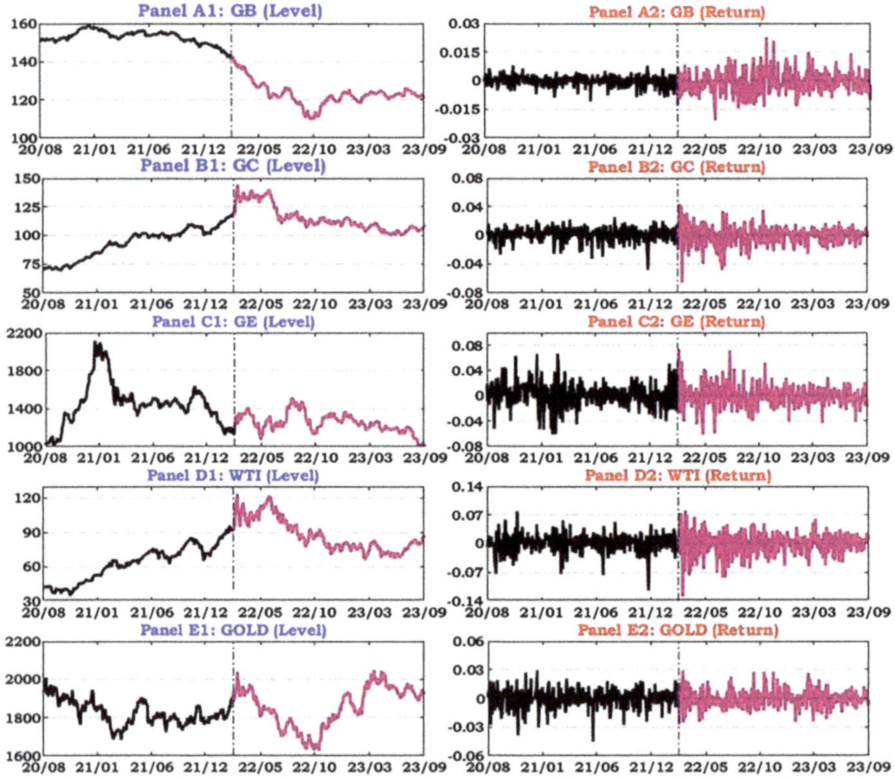

Fig. 1 Plots of time series. Note: The vertical lines denote the starting date of the invasion (February 24, 2022)

Table 1 Descriptive statistics and unconditional correlation estimations

	GB	GC	GE	WTI	GOLD		GB	GC	GE	WTI	GOLD
Panel A: Summary Statistics per Return Series											
Mean	−0.0002	0.0014	0.0003	0.002	0		−0.0004	−0.0003	−0.0003	−0.0001	0
Max	0.0081	0.0257	0.067	0.08	0.029		0.0227	0.0439	0.0741	0.0821	0.0289
Min	−0.0108	−0.0484	−0.0611	−0.114	−0.0451		−0.0207	−0.066	−0.0612	−0.1278	−0.027
SD	0.0029	0.0096	0.0192	0.0221	0.0097		0.0059	0.0132	0.0175	0.0276	0.0088
Skewness	−0.45***	−0.88***	0.14	−0.69***	−0.65***		0.23*	−0.41***	0.39***	−0.44***	0.19
Kurtosis	0.79**	2.34***	1.33***	2.69***	2***		0.76**	2.49***	2.04***	1.34***	1.18***
JB	23.03***	136.33***	29.65***	146.67***	90.95***		12.87***	110.56***	76.71***	41.53***	24.65***
Q	22.47***	6.87	26.45***	22.61***	9.18		19.22***	15.81***	16.97***	14.49**	8.99
Q2	7.8	13.04**	55.17***	21.41***	7.79		13.9**	138.06***	45.18***	71.48***	17.22***
ARCH	7.26	8.91	44.02***	15.76	7.71		22.38**	44.07***	17.08	55.15***	19.61*
ERS	−3.96***	−5.13***	−3.71***	−4.06***	−2.94***		−1.91*	−2.71***	−1.85*	−4.98***	−1.96*
Panel B: Pearson's Correlation Estimation											
GB											
GC	0.21***						0.17***				
GE	0.26***	0.15***					0.37***	0.28***			
WTI	−0.02	0.78***	0.08				0.03	0.84***	0.16***		
GOLD	0.49***	0.25***	0.14***	−0.02			0.47***	0.38***	0.21***	0.18***	

Notes: ***, **, and * denote the rejection of the null hypothesis at 10%, 5%, and 1% significance levels, respectively

the conflict period, as well as a significant safe haven function against green bond markets and vice versa during prolonged geopolitical uncertainties. Furthermore, gold has strong and moderate positive correlations with green assets, indicating that it is neither a hedge nor a safe haven before and during market volatility.

4 Methodology

In this chapter, we utilize three widely known portfolio construction approaches to shed insights on the diversification potential of green and black assets around the war in Ukraine. It is crucial to note that we assume that international investors can take short/long positions solely to invest in underlying assets using these techniques.

4.1 Hedge Ratio and Portfolio Weights

Following the financial literature on portfolio optimization (Chang et al., 2011; Gök et al., 2023), this chapter employs one of the most extensively used specifications, the DCC-GARCH approach, to estimate hedge ratios, hedging effectiveness, and portfolio weights. We use Kroner and Sultan's (1993) procedure to calculate a hedging ratio for green/black asset pairs, as follows:

$$\beta_{ij,t} = {}^{h_{ij,t}}\!/_{h_{jj,t}} \tag{1}$$

where $h_{ij,\,t}$ and $h_{jj,\,t}$, extracted from an optimal GARCH model (eGARCH or sGARCH), denotes, respectively, the conditional covariance between series i and j, and the conditional variance of series j at time t. It should be borne in mind, $\beta_{ij,\,t}$ implies, respectively, taking a long and short position in series i and j to hedge against the downside risk of series i. Afterward, we can estimate the optimal weights for a 1-dollar two-asset portfolio using the Kroner and Ng (1998) approach as given:

$$w_{ij,t} = \frac{\Sigma_{ii,t} - \Sigma_{jj,t}}{\Sigma_{ii,t} - 2\Sigma_{ij,t} + \Sigma_{jj,t}} \tag{2}$$

with $w_{ij,\,t}$ given by

$$w_{ij,t} = \begin{cases} 0, & \text{if } w_{ij,t} < 0 \\ w_{ij,t}, & \text{if } 0 \le w_{ij,t} \le 1 \\ 1, & \text{if } w_{ij,t} > 1 \end{cases} \tag{3}$$

where $w_{ij,\,t}$ stands for the weight of series i, whereas $1 - w_{ij,\,t}$ denotes the weight of series j at time t a 1-dollar two-asset portfolio.

4.2 Portfolio Allocation Approaches

Asset allocations for bivariate and multivariate portfolios can be calculated using the minimum variance portfolio (MVP) framework proposed by Markovitz (1959) as given:

$$w_t = \frac{H_t^{-1} I}{I H_t^{-1} I} \tag{4}$$

where w_t is the $m \times 1$ weights vector, I represents the identity matrix, and H_t denotes the $m \times m$ dimensional conditional variance-covariance matrix at time t.

We should note that while the objective of the aforementioned approach is to minimize the total variance, which is the sum of the variance of each asset and its covariance with each other, the minimum correlation portfolio suggested by Christoffersen et al. (2014) focuses on reducing conditional correlation. We can estimate portfolio weights under this method as given:

$$w_t = \frac{R_t^{-1} I}{I R_t^{-1} I} \tag{5}$$

with R_t given by

$$R_t = diag\left(H_t\right)^{-0.5} H_t diag\left(H_t\right)^{-0.5} \tag{6}$$

where R_t is the $m \times m$ dimensional correlation matrix.

Then, we apply the risk parity portfolio (RPP) proposed by Maillard et al. (2010), which is an easier portfolio construction approach compared to the other two techniques given that this method does not require estimation of expected returns. Under this technique, i.e., the risk-balanced portfolio, we can estimate portfolio weights after solving the minimization problem as described below:

$$\min \sum_{i,j=1}^{N} \left(w_{it} \left(H_t w_t\right)_i - w_{jt} \left(H_t w_t\right)_j \right)^2 \tag{7}$$

4.3 Portfolio Evaluation Approaches

First, we estimate Ederington's (1979) hedging effectiveness (HE) statistic to evaluate whether the effect of a single asset on portfolio volatility reduction is significant or not. We can define this metric by the following form:

$$\text{HE}_i = \frac{\text{var}_{\text{unhedged}} - \text{var}_{\text{hedged}}}{\text{var}_{\text{unhedged}}} \tag{8}$$

where HE_i stands for the percentage of risk reduction from the unhedged position of asset i, var_{hedged} signify the variance of the portfolio returns, and $var_{unhedged}$ is the variance of the unhedged asset, respectively. Note that, a higher HE statistic implies a higher risk reduction in a portfolio and vice versa.

To measure the expected excess return per unit of risk, we use the metric of the Sharpe ratio (Sharpe, 1994), which has the following form:

$$\mathrm{SR} = \left(\frac{\mu_p - R_f}{\sigma_p} \right) \tag{9}$$

where μ_p implies the portfolio return, R_f is the risk-free rate, and σ_p represent the portfolio standard deviation. Note that the risk-adjusted investment in a multivariate portfolio is more (less) appealing when the SR statistic is greater (lower).

4.4 *Wavelet Correlations*

This study utilizes wavelet analysis to measure co-dependence among green and black market instruments around the war period. Wavelet analysis is a refined version of Fourier analysis and offers localization in both time and frequency domains concurrently (Percival & Walden, 2000). The discrete wavelet transform (DWT) is preferred over the continuous wavelet transform (CWT) for analyzing financial time series with limited duration, as DWT assumes evenly spaced observations while CWT assumes signal continuity (Crowley, 2007). We prefer using the maximal overlap discrete wavelet transform (MODWT), despite discarding some properties of DWT, as it effectively decomposes series into coefficients associated with variations across scales and provides significant advantages over DWT. In this study, the MODWT method is applied to decompose a nondyadic series into scale-decomposed equivalent series, enabling the simultaneous examination of co-movements between black and green assets within different time scales and frequency bands. Using the MODWT with a least asymmetric wavelet filter of length 8 (hereafter, LA(8)), one can obtain a wavelet correlation estimator at scale τ_j as:

$$\hat{\rho}_{X,Y}\left(\tau_j\right) = \frac{\hat{\sigma}_{X,Y}\left(\tau_j\right)}{\hat{\sigma}_X\left(\tau_j\right)\hat{\sigma}_Y\left(\tau_j\right)} \tag{10}$$

with the unbiased estimator of wavelet variance at scale $\hat{\sigma}_{X,Y}\left(\tau_j\right)$ given by:

$$\hat{\sigma}_X^2\left(\tau_j\right) = \frac{1}{M_j} \sum_{t=L_j-1}^{N-1} \overset{\c{}}{W}{}_{j,t}^2 \tag{11}$$

and with the unbiased estimator of wavelet covariance at scale $\hat{\sigma}_{X,Y}\left(\tau_j\right)$ as follows:

$$\hat{\sigma}_{X,Y}(\tau_j) = \frac{1}{M_j} \sum_{t=L_j-1}^{N-1} \breve{W}_{j,t}^{(X)} \breve{W}_{j,t}^{(Y)} \quad (12)$$

where $\{\breve{W}_{j,t}\}$ represent the jth-level MODWT wavelet coefficients in Eq. (11) and $M_j \equiv N - L_j + 1 > 1$ denotes the number of coefficients in each scale τ_j, in Eq. (12), that are unaffected by boundary conditions.

Subsequently, we calculate an approximately 100(1 − 2p)% confidence interval as suggested by Percival and Walden (2000)) as shown below:

$$\left[\tanh\left\{h\left[\rho_{XY}(\tau_j)\right] - \frac{\theta^{-1}(1-p)}{\sqrt{N_j - 3}}\right\}, \tanh\left\{h\left[\rho_{XY}(\tau_j)\right] + \frac{\theta^{-1}(1-p)}{\sqrt{N_j - 3}}\right\}\right] \quad (13)$$

where $h(p) = \tanh h^{-1} \hat{\rho}$ corresponds to the Fisher's z-transformation and N_j is the number of wavelet coefficients derived by DWT at the j th level under the condition of uncorrelated Fisher's z-transformation.

5 Empirical Results and Discussions

To assess the integration order of green and black assets, we conducted two conventional unit root tests on the logarithmic level and return series. The results, shown in Table 2, indicate that GE is level stationary, while GB, GC, WTI, and GOLD become stationary in their first logarithmic differences.

5.1 *Portfolio-Back Testing Model Results*

We provide the summary statistics of weights for multivariate portfolios pre (Panel A) and during (Panel B) the war period using MVP, MCP, and RPP strategies in Table 3. The results show significant differences in average multivariate portfolio

Table 2 Stationarity test results

	Log-level series		First-differenced series	
	ADF (1979)	PP (1988)	ADF (1979)	PP (1988)
GB	−1.616	−1.611	−23.964***	−24.011***
GC	−1.537	−1.526	−25.14***	−25.037***
GE	−3.19*	−3.221*	−23.378***	−23.354***
WTI	−1.782	−1.61	−26.555***	−26.73***
GOLD	−2.812	−2.684	−28.825***	−28.95***

Note: The asterisks, *** and **, implies, respectively, the rejection of the null hypothesis that underlying series has a unit root at 5% and 1% levels of significance

weights and hedging effectiveness under various strategies and around the war conflict between Russia and Ukraine. We find that the green assets lose their popularity during the war periods and both gold and oil markets become more preferable assets under three portfolio construction strategies.

Despite making an insignificant and negative contribution to reducing volatility, GB emerges as the most preferred asset under both MVP and RPP techniques during both subperiods (Bai et al., 2023). In addition, the results of MCP technique show that investing around 20% of the fund in green bonds increases, rather than decreases, overall portfolio volatility. WTI and GE markets appear to be the most appealing investments under all techniques before and during the war periods, given their significant contribution to reducing overall volatility. Gold, however, shows comparable and meaningful performance in reducing volatility under the MVP and RPP approaches, but it exhibits a negative impact on the portfolio volatility during the war period under the MCP model. Figure 2 shows that the portfolio weights change substantially under the MVP strategy during the war period compared to the pre-crisis period but tend to remain stable under the other techniques during both periods. Throughout the investigation period, GB obtains the most allocation, followed by GOLD and GC under MVP and RPP, whereas WTI is preferred under MCP throughout both normal and stressed circumstances. We provide the cumulative portfolio returns under three strategies for both periods in Fig. 3. The findings show that establishing a multivariate portfolio containing both green and black assets produced positive returns prior to the military conflict in Ukraine, with the MVP and RPP generating the lowest and highest performance, respectively. However, given the rising uncertainty in financial markets, all strategies throughout the war slipped into negative territory, with the RPP and MVP delivering the highest and lowest poor performance, respectively. These findings imply that larger weights for green assets result in lower performance before the conflict but greater portfolio diversification benefits during the war period.

Table 4 presents average weights and hedging effectiveness scores for bivariate portfolios before and during the war periods. We find evidence of significant changes in portfolio weights and hedging effectiveness, with green bonds losing its popularity and gold increasing its proportion in two-asset portfolios for four cases due to changes in investor preferences. Furthermore, the results show that constructing a bivariate portfolio with green and/or black assets would make a positive contribution to reducing volatility in all cases, except for the two-asset portfolios GC/WTI and GB/GE. For both periods, the lower the asset allocation, the higher the risk reduction, and vice versa. Among assets, GB receives the largest weights in bivariate portfolios, yet its impact on volatility reduction is zero or close to zero while statistically significant before the war. During the war its contribution to reducing volatility strengthens against GC, WTI, and GOLD, nevertheless, it increases volatility for the bivariate portfolio GB/GE. Gold also obtains larger weights and significantly contributes to risk reduction in bivariate portfolios, particularly with GC, GE, and WTI as the second asset before the war. During the war period, gold remains popular among investors, but exhibits weakened hedging effectiveness, suggesting that gold acts as a safe-haven asset against other markets but GB when geopolitical

Table 3 Multivariate portfolio allocation analysis under different strategies

	Minimum variance portfolio			Minimum correlation portfolio			Risk parity portfolio		
	Average	SD	HE	Average	SD	HE	Average	SD	HE
Panel A: Before the military conflict in Ukraine									
GB	0.935	0.037	0.037	0.209	0.011	−9.666***	0.515	0.033	−1.391***
GC	0.012	0.032	0.922***	0	0	0.135	0.126	0.025	0.806***
GE	0.004	0.006	0.976***	0.236	0.01	0.729***	0.106	0.027	0.939***
WTI	0.043	0.025	0.983***	0.368	0.009	0.807***	0.082	0.019	0.957***
GOLD	0.006	0.027	0.897***	0.187	0.011	−0.146	0.171	0.027	0.743***
Panel B: During the War Period									
GB	0.729	0.095	0.071	0.193	0.022	−3.774***	0.384	0.044	−0.513***
GC	0.154	0.104	0.82***	0	0	0.072	0.146	0.022	0.706***
GE	0.001	0.004	0.897***	0.214	0.011	0.47***	0.139	0.018	0.832***
WTI	0.005	0.013	0.957***	0.402	0.013	0.778***	0.086	0.013	0.93***
GOLD	0.11	0.062	0.547***	0.191	0.013	−1.329***	0.245	0.019	0.262**

Notes: The asterisks, *** and **, indicate the rejection of the null hypothesis at 99% and 95% confidence levels, respectively

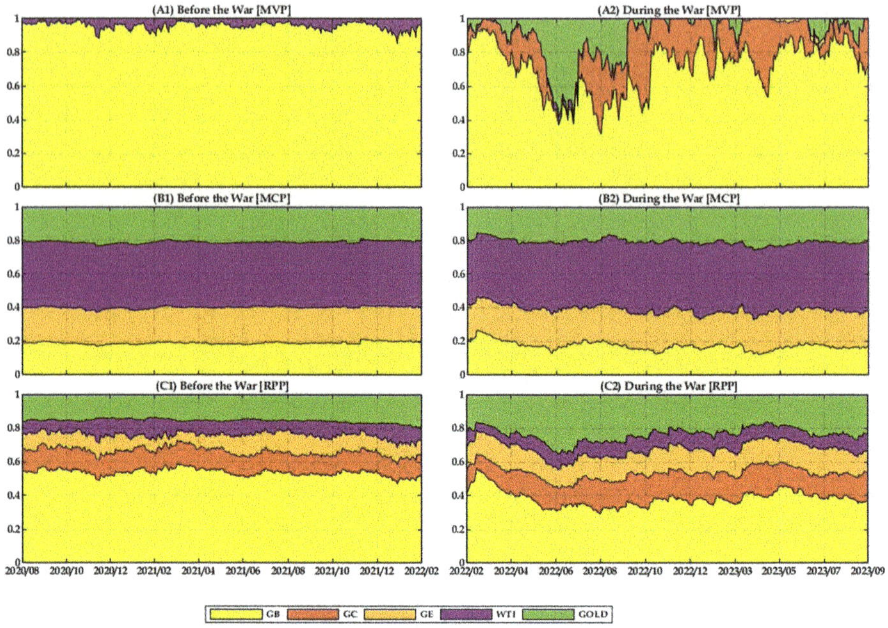

Fig. 2 Time-varying multivariate portfolio weights under various portfolio construction techniques before (Left) and during the war (Right) periods

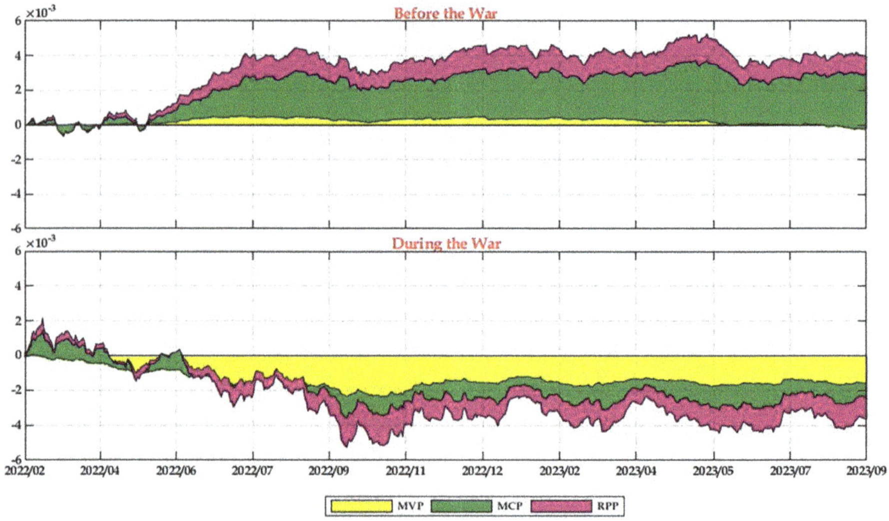

Fig. 3 Time-varying multivariate portfolio returns under various portfolio construction techniques before (top panel) and during the war periods (bottom panel)

Table 4 Bivariate portfolio weights and hedging effectiveness

	Panel A			Panel B		
	Mean	SD	HE	Mean	SD	HE
GB/GC	0.984	0.006	0.009***	0.856	0.067	0.065***
GB/GE	1	0	0***	0.979	0.016	−0.016***
GB/WTI	0.972	0.013	0.039***	0.957	0.018	0.029***
GB/GOLD	1	0	0***	0.834	0.164	0.047***
GC/GB	0.016	0.006	0.913	0.144	0.067	0.811
GC/GE	0.792	0.086	0.136***	0.685	0.121	0.207***
GC/WTI	0.996	0.03	−0.001***	1	0	0***
GC/GOLD	0.487	0.061	0.38***	0.301	0.153	0.611
GE/GB	0	0	0.978	0.021	0.016	0.884
GE/GC	0.208	0.086	0.783	0.316	0.121	0.553**
GE/WTI	0.561	0.161	0.452***	0.753	0.082	0.19***
GE/GOLD	0.2	0.079	0.784	0.196	0.093	0.773
WTI/GB	0.028	0.013	0.984	0.043	0.018	0.955
WTI/GC	0.004	0.03	0.81	0	0	0.772
WTI/GE	0.439	0.161	0.585***	0.247	0.082	0.672**
WTI/GOLD	0.203	0.085	0.853**	0.082	0.05	0.905
GOLD/GB	0	0	0.913	0.166	0.164	0.565
GOLD/GC	0.513	0.061	0.384***	0.699	0.153	0.124***
GOLD/GE	0.801	0.079	0.145***	0.804	0.093	0.091***
GOLD/WTI	0.797	0.085	0.229***	0.918	0.05	0.059***

Notes: The asterisks, *** and **, indicate the rejection of the null hypothesis at 99% and 95% confidence levels, respectively

risks intensify. A perfect match appears between crude oil and clean energy for both periods, with GE increasing its weight against WTI while decreasing its contribution to volatility reduction during the unstable period. Overall, portfolio construction using substitute assets emerges (WTI and GE) as the most effective way in decreasing portfolio risk during a military conflict.

Table 5 reveals descriptive statistics for hedge ratios and hedging effectiveness following Kroner and Sultan (1993), Ederington (1979) and Antonakakis et al. (2020), respectively, over both periods. Each positive hedge ratio (average) in the first column represents the amount of a short position in the second asset required to hedge a $1 long position in the first. In the case of a negative value, it may be helpful to state that a similar, i.e., long, position should be taken for both assets. The hedging effectiveness coefficients, shown in the third column, indicate the contribution of the second asset to risk reduction is significant or not. For instance, the positive hedge ratio of 0.14 (0.327) for the GB/GOLD pair suggests that a $1 long position in green bond markets could be hedged with a $0.14 ($0.327) investment in gold markets through holding a short position before (during) the conflict period. The results show that the cost of black assets is more expensive for GB and GC, while green assets are more expensive for GC, regardless of the sample period. Similarly, short positions in black assets tend to be less expensive than long positions in WTI

Table 5 Hedge ratio estimations for bivariate portfolios

	Panel A: Before the war period			Panel B: During the war period		
	Mean	SD	HE	Mean	SD	HE
GB/GC	0.065	0.005	0.042	0.095	0.033	0.048*
GB/GE	0.039	0.008	0.062	0.113	0.027	0.168
GB/WTI	−0.007	0.003	0	0.008	0.008	−0.003
GB/GOLD	0.14	0.012	0.229***	0.327	0.099	0.278***
GC/GB	0.814	0.066	0.043**	0.394	0.094	0.019***
GC/GE	0.089	0.019	0.024	0.212	0.043	0.067
GC/WTI	0.382	0.098	0.594	0.383	0.048	0.709
GC/GOLD	0.263	0.027	0.066***	0.476	0.154	0.176***
GE/GB	1.585	0.409	0.077**	0.87	0.156	0.131***
GE/GC	0.29	0.073	0.022	0.384	0.097	0.072*
GE/WTI	0.059	0.022	0.003	0.096	0.029	0.017
GE/GOLD	0.285	0.074	0.021***	0.379	0.131	0.05***
WTI/GB	−0.349	0.188	−0.002**	0.149	0.168	−0.007***
WTI/GC	1.606	0.474	0.641	1.817	0.247	0.71*
WTI/GE	0.078	0.033	0.014	0.254	0.07	0.019
WTI/GOLD	−0.026	0.028	0.001***	0.375	0.24	0.058***
GOLD/GB	1.683	0.176	0.253**	0.793	0.227	0.209***
GOLD/GC	0.252	0.028	0.064	0.258	0.06	0.139*
GOLD/GE	0.083	0.015	0.017	0.114	0.029	0.035
GOLD/WTI	−0.008	0.01	0.001	0.039	0.021	0.043

Notes: The asterisks, *** and **, indicate the rejection of the null hypothesis at 99% and 95% confidence levels, respectively

and gold. These results indicate that it is more reasonable and less expensive to use assets with similar characteristics to manage market risk, with the exception of clean energy assets. We, on the other hand, find that hedge ratios and effectiveness mostly increased during military conflict. This suggests that for all but a few cases, hedging costs increased during the crisis and the cost of using green bonds as a hedging instrument has been significantly reduced, thereby supporting the findings from Yahya et al. (2019). In addition, using GB as a hedging asset is not only more expensive but also less effective in both periods, in contrast to Jin et al. (2020), who found it more efficient and less expensive for hedging carbon risk during the crisis. It is obvious that the cost of hedging the risk of black assets is considerably higher than the cost of hedging the risk of green bonds. For example, to hedge a $1 long position in WTI, a $1.82 short position in GC is required during the war which is the highest hedging ratio in the two periods. It is worth noting that this strategy could significantly reduce the risk of a two-asset portfolio by 71%, which is the highest rate, amid heightened geopolitical risks. Despite the fact that the hedging cost of this portfolio emerged as the second highest after the GOLD/GB strategy in the pre-war period, the effectiveness coefficient indicates that this choice was worthless in terms of risk mitigation. Similarly, using gold as a hedge for green assets is effective in both periods but becomes more expensive during the war, confirming Raheem et al. (2023). Our findings, however, contradict Naeem et al. (2022) who observed weakening hedging effectiveness for WTI crude oil and gold during the financial crisis. Negative hedge ratios appear solely during the pre-war era; holding a long position in only black asset category significantly can reduce total portfolio risk. A $0.026 long position in GOLD against the risks of a $1.0 long position in WTI could only reduce portfolio risk by one thousandth. Despite the fact that the WTI and GB variables appear to be two distinct markets in which identical positions should be created to eliminate each other's risk, they cause the portfolio risk to increase rather than decrease. These results align with Ahmad et al. (2018) and Raheem et al. (2023), who show negative hedging effectiveness and conclude that hedged portfolios have less risk mitigation during the stressful periods.

Finally, we employ the annualized Sharpe Ratio (SR) to assess the performance of green and black assets in terms of their profitability relative to risks under three portfolio construction techniques before and during the war periods. The results, shown in Table 6, suggest that the expected return per unit of risk incurred varies greatly under various models, across markets, and over different periods; it is sometimes higher or lower than 1 and occasionally falls into negative territory, particularly during the war period (see Fig. 3). GB exhibits the worst performance under all strategies in both periods, whereas GC has the best reward-to-volatility ratio of 2.298 under the RPP model prior to the conflict period. WTI tends to be the most profitable asset due to the relatively higher risk-adjusted performance under three strategies only before the outbreak of war in Ukraine. However, it appears to be an unattractive and expensive asset like all other markets due to its negative score driven by heightened uncertainty. GOLD, on the other hand, seems to be a profitable asset during the conflict, achieving the RPP strategy's one and only positive SR ratio of 0.15, suggesting an extremely lower but positive anticipated extra return per unit of risk. Overall, RPP outperforms other methods by providing the highest ratio for green assets (GC) before the war period, whereas MCP outperforms other methods by providing the only positive ratio for black assets (GOLD) during heightened geopolitical risks. These findings contradict Cepni et al. (2022), who suggest that green bonds can enhance risk-adjusted returns during periods of high climate uncertainty, rather than solely focusing on risk reduction.

Table 6 Reward-to-volatility (Sharpe Ratios)

	Panel A: Before the war period			Panel B: During the war period		
	MVP	MCP	RPP	MVP	MCP	RPP
GB	−0.819	−0.899	−0.818	−1.021	−1.467	−1.237
GC	0.405	NA	2.298	−0.331	NA	−0.458
GE	NA	0.219	0.435	−0.328	−0.344	−0.408
WTI	1.416	1.452	1.449	−0.494	−0.062	−0.227
GOLD	−0.81	−0.068	−0.001	−0.993	0.15	−0.103

Notes: MVP, MCP, and RPP imply, respectively, minimum variance, minimum correlation; and risk parity portfolios. NA: No available

5.2 Rolling Wavelet Correlation Results

To assess the safe-haven characteristics between green and black assets during the Ukraine-Russian war across different investment horizons, we utilize the MODWT-based correlation estimation with Daubechies LA(8) filter. We select an optimal integer decomposition level of 4, although a higher level could be achieved (floor(log_2 (769)) $\geq$ 9). Our choice is motivated by the gradual decrease in the number of non-boundary affected wavelet coefficients as the scale increases. Consequently, we have 762, 748, 720, and 664 compatible wavelet coefficients for correlation estimation at four scales d1, d2, d3, and d4, respectively, representing [2–4), d2 [4–8), d3 [8–16), and d4 [16–32) daily time horizons. Following Gök and Çankal (2020) and Nguyen et al. (2021), we consider the first three scales (d1, d2, d3) for short-run and the last scale (d4) for long-run analysis, recognizing the diverse expectations and risk perceptions of international investors in financial markets. Figures 4 and 5 illustrate time-varying wavelet correlation estimations based on a rolling window of 150 observations across the four scales, with a color code representing the power range from dark blue to dark red. It is worth noting that darker (lighter) colors indicate stronger (weaker) correlations between market returns.

Panels A–D of Fig. 4 show time-varying estimations of correlations between green bond index returns and changes in other indices. The findings show positive and negative co-movements that vary in magnitude, direction, and significance across wavelet scales, with a decoupling pattern around the first months of the conflict and a recoupling positive trend for the rest of the sample period. Prior to the war period, GOLD ranked first with a strongly positive co-movement with GB, followed by GC and GE, however, a negative association, at varying degrees, emerged with the crude oil markets strengthening as wavelet scales increased. The appearance of comparatively lesser but positive correlations at scale d1 suggests that, during the tranquil period, all but GC markets act as a diversifier against green bonds (Baur & Lucey, 2010). On the other hand, the commencement of the war has caused a decoupling pattern among financial markets, with GB diverging considerably from other markets over some time. GB and the other markets are similarly, albeit to differing degrees, influenced by the heightened geopolitical risks across various time horizons, therefore, inducing negative associations. In fact, the conflict has had the most detrimental impact on the association between GB and GOLD returns, causing a flight-to-quality phenomenon for both short- and long-term investors in green bond markets. These findings imply that the interconnections between green bonds and other markets are driven by a variety of investor behavior and investment horizons during times of stress, indicating strong safe-haven role of green bonds for gold, oil, clean energy, and commodity markets, thus confirming earlier studies (Goodell et al., 2022; Nguyen et al., 2021). With the commencement of Ukraine's counter-offensive around August 2022, the GB-other markets correlation switched to its initial trajectory and the green bond market moved in tandem with the other markets, with the exception of the crude oil market, until May 2023. Afterward, we find

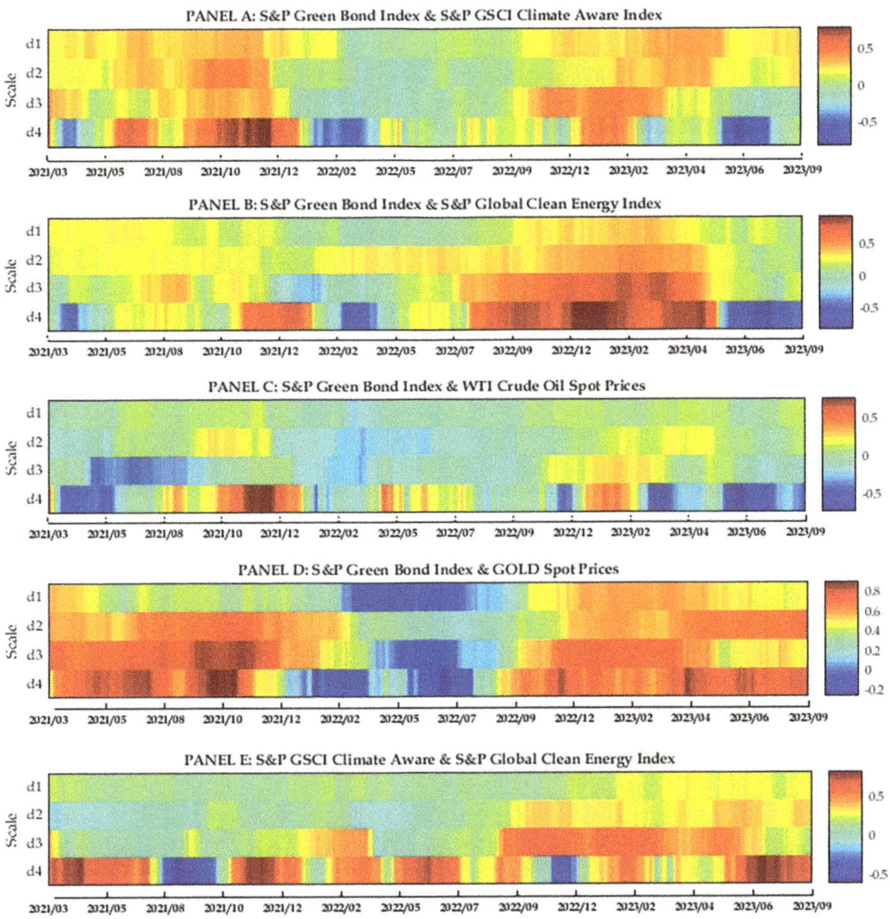

Fig. 4 Time-varying wavelet correlation estimations between market returns

that GB correlates positively with gold returns at all scales and with the rest of markets in the short run until the end of sample period but is highly negatively associated with all markets in the long run. We may conclude that green bond markets provide a safe haven for long-term investors against adverse fluctuations in commodities, clean energy, and oil markets in 2023 (Pham & Nguyen, 2021).

We also find significant evidence that the dynamics of correlation between the GSCI climate aware index and the others, including the clean energy, oil, and gold markets, fluctuate substantially across time and frequencies (See Panel E of Fig. 4 and Panels A–B of Fig. 5). Prior to and during the war periods, a weak relationship was observed between commodity (GC) and clean energy (GE) markets at scales d1, d2, and d3, indicating prominent diversification benefits of GC against GE for short-term horizons until the commencement of Ukraine's counteroffensives. Following that, a positive association between GC and GE emerges, with the

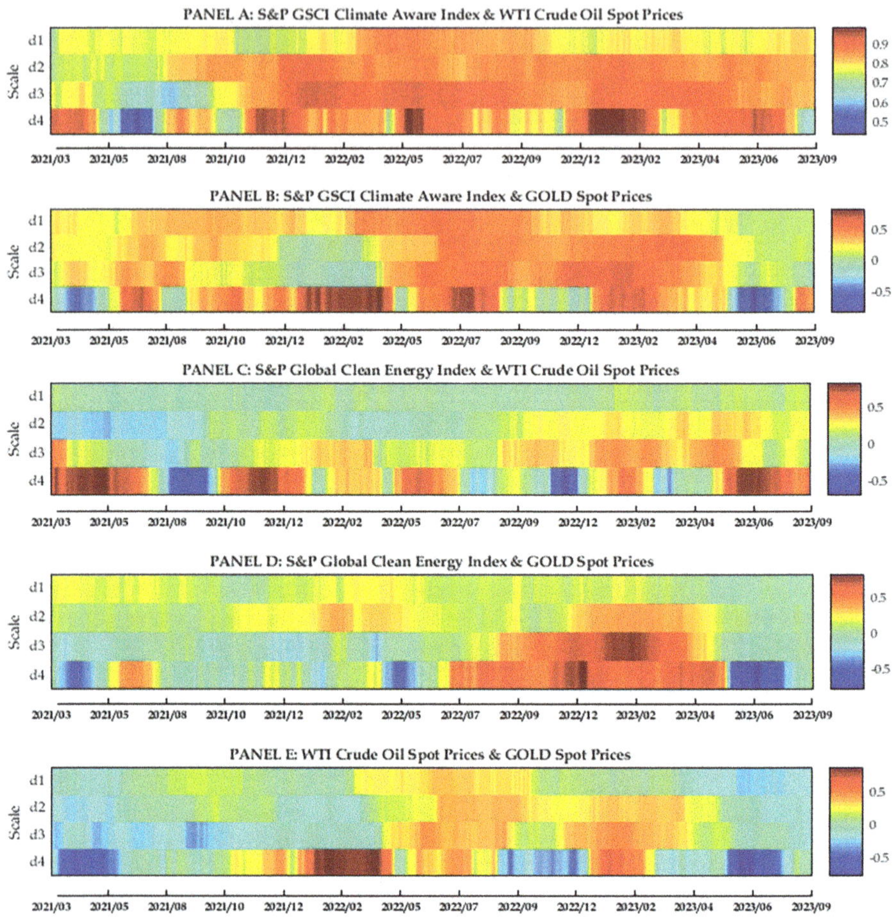

Fig. 5 Time-varying wavelet correlation estimations between market returns

magnitude being moderate at lower scales but strengthening at longer horizons. However, it had a substantial and positive correlation with GE on a scale of d4, providing minimal diversification advantages for long-term investors throughout the sample period, except for September to October 2021 and November 2022.

In Fig. 5, Panel A displays a strongly positive co-movement between GC and WTI, indicating that GC does not act as a safe haven or hedge against oil market fluctuations across various time horizons throughout the sample period. The magnitude of the correlation remained higher than 0.90 for the majority of the time, implying that the rising interaction between negative asset returns validates the positive correlation observed between assets as an inevitable consequence of the geopolitical crisis (Nguyen et al., 2021). Panel B depicts the mostly positive co-movements of GC and GOLD, which intensified during the conflict. GC acts as a weak safe haven for gold investors around May 2023 in the short run but offers substantial diversification benefits for long-term investors for a limited time. Panels C and D show weak correlations at lower scales but strong positive or negative correlations at higher scales between clean energy and oil market returns during the sample period. Before the onset of the war period, GE was weakly related to WTI and GOLD market returns at lower scales (d1 and d2), but the association strengthened at higher time horizons, experiencing a transition from negative to positive or vice versa. The appearance of warmer shades in Panel D from June 2022 to May 2023 shows a substantial increase in co-movement between clean energy equities and gold returns at scales d3 and d4, indicating that both market investors react completely similar, albeit at varying magnitude, due to changes in their preferences across different investment horizons. Panel E demonstrates that the interaction between gold and oil markets differs over time and at various scales. Prior to the start of the conflict, it appeared a scale-dependent negative co-movement, but a moderately negative and significantly positive relationship emerged in the short and long term, respectively, with the launch of the military invasion in Ukraine. This result suggests that short-term investors can benefit from diversification, but in the long run, both markets cannot serve as safe havens for each other. However, we can see that in the fourth quarter of 2022 and from May to June 2023, oil performs as a safe haven against gold for long-term investors.

6 Conclusions and Policy Implications

Using a DCC-GARCH model and a rolling wavelet correlation, we examine the hedging ability and safe-haven characteristics of three green assets against two major safe-haven instruments for a sample period covering August 13, 2020, to September 6, 2023. We can summarize the main results as follows. The results of the portfolio-back testing models, on the one hand, show substantial variations in average multivariate portfolio weights and hedging effectiveness under different techniques and in light of the Russo-Ukrainian military conflict. Despite having the highest weights under both MVP and RPP approaches in both subperiods, green bonds provide a negligible and negative contribution to volatility reduction. However, WTI and GE markets appear to be the most attractive investments with below-average portfolio weights under all techniques prior to and war periods, given their significant contribution to reducing overall volatility. Furthermore, while building a multivariate portfolio with both green and black assets yielded positive returns prior to the military war in Ukraine, the heightened uncertainty in financial markets caused all models to shift into negative territory due to poor performance. On the other hand, the results for the two-asset portfolios indicate that green bonds, followed by gold, had the highest asset allocations during the Russo-Ukraine war period, albeit with a slight decrease, and significantly but marginally reduced portfolio risk. This result implies that the lower the asset allocation, the higher the risk reduction and vice versa for bivariate portfolios. We find that hedge ratios and effectiveness mostly increased during military conflict, indicating that for all but a few cases, hedging costs increased during the crisis and the usage of green bonds as a hedging tool has become less costly while gold has become more expensive. GB exhibits the worst performance under all strategies in both periods, whereas GC, followed by WTI, has the best reward-to-volatility ratio under the RPP model prior to the conflict period. Time-varying wavelet correlation estimations reveal both negative and positive scale and event-dependent co-movements between all financial asset return pairs, with the exception of GC/WTI moving in tandem, throughout the sample period. The outbreak of the war in February 2022, alongside Ukraine's counterattacks starting in August 2023 exerts a strong impact on the direction of co-movements. During the initial stages of the war, green bonds acted as a safe haven for both short- and long-term gold and oil market investors. The long-term safe-haven properties of green bonds for all but gold markets during Ukraine's counterattack period and in 2023 are quite apparent. Furthermore, oil is a strong safe haven for the gold market in the short and long run, while clean energy index returns play a critical safe-haven role for long-term gold investors.

Our results are crucial for politicians and investors. From the policy-making perspective, on the one hand, the diversification benefits of green markets against conventional market instruments, such as oil and gold, prior to or during stressful periods, are of high importance for the attainment of financial stability and strengthening the functionality of those markets. Furthermore, the divergent effects in both magnitude and direction over various time frames indicate that policymakers ought to devise policies that are coherent with the maturity of the market and account for unanticipated events. On the other hand, it is quite evident that efficient diversification in portfolio formation is heavily influenced by asset type, market circumstances, and portfolio construction strategy. Investors are strongly encouraged to not only evaluate the cost of including an asset in the portfolio but also its contribution to reducing the risk of the portfolio. Furthermore, while allocating funds, it is critical to consider the sensitivity of assets to heightened uncertainty, given portfolio weight, hedging ratio, and degree of effectiveness all change significantly. The results propose that, with the exception of clean energy assets, using assets with similar characteristics to mitigate market risk makes more sense and is less expensive for bivariate portfolios. Conducting portfolio analysis during stressful periods with the application of quantile-based models, cross quantilogram approach, etc. might be an option for future studies.

References

Abakah, E. J. A., Tiwari, A. K., Sharma, A., & Mwamtambulo, D. J. (2022). Extreme connectedness between green bonds, government bonds, corporate bonds and other asset classes: Insights for Portfolio investors. *Journal of Risk and Financial Management, 15*(10), 10. https://doi.org/10.3390/jrfm15100477

Agoraki, M.-E. K., Aslanidis, N., & Kouretas, G. P. (2023). How has COVID-19 affected the performance of green investment funds? *Journal of International Money and Finance, 131*, 102792. https://doi.org/10.1016/j.jimonfin.2022.102792

Ahmad, W., Sadorsky, P., & Sharma, A. (2018). Optimal hedge ratios for clean energy equities. *Economic Modelling, 72*, 278–295.

Antonakakis, N., Cunado, J., Filis, G., Gabauer, D., & de Gracia, F. P. (2020). Oil and asset classes implied volatilities: Investment strategies and hedging effectiveness. *Energy Economics, 91*, 104762.

Arif, M., Hasan, M., Alawi, S. M., & Naeem, M. A. (2021). COVID-19 and time-frequency connectedness between green and conventional financial markets. *Global Finance Journal, 49*, 100650. https://doi.org/10.1016/j.gfj.2021.100650

Arif, M., Naeem, M. A., Farid, S., Nepal, R., & Jamasb, T. (2022). Diversifier or more? Hedge and safe haven properties of green bonds during COVID-19. *Energy Policy, 168*, 113102. https://doi.org/10.1016/j.enpol.2022.113102

Bai, L., Wei, Y., Zhang, J., Wang, Y., & Lucey, B. M. (2023). Diversification effects of China's carbon neutral bond on renewable energy stock markets: A minimum connectedness portfolio approach. *Energy Economics, 123*, 106727.

Baur, D. G., & Lucey, B. M. (2010). Is gold a hedge or a safe haven? An analysis of stocks, bonds and gold. *Financial Review, 45*(2), 217–229.

Będowska-Sójka, B., Demir, E., & Zaremba, A. (2022). Hedging geopolitical risks with different asset classes: A focus on the Russian Invasion of Ukraine. *Finance Research Letters, 50*, 103192. https://doi.org/10.1016/j.frl.2022.103192

Cepni, O., Demirer, R., & Rognone, L. (2022). Hedging climate risks with green assets. *Economics Letters, 212*, 110312.

Chang, C.-L., McAleer, M., & Tansuchat, R. (2011). Crude oil hedging strategies using dynamic multivariate GARCH. *Energy Economics, 33*(5), 912–923. https://doi.org/10.1016/j.eneco.2011.01.009

Christoffersen, P., Errunza, V., Jacobs, K., & Jin, X. (2014). Correlation dynamics and international diversification benefits. *International Journal of Forecasting, 30*(3), 807–824.

Crowley, P. M. (2007). A guide to wavelets for economists. *Journal of Economic Surveys, 21*(2), 207–267.

Cui, T., Suleman, M. T., & Zhang, H. (2022). Do the green bonds overreact to the COVID-19 pandemic? *Finance Research Letters, 49*, 103095. https://doi.org/10.1016/j.frl.2022.103095

Deng, J., Guan, S., Zheng, H., Xing, X., & Liu, C. (2022). Dynamic spillovers and asymmetric connectedness between fossil energy and green financial markets: Evidence from China. *Frontiers in Energy Research, 10*. https://www.frontiersin.org/articles/10.3389/fenrg.2022.986341

Dickey, D. A., & Fuller, W. A. (1979). Distribution of the estimators for autoregressive time series with a unit root. *Journal of the American Statistical Association, 74*, 427–431.

Dong, X., Xiong, Y., Nie, S., & Yoon, S.-M. (2023). Can bonds hedge stock market risks? Green bonds vs conventional bonds. *Finance Research Letters, 52*, 103367. https://doi.org/10.1016/j.frl.2022.103367

Duchêne, S. (2020). Review of handbook of green finance. *Ecological Economics, 177*, 106766. https://doi.org/10.1016/j.ecolecon.2020.106766

Dutta, A., Jana, R. K., & Das, D. (2020). Do green investments react to oil price shocks? Implications for sustainable development. *Journal of Cleaner Production, 266*, 121956. https://doi.org/10.1016/j.jclepro.2020.121956

Ederington, L. H. (1979). The hedging performance of the new futures markets. *The Journal of Finance, 34*(1), 157–170.

Elsayed, A. H., Naifar, N., Nasreen, S., & Tiwari, A. K. (2022). Dependence structure and dynamic connectedness between green bonds and financial markets: Fresh insights from time-frequency analysis before and during COVID-19 pandemic. *Energy Economics, 107*, 105842. https://doi.org/10.1016/j.eneco.2022.105842

Ferrer, R., Benítez, R., & Bolós, V. J. (2021). Interdependence between green financial instruments and major conventional assets: A wavelet-based network analysis. *Mathematics, 9*(8), 8. https://doi.org/10.3390/math9080900

Gök, R., & Çankal, E. (2020). Analyzing time-frequency nexus between stock returns and bond yields through wavelets: the case of Turkey-Part I. *Journal of Financial Researches & Studies, 12*(23), 45–494.

Gök, R., Bouri, E., & Gemici, E. (2023). Volatility spillovers between sovereign CDS and futures markets in various volatility states: Evidence from an emerging economy around the pandemic. *Research in International Business and Finance, 66*, 102023.

Goodell, J. W., Corbet, S., Yadav, M. P., Kumar, S., Sharma, S., & Malik, K. (2022). Time and frequency connectedness of green equity indices: Uncovering a socially important link to Bitcoin. *International Review of Financial Analysis, 84*, 102379.

Guo, D., & Zhou, P. (2021). Green bonds as hedging assets before and after COVID: A comparative study between the US and China. *Energy Economics, 104*, 105696. https://doi.org/10.1016/j.eneco.2021.105696

Hammoudeh, S., Ajmi, A. N., & Mokni, K. (2020). Relationship between green bonds and financial and environmental variables: A novel time-varying causality. *Energy Economics, 92*, 104941. https://doi.org/10.1016/j.eneco.2020.104941

Huang, J., Cao, Y., & Zhong, P. (2022). Searching for a safe haven to crude oil: Green bond or precious metals? *Finance Research Letters, 50*, 103303. https://doi.org/10.1016/j.frl.2022.103303

Huang, Y., Duan, K., & Urquhart, A. (2023). Time-varying dependence between Bitcoin and green financial assets: A comparison between pre- and post-COVID-19 periods. *Journal of International Financial Markets, Institutions and Money, 82*, 101687. https://doi.org/10.1016/j.intfin.2022.101687

Jin, J., Han, L., Wu, L., & Zeng, H. (2020). The hedging effect of green bonds on carbon market risk. *International Review of Financial Analysis, 71*, 101509.

Kamal, J. B., & Hassan, M. K. (2022). Asymmetric connectedness between cryptocurrency environment attention index and green assets. *The Journal of Economic Asymmetries, 25*, e00240. https://doi.org/10.1016/j.jeca.2022.e00240

Karim, S., & Naeem, M. A. (2022). Do global factors drive the interconnectedness among green, Islamic and conventional financial markets? *International Journal of Managerial Finance, 18*(4), 639–660. https://doi.org/10.1108/IJMF-09-2021-0407

Khamis, M., & Aassouli, D. (2023). The eligibility of green bonds as safe haven assets: A systematic review. *Sustainability, 15*(8), 8. https://doi.org/10.3390/su15086841

Kroner, K. F., & Sultan, J. (1993). Time-varying distributions and dynamic hedging with foreign currency futures. *Journal of Financial and Quantitative Analysis, 28*(04), 535–551.

Kroner, K., & Ng, V. (1998). Modeling asymmetric movement of asset prices. *Review of Financial Studies, 11*(4), 844–871.

Kumar, S., Jain, R., Narain, Balli, F., & Billah, M. (2023). Interconnectivity and investment strategies among commodity prices, cryptocurrencies, and G-20 capital markets: A comparative analysis during COVID-19 and Russian-Ukraine war. *International Review of Economics & Finance, 88*, 547–593. https://doi.org/10.1016/j.iref.2023.06.039

Maillard, S., Roncalli, T., & Teïletche, J. (2010). The properties of equally weighted risk contribution portfolios. *The Journal of Portfolio Management, 36*(4), 60–70.

Markovitz, H. M. (1959). *Portfolio selection: Efficient diversification of investments*. Wiley.

Naeem, M. A., Farid, S., Ferrer, R., & Shahzad, S. J. H. (2021). Comparative efficiency of green and conventional bonds pre- and during COVID-19: An asymmetric multifractal detrended fluctuation analysis. *Energy Policy, 153*, 112285. https://doi.org/10.1016/j.enpol.2021.112285

Naeem, M. A., Hasan, M., Arif, M., Suleman, M. T., & Kang, S. H. (2022). Oil and gold as a hedge and safe-haven for metals and agricultural commodities with portfolio implications. *Energy Economics, 105*, 105758.

Nguyen, T. T. H., Naeem, M. A., Balli, F., Balli, H. O., & Vo, X. V. (2021). Time-frequency comovement among green bonds, stocks, commodities, clean energy, and conventional bonds. *Finance Research Letters, 40*, 101739.

Percival, D. P., & Walden, A. T. (2000). *Wavelet methods for time series analysis (Volume 4)*. Cambridge University Press.

Pham, L., & Nguyen, C. P. (2021). Asymmetric tail dependence between green bonds and other asset classes. *Global Finance Journal, 50*, 100669.

Phillips, P. C., & Perron, P. (1988). Testing for a unit root in time series regression. *Biometrika, 75*(2), 335–346.

Raheem, I. D., Akinkugbe, O., Yusuf, A. H., & Asl, M. G. (2023). Hedging strategies among financial markets: the case of green and brown assets. *Empirical Economics*, 1–43.

Rehman, M. U., Zeitun, R., Vo, X. V., Ahmad, N., & Al-Faryan, M. A. S. (2023). Green bonds' connectedness with hedging and conditional diversification performance. *Journal of International Financial Markets, Institutions and Money, 86*, 101802. https://doi.org/10.1016/j.intfin.2023.101802

Reboredo, J. C., Ugolini, A., & Aiube, F. A. L. (2020). Network connectedness of green bonds and asset classes. *Energy Economics, 86*, 104629. https://doi.org/10.1016/j.eneco.2019.104629

Reboredo, J. C., & Ugolini, A. (2020). Price connectedness between green bond and financial markets. *Economic Modelling, 88*, 25–38. https://doi.org/10.1016/j.econmod.2019.09.004

Ren, B., & Lucey, B. (2022). A clean, green haven?—Examining the relationship between clean energy, clean and dirty cryptocurrencies. *Energy Economics, 109*, 105951. https://doi.org/10.1016/j.eneco.2022.105951

Rizvi, S. K. A., Naqvi, B., Mirza, N., & Umar, M. (2022). Safe haven properties of green, Islamic, and crypto assets and investor's proclivity towards treasury and gold. *Energy Economics, 115*, 106396. https://doi.org/10.1016/j.eneco.2022.106396

Saeed, T., Bouri, E., & Tran, D. K. (2020). Hedging strategies of green assets against dirty energy assets. *Energies, 13*(12), 12. https://doi.org/10.3390/en13123141

Sharpe, W. F. (1994). The Sharpe ratio. *Journal of Portfolio Management, 21*(1), 49–58.

Su, T., Zhang, Z., & (Justin), & Lin, B. (2022). Green bonds and conventional financial markets in China: A tale of three transmission modes. *Energy Economics, 113*, 106200. https://doi.org/10.1016/j.eneco.2022.106200

Tiwari, A. K., Abakah, E. J. A., Yaya, O. S., & Appiah, K. O. (2023). Tail risk dependence, co-movement and predictability between green bond and green stocks. *Applied Economics, 55*(2), 201–222. https://doi.org/10.1080/00036846.2022.2085869

Xia, Y., Shi, Z., Du, X., Niu, M., & Cai, R. (2023). Can green assets hedge against economic policy uncertainty? Evidence from China with portfolio implications. *Finance Research Letters, 55*, 103874. https://doi.org/10.1016/j.frl.2023.103874

Xiang, S., & Cao, Y. (2023). Green finance and natural resources commodities prices: Evidence from COVID-19 period. *Resources Policy, 80*, 103200. https://doi.org/10.1016/j.resourpol.2022.103200

Yahya, M., Oglend, A., & Dahl, R. E. (2019). Temporal and spectral dependence between crude oil and agricultural commodities: A wavelet-based copula approach. *Energy Economics, 80*, 277–296.

Yousaf, I., Suleman, M. T., & Demirer, R. (2022). Green investments: A luxury good or a financial necessity? *Energy Economics, 105*, 105745. https://doi.org/10.1016/j.eneco.2021.105745

Does Foreign Green Finance Inspire Environmental, Social, and Governance Sustainability Metrics? Evidence from MENA Countries

Faris Alshubiri

Abstract The chapter aims to investigate the relationship between foreign green finance and environmental, social, and governance sustainability in MENA countries over the period from 2011 to 2021. The chapter used diagnostic tests as tests for selecting appropriate estimators. The chapter used Feasible Generalized Least Squares (FGLS), instrumental variables (2SLS) regression, and instrumental variables (IV-GMM) regression for more robustness. The main conclusion shows that there is a significant negative relationship between foreign green finance and environmental, sustainability. Meanwhile, there is a significant positive relationship between foreign green finance and governance and social sustainability in the MENA countries. The main recommendations of this chapter are focused on the importance of formalized strategies and policies to improve transparency, accountability, capacity building, and community involvement in long-term planning to increase foreign green finance and promote positive environmental, social, and governance sustainability.

1 Introduction

In recent years, interest has increased in the need for coordination between countries in finding radical solutions to economic and social problems such as global warming and climate liberalization, to achieve sustainable development (Zubair Mumtaz & Alexander Smith, 2019; Shafique & Majeed, 2020). Achieving this requires a diagnosis of economic indicators at the local and international levels and interest in the green financing agenda to increase opportunities to enhance strengths, mitigate weaknesses, and avoid challenges (Drago & Gatto, 2022). Therefore, countries have formulated strategies to overcome the challenges of sustainability by

F. Alshubiri (✉)
Department of Finance and Economics, College of Commerce and Business Administration, Dhofar University, Salalah, Oman
e-mail: falshubiri@du.edu.om

N. Naifar (ed.), *Climate Change and Finance*, Sustainable Finance,
https://doi.org/10.1007/978-3-031-56419-2_3

identifying the right investment projects within the framework of structured financing in the public and private sectors.

Achieving the required successes from countries within this context requires governments to prepare effectively within the long-term strategic plans and allocate a budget to implement projects and achieve green investment Dikau and Volz (2018). Therefore, governments have encouraged green projects in various sectors such as water, sanitation, education, energy, and others. Interest in infrastructure development has increased in developing Asian countries, and World Bank reports indicate that countries need $26 trillion from 2016 to 2030 if the region wants to maintain economic growth, reduce poverty, and respond to rapid climate changes (World Bank, 2023).

In-depth, green finance plays a vital role in environmental, social, and governance (ESG) sustainability as green finance refers to financial products, services, and investment activities that support sustainable projects and initiatives. However, green finance involves mobilizing financial resources toward activities that contribute to the transition to a low-carbon, climate-resilient, and economic environment friendly. Through integrating sustainability pillars into financial decisions, green finance aims to align economic activities with long-run environmental goals and enhance the transition to an increase in the level of a sustainable and resilient economy. The main objective of green finance is to encourage investment projects that address climate change, environmental degradation, and other sustainability challenges. Therefore, green finance redirects capital flows from activities that harm the environment toward projects that promote renewable energy, in an efficient manner leading to sustainable transportation, waste management, and green infrastructure sectors. Green finance encompasses a range of financial instruments including green finance such as green bonds and loans, green investment funds, and other innovative tools. However, these financial tools promote investors, and governments to consider the opportunities to participate in sustainable development activities and generate a positive environmental and social impact.

Due to the many environmental, economic, and social impacts on many MENA countries, the governments increase the importance to work in investing in green products by using innovation models and sustainable financing to mitigate any environmentally damaging forms of economic growth (Khan et al., 2022).

Yin et al. (2019) and Drago and Gatto (2022) indicated that environmental sustainability needs attractive policies and regulations. However, Zubair Mumtaz and Alexander Smith (2019), Shafique and Majeed (2020), and Malik et al. (2018) focused on how the depth of green finance can be changed between provinces under the same level of business. The development of a green finance mechanism solves the challenges faced by industrial stakeholders and financial institutions by including an awareness of the financial sector and effective tools for environmental-based credit risk assessment within financing procedures (Shafique & Majeed, 2020). Furthermore, renewable resources enhance SMEs' capacity and support the shortage of budget to prepare appropriate strategies for effective management resources (Zubair Mumtaz & Alexander Smith, 2019). Dikau and Volz (2018) indicated that central banks should address the financial governance policies to support the growth of green finance models and determine the pricing of environmental products and related risks by financial institutions. Therefore, Dikau and Volz (2018) support the importance of macroeconomic stability in filling the green financing gap and managing climate issues and environmental risks on a systemic country level.

The financial policies used can affect renewable energy and its investments and reflect positively on sustainable development, by increasing tax revenues for local governments in the light of financing private sector projects (Yoshino & Taghizadeh-Hesary, 2018). Furthermore, Internet sales are heavily used to finance vehicles which helps risky sectors to grow (Yoshino et al., 2019). Finally, new financial technologies (FinTech) can develop many economic sectors which reduce operational costs. However, the following question remains: Does foreign green finance inspire environmental, social, and governance sustainability metrics of MENA Countries?

Through explaining the MENA countries' future trends and challenges, this chapter aims to (a) determine a gap analysis between foreign green finance and ESG matrix; (b) formalizing guidelines to manage green finance and their effect on ESG and (c) examine the statistical relationship between foreign green finance and ESG for MENA countries. For this purpose, the chapter aims to contribute to the analysis of the theories related to green finance and ESG to strengthen the debate. The MENA countries are rich in natural resources, particularly oil and gas depending on fossil fuels, and transferring toward green finance and sustainable practices plays a vital role in economic diversification and geopolitical challenges, to ensure long-term economic sustainability, regional stability, mitigating the effects of climate change, significant potential for renewable energy and enhancement of market competitiveness between green finance and ESG sustainability of the MENA region.

The main motivations for this study are to explain the environmental challenges such as water Scarcity to make sustainable water management crucial. MENA region is prone to desertification, and sustainable practices of climate change. Green finance and sustainable practices can help diversify MENA economies, investments in renewable energy projects supporting the transition to a low-carbon economy, and seek foreign green finance to support these efforts and demonstrate commitment to responsible business conduct. Implementing sustainable practices and attracting foreign green finance can help mitigate climate-related risks and environmental degradation. Foreign green finance aligns with evolving regulations and avoids potential trade barriers.

The remaining sections of the chapter are as follows: Sect. 2 presents the literature review, the methodology, the main model, and the variables presented in Sect. 3. Section 4 presented results and discussion. Section 5 presented the conclusion and policy implications.

2 Literature Review

Most of the developing economies have significant impacts on the sustainable environment compared to the developed economies. Therefore, the deployment of green financing and green infrastructure projects is difficult in developing countries, and they still have problems that limit growth, especially the large costs associated with green growth. A combination of China's existing green financial product composition, green finance assessment according to five criteria, and the use of both expert rating and capital size across financial sectors was used. SMEs play a huge role in job creation and are the main contributors to the country's GDP, growth, and exports (Loewe et al., 2013). Small and medium enterprises are active in developing entrepreneurial potential and maintaining sustainable industrial growth. Emerging economies have a higher impact than developed economies in evaluating green finance (Khan et al., 2022). Indian industries explain the offset of environmental degradation which leads to dioxide absorption, this facilitates environmentally sustainable financing (Soundarrajan & Vivek, 2016). Infrastructure projects, green economy, and green financing decisions; have been developed in Bangladesh m in recent years but not at an up level due to limiting growth causes obstacles to green products and the substantial transaction costs associated with green projects (Zheng et al., 2021). Khairunnessa et al. (2021) evaluate current green financial products based on expert rating, capital volume across financial sectors, and eco-friendly economic literacy, green business, and eco-friendly culture (Wang et al., 2021). Sassanelli and Terzi (2022) and (Loewe et al., 2013) indicated that many small and medium enterprises (SMEs) sector consider green finance as an integral component of each economy by improving human economic and cultural quality, this leads to employment creation which allows for an increase in a country's gross domestic production, growth, and exports.

Flaherty et al. (2017) and Sinha et al. (2021) found a positive impact of green finance on sustainable development, but sometimes sustainable development hinders the development of the public. The countries try to manage industrial strategy, fiscal taxation, environmental protection, and other designs to direct green financing effectively and promote sustainable development (Shen et al., 2022). Some characteristics limit the relationship between green finance and sustainable development (Su et al., 2021). Yoshino et al. (2017) found that raising energy self-sufficiency promotes sustainable development and energy security by diversification of energy resources. The limited resources of energy (coal, oil, or gas) lead to a reduction in the resiliency of the economy and price fluctuations. Several studies (Hamilton, 1983; Barsky & Kilian, 2004; Taghizadeh-Hesary & Yoshino, 2016) found that increases in oil prices lead to economic growth and create inflation for most oil-importing countries. Taghizadeh-Hesary et al. (2017) Japan needs to diversify its energy supplies to increase energy self-dependency and energy security. Meanwhile, Lucas et al. (2016) focused on the diversification of energy sources as a coherent strategy that reduces emissions and promotes sustainable development.

There is a strong relationship between environmental sustainability and green (Hall et al., 2018; Ji et al., 2021). This situation needs renewable energy trade for long-term environmental sustainability and to minimize climate change (Sun et al., 2020; Taghizadeh-Hesary et al., 2021; Ferrat et al., 2022). Renewable alternatives give a better understanding of the driving forces behind CO_2 emissions (Murshed et al., 2021). Pakrooh et al. (2020) and Saravia-Matus et al. (2019) indicated a performance ratio such as greenhouse gas emissions is needed to analyze the relationship between agricultural production and greenhouse gas emissions. Nasim et al. (2022) explore the effectiveness of environment quality and trade along with human capital. Wang et al. (2019) indicated that OECD economies have implemented a green economic growth strategy by updating environmental rules and regulations. As well as reduced CO_2 emissions (Ferrat et al., 2022).

Based on a review of the literature, which investigates the relationship between green finance and environmental sustainability, highlighting the importance of public–private partnership investment for renewable and efficient energy projects and economic growth, as a result, various factors not previously considered in foreign green finance, and effect on environmental, social, and governance sustainability on MENA countries.

This chapter is based on the following hypotheses:

H1: Foreign green finance has a significant positive impact on the environmental sustainability of MENA countries.
H2: Foreign green finance has a significant positive impact on the social sustainability of MENA countries.
H3: Foreign green finance has a significant positive impact on the governance sustainability of MENA countries.

3 Methodology, Main Model, and Variables

The current study aimed to investigate whether foreign green finance can inspire environmental, social, and governance sustainability in MENA countries. The study population included 13 MENA countries and covered the period of 2011–2021. The study excluded 6 countries in MENA due to unavailable data as this is considered the limitation of the study. The dataset included 143 observations for the studied MENA countries.

The panel data will be used to test the main model of the chapter which is categorized into three dependent variables of ESG, i.e., environmental proxy measured by CO_2 emissions (kg per 2017 PPP \$ of GDP); (CO2E), social proxy measured by GNI per capita growth (annual %) (GNIPCG) and governance measured by corruption perception index (CORR). The independent variable of foreign green finance will be measured by the value of announced greenfield FDI projects (LOGVGFFDIP). Four control variables will use trade openness, calculated as the log of exports plus imports of commodities (% of GDP); (LOGTO), adjusted net national income per capita; (LOGANNIPC), inflation, calculated as the log of the consumer price index (% of annual) (INFCP); GDP growth rate (annual %) (GDPG). Furthermore, the chapter will use diagnostic tests to check the appropriate models and variables for more robustness.

The main justification for selecting the studied MENA countries due to their geographic importance, wealth of natural resources, environmental challenges, goals of economic diversification, and the broader impact of their sustainability efforts on global and regional dynamics. Studying these countries can provide critical roles for the potential of foreign green finance to manage positive change in diverse contexts. The panel data for the main model variables were collected from World Development Indicators (World Bank, 2022).

The chapter used descriptive statistics to explain the characteristics of the variables of the study. Then, a correlation matrix is used to show the relationship between variables and levels of significance. Some diagnostic tests were used such as variance inflation factor (VIF) to test the collinearity, Breusch-Pagan/Cook-Weisberg test for heteroscedasticity, and Wooldridge test for autocorrelation. The main estimator was Feasible Generalized Least Squares (FGLS) as this estimator deals with violations of the assumptions underlying ordinary least squares (OLS), such as heteroscedasticity and autocorrelation (Hansen, 2007; Bickel & Levina, 2008). This estimator provides the results with a more efficient, robust, and unbiased estimation of regression parameters making these issues exit. To overcome the endogeneity issues which was in panel data, the study used instrumental variables (2SLS) regression and instrumental variables (IV-GMM) regression for more robustness. The 2SLS addresses issues related to endogeneity, causality, and measurement error (Benda & Corwyn, 1997). Meanwhile, IV-GMM extends the capabilities of traditional 2SLS in addressing endogeneity. It is important and valuable with dynamic panel data, especially if it has weak instruments, nonlinear models, and nonstationary data (Staiger & Stock, 1997; Hansen et al., 1996).

4 Results and Discussions

The results of the descriptive statistics, correlation matrix, and diagnostic test were used to select the appropriate estimators. Therefore, the study used FGLS, 2SLS, and IV-GMM estimators to overcome the heterogeneity and endogeneity issues for the panel data.

4.1 Descriptive Statistics

Table 1 presents the analysis of 8 variables and 143 observations of 13 MENA countries throughout 2011–2021. The preliminary descriptive statistics revealed the standard deviation (SD), mean, median, skewness, kurtosis, and Jarque-Bera values.

Table 1 Descriptive statistics

	LOGVGFFDIP	CORR	CO2E	GNIPCG	INFCP	LOGTO	LOGANNIPC	GDPG
Mean	2.726	1.508	0.337	0.272	4.538	1.899	3.908	2.307
Median	2.668	1.612	0.329	0.415	2.647	1.914	3.669	2.699
Maximum	4.459	1.949	0.536	16.67	39.90	2.276	4.654	13.93
Minimum	0.778	0.255	0.142	−13.79	−2.317	1.480	3.351	−11.32
Std. Dev.	0.879	0.343	0.101	4.123	7.125	0.186	0.425	3.854
Skewness	−0.094	−2.032	0.196	0.432	2.973	0.106	0.366	−0.662
Kurtosis	2.349	6.511	1.775	5.915	12.43	2.470	1.508	5.262
Jarque-Bera	2.736	171.9	9.855	55.10	741.25	1.938	16.45	40.98
Probability	0.254	0.000	0.007	0.000	0.000	0.379	0.000	0.000
Sum	389.8	215.7	48.32	38.99	649.05	271.58	558.90	329.97
Sum Sq. Dev.	109.7	16.70	1.452	2414.3	7210.01	4.961	25.76	2110.15
Observations	143	143	143	143	143	143	143	143

Source: The data collected from the World Bank (2022)

The mean and median LOGANNIPC values were the highest, and the GNIPCG and CO2E values were low. SD represents the average distance of the variable from the mean. INFCP had the highest SD, and CO2E had the lowest SD. Skewness in the study reflects two issues between right (positive) and left (negative). The distribution of tails was close to the mean, reflecting positive kurtosis values as a leptokurtic distribution. The coefficient value of the Jarque-Bera test statistic was significant for all variables of the model of the MENA countries, except VGFFDIP and LOGTO indicating that the variables followed the normal distribution.

4.2 Correlation Matrix

Table 2 shows the correlation matrix coefficients for all 8 study variables as well as the significance levels. The results were consistent with the main model estimators of FGLS, 2SLS, and IV-GMM. The coefficient of values of most variables was less than 40%; this ratio represents the strength of the coefficients and indicates no multicollinearity between predictor and response.

4.3 Diagnostics Tests

The study used diagnostic tests to show the robustness of the models and appropriate estimators. The study used the variance inflation factor for all variables and findings show in Table 3 that VIF is less than 3 explaining that no multicollinearity between two and more independent variables in a regression model. Then Table 4 shows that, the study explains the heteroscedasticity by using Breusch-Pagan/Cook-Weisberg test for heteroscedasticity test and the findings show that the probability for model 1 above 5%, models 2 and 3 were less than 5%, which explains that no heteroscedasticity issues in model 1 while models 2 and 3, there were heteroscedasticity issues.

Additionally, the Wooldridge test for autocorrelation was used and the findings show there is a serial correlation in all models because the significant level was less than 5%.

4.4 FGLS, 2SLA, and IV-GMM Dynamic Panel Data Estimation

The FGLS test was run as the a pre estimator, for regression and presented in Table 6. The FGLS results are presented in Table 5. The FGLS, estimator in model 1 (CO2E) revealed a significant negative relationship between LOGVGFFDIP and

Table 2 Correlation coefficient matrix

	MENA countries							
Correlation-Probability	LOGVGFFDIP	CORR	CO2E	GNIPCG	INFCP	LOGTO	LOGANNIPC	GDPG
VGFFDIP	1.000							
Prob. level	—							
CORR	0.346	1.000						
Prob. level	0.000	—						
CO2E	−0.064	−0.124	1.000					
Prob. level	0.446	0.138	—					
GNIPCG	0.098	−0.065	−0.145	1.000				
Prob. level	0.244	0.438	0.082	—				
INFCP	−0.123	−0.218	0.281	0.055	1.000			
Prob. level	0.141	0.008	0.000	0.512	—			
LOGTO	0.205	0.159	0.124	−0.017	−0.461	1.000		
Prob. level	0.014	0.057	0.137	0.840	0.000	—		
LOGANNIPC	0.646	0.303	0.173	−0.024	−0.315	0.473	1.000	
Prob. level	0.000	0.000	0.038	0.773	0.000	0.000	—	
GDPG	0.061	−0.186	−0.085	0.685	−0.025	−0.013	−0.002	1.000
Prob. level	0.464	0.026	0.311	0.000	0.761	0.874	0.977	—

Note: The table shows the coefficient and significant values based on data (World Bank 2022)

Table 3 Collinearity diagnostics

Variable	VIF	SQRT-VIF	R-Tolerance	Squared
LOGVGFFDIP	2.02	1.42	0.495	0.504
CO2E	1.39	1.18	0.717	0.282
CORR	1.27	1.13	0.788	0.211
GNIPCG	1.99	1.41	0.501	0.498
INFCP	1.66	1.29	0.603	0.396
LOGTO	1.58	1.26	0.631	0.369
LOGANNIPC	2.50	1.58	0.400	0.599
GDPG	2.01	1.42	0.496	0.503

Note: Mean VIF 1.80

Table 4 Diagnostic tests for estimators

	Panel least squares		
	Model (1)	Model (2)	Model (3)
Estimators	CO2E	GNIPCG	CORR
Prob. Breusch-Pagan/Cook-Weisberg test for heteroskedasticity	0.318	0.000	0.000
Prob. Wooldridge test for autocorrelation	0.008	0.000	0.000

Note: The significance levels refer to $p < 0.01$ (***), $p < 0.05$ (**) and $p < 0.1$ (*)

Table 5 Feasible generalized least squares

	Feasible generalized least squares		
	Model-1	Model-2	Model-3
Variables	CO2E	GNIPCG	CORR
LOGVGFFDIP	−0.037***	0.559	0.117***
	(0.011)	(0.375)	(0.038)
INFCP	0.006***	0.046	−0.008**
	(0.001)	(0.039)	(0.004)
LOGTO	0.108**	1.12	−0.025
	(0.047)	(1.63)	(0.169)
LOGANNIPC	0.104***	−0.956	0.048
	(0.025)	(0.865)	(0.089)
GDPG	−0.001	0.727***	0.018***
	(0.001)	(0.064)	(0.006)
Constant	0.200***	1.54	1.12***
	(0.098)	(3.36)	(0.349)
Observations	143	143	143
Number of ID	13	13	13
Wald $chi^2(5)$	50.11	134.11	35.22
Prob > chi^2	0.000	0.000	0.000
Panels	Homoskedastic	Homoskedastic	Homoskedastic
Correlation	No autocorrelation	No autocorrelation	No autocorrelation

Note: The significance levels refer to $p < 0.01$ (***), $p < 0.05$ (**), and $p < 0.1$ (*)

Table 6 Instrumental variables (2SLS) regression and Instrumental variables (IV-GMM) regression

	Instrumental variables (2SLS) regression			Instrumental variables (GMM) regression		
	Model-1	Model-2	Model-3	Model-1	Model-2	Model-3
Variables	CO2E	GNIPCG	CORR	CO2E	GNIPCG	CORR
LOGVGFFDIP	−0.041***	0.814	0.084***	−0.041***	0.814*	0.084***
	(0.012)	(0.500)	(0.018)	(0.009)	(0.476)	(0.022)
INFCP	0.008***	−0.003	−0.004*	0.008***	−0.003	−0.004**
	(0.001)	(0.062)	(0.002)	(0.001)	(0.052)	(0.001)
LOGTO	0.131**	0.614	0.141*	0.131**	0.614	0.141**
	(0.055)	(2.21)	(0.081)	(0.062)	(2.32)	(0.068)
LOGANNIPC	0.109***	−1.94	0.069	0.109***	−1.94	0.069*
	(0.029)	(1.20)	(0.043)	(0.025)	(1.20)	(0.035)
GDPG	−0.001	0.228	−0.012	−0.001	0.228	−0.102
	(0.009)	(0.366)	(0.013)	(0.008)	(0.305)	(0.016)
Constant	−0.259**	3.95	0.874***	−0.259**	3.95	0.874***
	(0.131)	(5.29)	(0.193)	(0.114)	(4.31)	(0.138)
Wald $chi^2(5)$	23.47	6.08	108.74	97.19	6.20	202.01
Prob > chi^2	(0.000)	(0.298)	(0.000)	(0.000)	(0.287)	(0.000)
R-squared	0.247	0.259	0.441	0.247	0.259	0.441
GMM C statistic chi2(4)	—	—	—	5.44(0.244)	9.10(0.058)	3.23(0.519)
Tests of endogeneity						
Durbin (score) $chi^2(4)$	5.56(0.234)	16.87(0.002)	3.53(0.472)	—	—	—
Wu-Hausman F (4,120)	1.34(0.258)	4.47(0.002)	0.838(0.503)	—	—	—

Note: The significance levels refer to $p < 0.01$ (***), $p < 0.05$ (**), and $p < 0.1$ (*). Model 1, Model 2, and Model 3: Instrumented: logannipc gdpg logto infcp, Instruments: logvgffdip L.logannipc L.gdpg L.logto L.infcp

CO2E, indicating that foreign green finance had a significant negative effect on environmental, sustainability in the MENA countries. Specifically, a 1% increase in LOGVGFFDIP led to −0.037 decreases in CO2E. Meanwhile, model 3 (CORR) revealed a significant positive relationship between LOGVGFFDIP and CORR, indicating that foreign green finance had a significant positive effect on governance, and sustainability in the MENA countries. Specifically, a 1% increase in LOGVGFFDIP led to a 0.117 increase in CORR. Finally, there is no significant relationship between LOGVGFFDIP and GNIPCG as a measure of social sustainability in Model 2. The control variables result shows that there is a positive significant relationship between INFCP, LOGTO, and LOGANNIPC on CO2E for model 1. Meanwhile, the control variable for GDPG in model 2 shows a positive significant relationship between GDPG and GNIPCG. Finally, the control variables for model 3 show that there is a negative significant relationship between INFCP and CORR. The diagnostic test results show that the panel has no autocorrelation and is homoscedastic.

The 2SLS test was run as the advance estimator. The 2SLS results are presented in Table 6. The 2SLS, estimator in model 1 (CO2E) revealed a significant negative relationship between LOGVGFFDIP and CO2E, indicating that foreign green finance had a significant negative effect on environmental sustainability in the MENA countries. Specifically, a 1% increase in LOGVGFFDIP led to −0.041 decreases in CO2E. These results are confirmed by IV-GMM. Meanwhile, model 3 (CORR) revealed a significant positive relationship between LOGVGFFDIP and CORR, indicating that foreign green finance had a significant positive effect on governance sustainability in the MENA countries. Specifically, a 1% increase in LOGVGFFDIP led to 0.084 increases in CORR. These results are confirmed by IV-GMM. Finally, there is no significant relationship between LOGVGFFDIP and GNIPCG as a measure of social sustainability for model 2 in the 2SLS estimator while there is a significant positive relationship between LOGVGFFDIP and GNIPCG, indicating that foreign green finance had a significant positive effect on social sustainability in the MENA countries in the IV-GMM estimator. Specifically, a 1% increase in LOGVGFFDIP led to 0.814 increases in CORR.

The control variables consistent in both 2SLS and IV-GMM estimators results show that there is a positive significant relationship between INFCP, LOGTO, and LOGANNIPC on CO2E for model 1. Meanwhile, the control variables for model 3, show that there is a negative significant relationship between INFCP and CORR, and a significant positive relationship between LOGTO and LOGANNIPC. Furthermore, for robustness, all these results are consistent with FGLS results. The diagnostic test results show that the tests of endogeneity of both Durbin (score) and Wu-Hausman findings present a significant level of more than 5%, explaining that models 1 and 3 are appropriate and fit.

5 Discussion of Results

The findings of results are consistent in three estimators FGLS, 2SLS, and IV-GMM, and showed that model 1 (CO2E) revealed a significant negative relationship between LOGVGFFDIP and CO2E, indicating that foreign green finance had a significant negative effect on environmental sustainability in the MENA countries. This result can be interpreted as a lack of laws or inadequate regulations in MENA countries which leads to unsustainable practices. MENA countries may not meet the necessary environmental standards, creating a negative effect on environmental sustainability. The diligence when selecting projects to finance is weak, this leads to harm to projects and the environment. They prioritized financial returns in the short run over long-term environmental sustainability. MENA countries could depend heavily on foreign capital, which discourages local governments from investing in sustainable practices. Finally, economic and political instability in some MENA countries affected environmental sustainability. These results are inconsistent with Hall et al. (2018), Zheng et al. (2021), Ji et al. (2021) which indicated that environmental sustainability can be green. According to Sun et al. (2020), Taghizadeh-Hesary et al. (2021), Ferrat et al. (2022) that environmental sustainability can minimize climate change. Nasim et al. (2022) explore the effectiveness of environment quality and trade along with human capital. Updating environmental rules and regulations can green finance and economics Wang et al. (2019), then reduce CO_2 emissions (Ferrat et al., 2022) or diversification of energy sources (Lucas et al., 2016). In contrast, some studies consistent with the current study by Flaherty et al. (2017), Su et al. (2021), and Sinha et al. (2021) found a positive impact of green finance on sustainable development, but sometimes sustainable development hinders the development of the public.

Meanwhile, model 3 (CORR) revealed a significant positive relationship between LOGVGFFDIP and CORR, indicating that foreign green finance had a significant positive effect on governance sustainability in the MENA countries. This result can be interpreted as the governments have effective and accountable systems over the long term. This result is because of increased transparency and accountability; the attraction of foreign green finance is due to strong regulations which leads to strong governance structures. Mena countries have such training and development programs to invest in capacity building, which facilitates managing sustainable projects effectively. Sharing the capital and experiences between Public–Private sectors through joint decision-making leads to highly effective project implementation. Furthermore, adapt the Sustainable Development Goals (SDGs) by including reducing inequality, justice, and promoting good governance. Finally, the Safeguards of social and environmental projects lead to better governance in long-term planning as well as an increased international collaboration with an enhanced a good reputation to address global environmental challenges. These results are consistent with Khan et al. (2022), Wang et al. (2021), Khairunnessa et al. (2021) indicated that maintaining sustainable industrial growth can evaluate green finance. Khairunnessa et al. (2021) evaluate current green financial products based on expert rating, capital volume across financial sectors, eco-friendly economic literacy, green business, and eco-friendly culture. Meanwhile, Shen et al. (2022) focused on industrial strategy, fiscal taxation, and environmental protection in green sustainable sectors.

Finally, there is no significant relationship between LOGVGFFDIP and GNIPCG as a measure of social sustainability for model 2 in the FGLS and 2SLS estimator while there is a significant positive relationship between LOGVGFFDIP and GNIPCG, indicating that foreign green finance had a significant positive effect on social sustainability in the MENA countries in the IV-GMM estimator. This result can be interpreted because fair, improving well-being leads to a positive effect on foreign green finance on social sustainability. More deeply, because of green projects, jobs will be created, which will reduce unemployment and improve economic well-being, and then social sustainability will increase. Furthermore, overall quality regarding poverty alleviation presents access to clean energy, quality of education, and a strong healthcare sector. These issues will improve community development and promote green projects, which lead to better social cohesion. Developing the health sector, education skills, technology and access to services, inequality, and culture, all these facts improved social sustainability. Furthermore, resilience to climate change leads to improved foreign green finance and projects by building infrastructure and enhancing social sustainability. Finally, collaboration with international institutions can facilitate knowledge transfer, which; leads to improving local communities to solve sustainability challenges independently. These results are consistent with Loewe et al. (2013) explained that huge opportunities for jobs are because of GDP, growth, and exports of SMEs which promote sustainable products. Improving human economic and cultural quality leads to employment creation which allows for GDP increase and green economies (Sassanelli & Terzi, 2022).

6 Conclusion and Policy Implications

The chapter aims to investigate the relationship between foreign green finance and environmental, social, and governance sustainability in MENA countries over the period from 2011 to 2021. The findings used FGLS, 2SLS, and IV-GMM estimators for more robustness. The main first conclusion shows that there is a significant negative relationship between foreign green finance and environmental, sustainability. Meanwhile, there is a significant positive relationship between foreign green finance and governance and social sustainability in the MENA countries. The policy implications highlighted that once there confirmed that foreign green finance has a significant negative effect on environmental sustainability, this requires to reassessment of policies and strategies to mitigate these issues, such as strength regulations, and implementing restrictive control of finance projects, to attract investors in a green and friendly environment. Available transparency to promote environmental credentials of green finance investments. Select an appropriate allocation of foreign green finance and assess the environmental sustainability. Community engagement ensures that local environmental concerns are addressed by applying environmental safeguards, to protect natural resources, safety ecosystems, and biodiversity in financed projects. The MENA countries should promote long-term sustainability planning and facilitate technology transfer to prioritize environmental protection and diversification of green finance sources. Finally, the MENA countries should re-evaluate investment priorities to reassess the allocation of foreign green.

The second main concluded that foreign green finance has a positive significant effect on governance sustainability, which should MENA countries strengthen governance practices. In-depth, development of institutional capacity, strengthening transparency and accountability, and improved regulatory issues, lead to anticorruption. Furthermore, green finance enhances, community participation and engagement, and enhances local ownership and governance. Finally, developing robust data information systems will lead to foreign green finance on governance sustainability and promote long-term planning to ensure that the advantages of foreign green finance are sustained over time.

Finally, the last conclusion shows that foreign green finance has a significant positive effect on social sustainability, which encourages policymakers to maximize these benefits to enhance the well-being of societies. This result can encourage policymakers to distribute equally the benefits of foreign green finance among different social groups, by planning, implementing, and monitoring green finance projects. Sustainable initiatives can be developed through education skills to participate in the green economy. Providing healthcare and well-being to enhance a good population with less poverty, this issue should be monitored, and green projects evaluated continuously. Finally, cultural preservation and resilience building are required to promote social sustainability.

The main recommendations of this chapter are focused on the importance of transparency, accountability, capacity building, and community involvement. These facts should be built into long-term planning to ensure that foreign green finance promotes positive environmental, social, and governance sustainability. Additionally, such strategies should be formalized for each MENA country to overcome challenges. Future studies can explain the factors of effects on sustainability such as corruption, and politics to understand how foreign green finance behavior is realized.

References

Bickel, P. J., & Levina, E. (2008). Regularized estimation of large covariance matrices. *The Annals of Statistics, 36*(1), 199–227.

Benda, B. B., & Corwyn, R. F. (1997). A test of a model with reciprocal effects between religiosity and various forms of delinquency using 2-stage least squares regression. *Journal of Social Service Research, 22*(3), 27–52.

Barsky, R. B., & Kilian, L. (2004). Oil and the macroeconomy since the 1970s. *Journal of Economic Perspectives, 18*(4), 115–134.

Dikau, S., & Volz, U. (2018). *Central banking, climate change, and green finance*. ADBI working paper 867. Asian Development Bank Institute, Tokyo.

Drago, C., & Gatto, A. (2022). Policy, regulation effectiveness, and sustainability in the energy sector: A worldwide interval-based composite indicator. *Energy Policy, 167*, 112889.

Flaherty, M., Gevorkyan, A., Radpour, S., & Semmler, W. (2017). Financing climate policies through climate bonds – A three-stage model and empirics. *Research in International Business and Finance, 42*, 468–479.

Ferrat, Y., Daty, F., & Burlacu, R. (2022). Short- and long-term effects of responsible investment growth on equity returns. *Journal of Risk Finance, 23*(1), 1–13.

Hamilton, J. D. (1983). Oil and the macroeconomy since World War II. *Journal of Political Economy, 91*(2), 228–248.

Hall, S., Roelich, K. E., Davis, M. E., & Holstenkamp, L. (2018). Finance and justice in low-carbon energy transitions. *Applied Energy, 222*, 772–780.

Hansen, C. B. (2007). Generalized least squares inference in panel and multilevel models with serial correlation and fixed effects. *Journal of Econometrics, 140*(2), 670–694.

Hansen, L., Heaton, J., & Yaron, A. (1996). Finite sample properties of some alternative GMM estimators. *Journal of Business and Economic Statistics, 14*(3), 262–280.

Ji, X., Chen, X., Mirza, N., & Umar, M. (2021). Sustainable energy goals and investment premium: Evidence from renewable and conventional equity mutual funds in the Eurozone. *Resources Policy, 74*, 102387.

Khan, M. A., Riaz, H., Ahmed, M., & Saeed, A. (2022). Does green finance deliver what is expected? An empirical perspective. *Borsa İstanbul Review, 22*, 586–593.

Khairunnessa, F., Vazquez-Brust, D. A., & Yakovleva, N. A. (2021). Review of the recent developments of green banking in Bangladesh. *Sustainability, 13*, 1904.

Loewe, M., Al-Ayouty, I., Altpeter, A., Borbein, L., Chantelauze, M., Kern, M., Niendorf, E., & Reda, M. (2013). Which factors determine the upgrading of Small and Medium-Sized Enterprises (SMEs)? In *The Case of Egypt; Co-Operation with the Egyptian Center for Economic Studies (ECES)*: Cairo, Egypt.

Lucas, J. N. V., Francés, G. E., & González, E. S. M. (2016). Energy security and renewable energy deployment in the EU: Liaisons dangereuses or virtuous circle? *Renewable and Sustainable Energy Reviews, 62*(C), 1032–1046.

Murshed, M., Elheddad, M., Ahmed, R., Bassim, M., & Then, E. T. (2021). Foreign direct investments, renewable electricity output, and ecological footprints: Do financial globalization facilitate renewable energy transition and environmental welfare in Bangladesh? *Asia-Pacific Financial Markets, 29*, 33–78.

Malik, S., Qasim, M., & Saeed, H. (2018). *ADBI Working Paper Series Green Finance in Pakistan: Barriers and Solutions Asian Development Bank Institute*. Available online: https://www.adb.org/publications/green-finance-pakistan-barriers-and-solutions (accessed on 14 July 2022).

Nasim, I., Ahmad, R., Bashir, F., & Noreen, S. (2022). Trade openness as a determinant of sectoral growth in Pakistan: A time series analysis. *Review of Education, Administration & Law, 5*(2), 73–83.

Pakrooh, P., Hayati, B., Pishbahar, E., Nematian, J., & Brännlund, E. R. (2020). Focus on the provincial inequalities in energy consumption and CO_2 emissions of Iran's agriculture sector. *Science of the Total Environment, 715*, 137029.

Shafique, O., & Majeed, A. (2020). Factors influencing bankers' intention to adopt green finance in Pakistan. *Review of Economics and Development Studies, 6*, 773–785.

Staiger, D., & Stock, J. H. (1997). Instrumental variables regression with weak instruments. *Econometrica, 65*(3), 557–586.

Soundarrajan, P., & Vivek, N. (2016). Green finance for sustainable green economic growth in India. *Journal of Agricultural Economics, 62*, 35–44.

Sassanelli, C., & Terzi, S. (2022). The D-BEST reference model: A flexible and sustainable support for the digital transformation of small and medium enterprises. *Global Journal of Flexible Systems Management, 23*, 345–370.

Sinha, A., Mishra, S., Sharif, A., & Yarovaya., L. (2021). Does green financing help to improve environmental & social responsibility? Designing SDG framework through advanced quantile modeling. *Journal of Environmental Management, 292*(15), 112751.
Shen, J. H., Long, Z., Lee, C. C., & Zhang, J. (2022). Comparative advantage, endowment structure, and trade imbalances. *Structural Change and Economic Dynamics, 60*, 365–375.
Su, C. W., Khan, K., Umar, M., & Zhang, W. (2021). Does renewable energy redefine geopolitical risks? *Energy Policy, 158*, 112566.
Sun, H., Mohsin, M., Alharthi, M., & Abbas, Q. (2020). Measuring environmental sustainability performance of South Asia. *Journal of Cleaner Production, 251*, 119519.
Saravia-Matus, S. L., Hörmann, P. A., & Berdegué, J. A. (2019). Environmental efficiency in the agricultural sector of Latin America and the Caribbean 1990–2015: Are greenhouse gas emissions reducing while agricultural production is increasing? *Ecological Indicators, 102*, 338–348.
Taghizadeh-Hesary, F., Rasoulinezhad, E., Yoshino, N., Sarker, T., & Mirza, N. (2021). Determinants of the Russia and Asia–Pacific energy trade. *Energy Strategy Review, 38*, 100681.
Taghizadeh-Hesary, F., & Yoshino, N. (2016). Monetary policy, oil prices and the real macroeconomic variables: An empirical survey on China, Japan, and the United States. *China: An International Journal, 14*(4), 46–69.
Taghizadeh-Hesary, F., Yoshino, N., & Rasoulinezhad, E. (2017). Impact of the Fukushima nuclear disaster on the oil-consuming sectors of Japan. *Journal of Comparative Asian Development, 16*(2), 113–134.
Wang, X., Zhao, H., & Bi, K. (2021). The measurement of green finance index and the development forecast of green finance in China. *Environmental and Ecological Statistics, 28*, 263–285.
Wang, Y., Sun, X., & Guo, X. (2019). Environmental regulation and green productivity growth: Empirical evidence on the Porter Hypothesis from OECD industrial sectors. *Energy Policy, 132*, 611–619.
Yoshino, N., Taghizadeh-Hesary, F., & Tawk, N. (2017). Decline of oil prices and the negative interest rate policy in Japan. *Economic and Political Studies, 5*(2), 233–250.
Yoshino, N., & Taghizadeh-Hesary, F. (2018). Alternatives to private finance: Role of fiscal policy reforms and energy taxation in development of renewable energy projects. In V. Anbumozhi, K. Kalirajan, & F. Kimura (Eds.), *Financing for low-carbon energy transition: Unlocking the potential of private capital*. Springer.
Yin, W., Kirkulak-Uludag, B., & Zhang, S. (2019). Is financial development in China green? Evidence from city-level data. *Journal of Cleaner Production, 211*, 247–256.
Yoshino, N., Taghizadeh-Hesary, F., & Nakahigashi, M. (2019). Modelling the social funding and spillover tax for addressing the green energy financing gap. *Economic Modelling, 77*, 34–41.
Zubair Mumtaz, M., & Alexander Smith, Z. (2019). Green finance for sustainable development in Pakistan. *IPRI Journal, 19*, 1–34.
Zheng, G. W., Siddik, A. B., Masukujjaman, M., Fatema, N., & Alam, S. S. (2021). Green finance development in Bangladesh: The role of Private Commercial Banks (PCBs). *Sustainability, 13*, 795.

Investor Attitudes and Preferences Toward Climate Change and Sustainable Investments: A Bibliometric Analysis

Juan Infante and Carlos Estevez-Mendoza

Abstract This chapter explores the evolving academic perspective on the nexus between climate change, finance, and responsible investment. Within the global academic landscape, there is an intensified emphasis on the intricate relationship between business practices and the urgent issue of climate change. Specifically, the finance domain is at the forefront, delving into the financial implications of environmentally conscious decisions and how they shape the modern investment arena. As the discourse progresses, studies have underscored the growing significance of climate-related risks for institutional investors, suggesting that contemporary financial strategies are increasingly attuned to climate considerations. There is also a marked rise in publications illustrating the link between climate concerns and investment behaviors, emphasizing the synthesis of ethical considerations with financial strategies. Echoing broader societal shifts, the academic sphere presents climate change as an immediate concern, necessitating a proactive response from the influential investment community. This chapter employs bibliometric techniques to examine the research literature evolution methodically. It identifies the most relevant authors, the theoretical framework they built to support their contributions, and the themes they developed around the topic. This way, we provide objective grounds for further studies and identify potential gaps that could be addressed.

This chapter was carried out within the framework of the activities of the "Impact Investments" Research Group at Universidad Villanueva.

J. Infante (✉)
Universidad Villanueva, Madrid, Spain
e-mail: juan.infante@villanueva.edu

C. Estevez-Mendoza
Complutense University of Madrid; CUNEF University, Madrid, Spain

N. Naifar (ed.), *Climate Change and Finance*, Sustainable Finance,
https://doi.org/10.1007/978-3-031-56419-2_4

1 Introduction

The global academic landscape has seen a dramatic shift in recent years, emphasizing the critical nature of climate change and its intertwining relationship with business practices. Within the broad spectrum of research, there is a burgeoning interest in the crossroads of finance, investment, and socially responsible behaviors, particularly as they relate to climate change.

Leading the charge in this dialogue is the field of finance, where researchers are increasingly examining the financial implications of ethical and environmentally conscious decisions. Renneboog et al. (2008) stand as a testament to this, offering an in-depth exploration into the performance of socially responsible mutual funds. Their findings suggest that environmental considerations, among other ethical concerns, play a pivotal role in shaping the modern investment landscape.

Another pivotal piece comes from Krueger et al. (2020). In their seminal study, they underscored the growing importance of climate-related risks for institutional investors. Their findings reflect a paradigm shift, suggesting that contemporary financial strategies are more than ever attuned to the potential risks and rewards associated with climate-conscious decisions.

Further underpinning the importance of this discourse is the surge in publications addressing the intricate link between climate considerations and investment behaviors. Hartzmark and Sussman (2019) research provides a compelling perspective on this, elucidating how investor reactions to sustainability rankings translate into tangible shifts in fund flows. Their work, among others, serves as evidence of the increasing confluence between sound financial management and a proactive approach to climate change.

The findings from a bibliometric review of recent literature are revealing. We can observe a notable increase in publications in this field. Moreover, the evolving themes suggest a comprehensive and integrated strategy toward addressing climate change in investment. Ethical considerations, which were once regarded as less important, are now at the forefront, driven by both moral responsibility and today's financial demands.

It is evident that the academic sphere is echoing the broader societal shifts. Climate change is no longer a distant concern but a pressing reality that demands a response. The investment community, with its substantial influence, is increasingly playing a proactive role, underpinned by a synthesis of ethics and strategy. This blend of social responsibility and financial acumen is likely to shape the future trajectory of both academia and the global economy. Then, it is convenient to delve into this topic based on the study of the scientific production already in place, setting an objective ground on the relevance of the authors, topics, and the extant intellectual framework. This, in turn, will enlighten potential gaps for future research. To accomplish this aim, we utilize bibliometric techniques (Groos & Pritchard, 1969; McCain, 1990) and the following sections of this chapter address the methodology, the results, discussion of the obtained results, and, finally, the conclusions and a prospect of avenues for future research.

2 Methodology

Bibliometric techniques utilize mathematical and statistical tools (McCain, 1990) to analyze data from published works. These methods conform to an objective base to study the structure of the research literature, allowing to establish relevant grounds to justify and drive a proper discussion on current and new approaches (Durisin & Puzone, 2009). Several disciplines and research fields have been analyzed using them. It is the case of management (Tahai & Meyer, 1999), finance (Alexander & Mabry, 1994; Merigó et al., 2015), strategic management (Nerur et al., 2008; Ramos-Rodríguez & Ruíz-Navarro, 2004), operations management (Pilkington & Liston-Heyes, 1999; Pilkington & Meredith, 2009), organizational behavior (Culnan et al., 1990), human resources management (Fernandez-Alles & Ramos-Rodríguez, 2009), or innovation (Fagerberg et al., 2012). Besides, they are also employed in the analysis of topics such as family business (Casillas & Acedo, 2007), entrepreneurship (Etemad & Lee, 2003; Landström et al., 2012; Schildt et al., 2006), human resources training (Danvila-del-Valle et al., 2019), or diversity (Estevez-Mendoza, 2022). Related fields like corporate social responsibility (De Bakker et al., 2005) and socially responsible investments have been addressed (Estevez-Mendoza, 2023).

In this chapter, we will utilize impact indicators and co-occurrence analysis for the case of the documents, authors, references, and the topics they refer. Thus, relevant contributions, similarities among topics, and intellectual frameworks are more easily identified (Culnan, 1987; Pilkington & Liston-Heyes, 1999; Ramos-Rodríguez & Ruíz-Navarro, 2004). We use the visualization of mapped networks with bibliometrix R package and VOSviewer to help understand and interpret the results of the analysis (Aria & Cuccurullo, 2017; van Eck et al., 2010).

2.1 Methodology Steps

Bibliometric analyses require to follow procedural steps that include data retrieving and cleaning, descriptive and relational analysis, and visualization (Zupic & Cater, 2015). Relevant databases and appropriate search terms are key for data retrieval. Since inconsistencies and wrong documents may appear, cleaning follows to secure the contributions that are adequate for the study.

In the phase of analysis, descriptive and relational bibliometric indicators are established. Thus, the frequency of publications helps to understand the evolution of the research and to set different stages if that is the case. Main authors are defined by the number of citations and impact indicators and depict those topics the Academia is giving more importance. The study of the keywords utilized for the categorization of the documents identifies related topics and, consequently, gaps for future research.

To identify their evolution and behavior, we created co-occurrence networks and applied clustering algorithms (Cobo et al., 2011, 2012), helped by two analytical elements: the strategic diagram and the thematic network (Bailón-Moreno et al., 2005; Callon et al., 1991; Cobo et al., 2011; López-Herrera et al., 2009). Using density and centrality measures, they allow classification into motor, basic, highly developed/isolated, and emerging/declining themes. Each theme has an associated network of related keywords (see Fig. 1). Since we split the analysis into two stages, the evolution of topics is visualized, and the process is also summarized computing the stability between periods (see Fig. 2).

We analyze these themes alongside the works of pertinent authors and their references. These references provide the foundational intellectual context for the research. Co-occurrence maps can assist in identifying and examining them.

The sample of documents was obtained from the Web of Science Main Collection Database. The applied criteria of selection and the cleaning steps brought 378 documents from 146 different journals. There were 956 authors and they accounted for 18,943 references in their documents.

2.2 *Evolution of Research Works*

Research evolves over the 30-year period and shows a growing pattern, especially noticeable in the last decade (Fig. 3). There is a seeding stage which is characterized by a low number of highly cited publications. Then, it appears a sharp increase in the last 7 years, yielding two identifiable stages that might be used in further steps of the analysis.

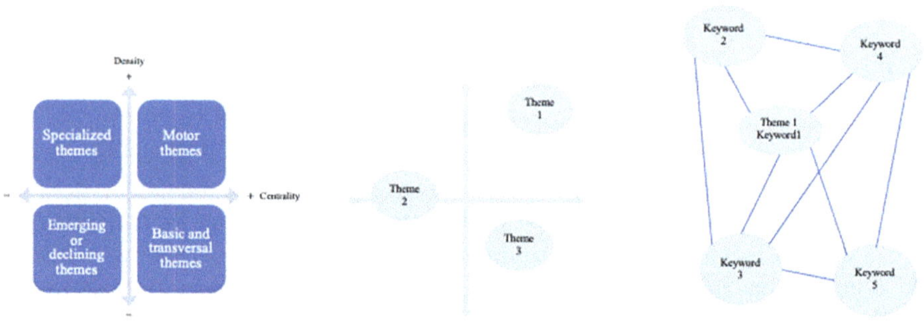

(a) Strategic Diagram (b) Themes placed in the strategic diagram (c) Thematic network of the themes placed in the strategic diagram

Fig. 1 Strategic diagram and thematic network layouts examples. Source: Adapted from Cobo et al. (2012))

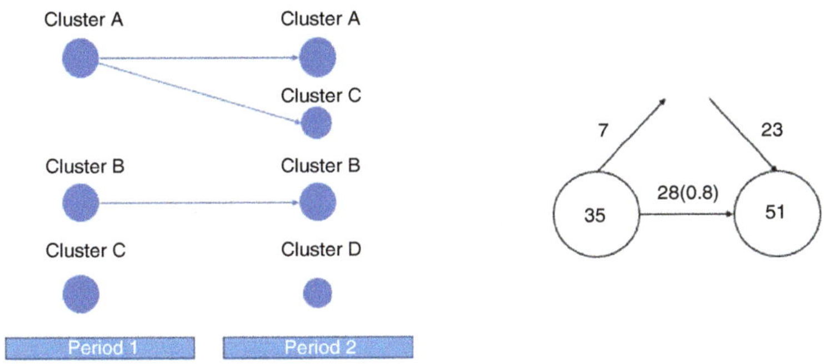

(a) Evolution of areas

(b) Stability between periods

Fig. 2 Evolution of areas and stability between periods. Examples. Source: Adapted from Cobo et al. (2012))

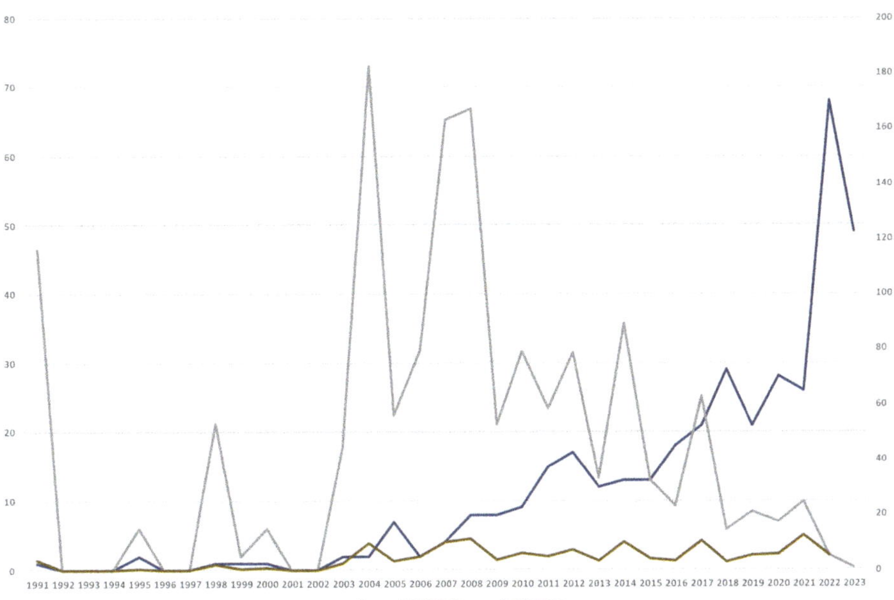

Fig. 3 Evolution of published works

2.3 *Impact and Relevance of Journals*

We ranked the journals behind the documents of our sample according to h-index and the total number of citations (Table 1). A journal dedicated to ethics, the *Journal of Business Ethics*, heads both classifications and shows the increasing interest inside business ethics toward the topic of our analysis. Journals in the field of finance such as the *Journal of Banking and Finance*, the *Journal of Financial Economics*,

Table 1 Impact and relevance of journals by indices and citations

Journal	H-Index	G-Index	M-Index	Journal	TC	NP
Journal of Business Ethics	19	32	0.90	Journal of Business Ethics	1997	32
Energy Policy	18	27	0.62	Energy Policy	1332	27
Ecological Economics	11	21	0.38	Journal of Banking & Finance	847	7
Business Strategy and the Environment	8	16	0.62	Ecological Economics	749	21
Journal of Banking & Finance	7	7	0.37	Journal of Economic Perspectives	708	1
Energy Economics	6	6	0.50	Academy of Management Review	567	1
European Journal of Operational Research	6	6	0.50	Journal of Financial Economics	412	3
Technological Forecasting and Social Change	6	10	0.40	Journal of Risk and Uncertainty	317	2
Business Ethics-A European Review	5	5	0.33	Journal of Finance	278	1
Corporate Social Responsibility and Environmental Management	5	5	0.33	Business Strategy and the Environment	273	16
Environmental & Resource Economics	5	6	0.26	Accounting Auditing & Accountability Journal	254	3
Business & Society	4	4	0.44	Strategic Management Journal	254	1
Journal of Economic Behavior & Organization	4	6	0.44	Resource and Energy Economics	252	4
Resource and Energy Economics	4	4	0.21	Business Ethics-A European Review	212	5
Sustainability Accounting Management and Policy Journal	4	5	0.44	European Journal of Operational Research	186	6
Transportation Research Part B-Methodological	4	4	0.67	World Development	182	4
Accounting Auditing & Accountability Journal	3	3	0.30	Technological Forecasting and Social Change	169	10
Australian Journal of Management	3	3	0.25	Energy Economics	166	6
Economic Analysis and Policy	3	3	0.75	American Economic Review	153	1
Economics Letters	3	3	0.60	Environmental & Resource Economics	151	6
Energy Journal	3	4	0.23	Organization Studies	148	1

(continued)

Table 1 (continued)

Journal	H-Index	G-Index	M-Index	Journal	TC	NP
Journal of Behavioral and Experimental Finance	3	3	0.75	Australian Journal of Management	144	3
Journal of Environmental Economics and Management	3	5	0.25	European Journal of Finance	144	1
Journal of Financial Economics	3	3	0.50	Journal of Environmental Economics and Management	126	5
World Development	3	4	0.27	Transportation Research Part A-Policy and Practice	122	2
Business Ethics Quarterly	2	2	0.11	Review of Financial Studies	120	2
California Management Review	2	2	0.08	Energy Journal	116	4
Corporate Governance-An International Review	2	4	0.12	Journal of Consumer Affairs	116	1
Economic Systems Research	2	2	0.29	Journal of Economic Behavior & Organization	96	6
Emerging Markets Review	2	2	0.40	Tourism Management	92	2
Finance Research Letters	2	3	0.67	Omega-International Journal of Management Science	89	1
International Review of Economics & Finance	2	3	0.20	Transportation Research Part B-Methodological	88	4
Journal of Corporate Finance	2	2	0.40	Corporate Social Responsibility and Environmental Management	84	5

or the *Journal of Finance* also have an important role. But the contributions are also boosted by journals in the field of management (*Academy of Management Review, Strategic Management Journal*) and others that are more explicitly devoted to sustainability (*Business Strategy and the Environment, Technological Forecasting and Social Change, Corporate Social Responsibility and Environmental Management, Environmental & Resource Economics, Ecological Economics*). The recent surge in publications over the past decade can be largely attributed to this development, laying the groundwork for an expansion in research endeavors. This evolution not only underscores the value of specialized resources but also highlights the potential they have for fostering cross-disciplinary impacts.

2.4 Relevant Authors and Their Topics

After computing the h-index and the total citations received by authors in the field we ranked them and show a summary in Table 2.

The most relevant authors inside the research community address different elements. We summarize first those among the top-ranked authors according to the Hirsch index and then we complete the landscape with those receiving more citations. Bauer et al. (2021) emphasize the significance of social preferences in sustainable investing, where investors are willing to sacrifice financial returns for sustainability. The study also explores how pension funds should address their clients' social preferences but notes that investment managers often find it challenging and costly to elicit such preferences, resulting in their neglect within pension funds' decision-making processes. On the other hand, Bauer and Smeets (2015) focus on the investment decisions of clients at socially responsible banks, revealing that social identification predominantly influences their allocations.

Other investigations have contributed to portfolio selection and sustainable investment. Bilbao-Terol et al. (2013) present a portfolio selection approach that incorporates socially responsible financial assets, multi-objective programming, and a new SR measure. The empirical results indicate that socially responsible investments have a relatively small financial sacrifice for "cautious" investors, encouraging SR adoption without significantly impacting financial returns. In their subsequent work, Bilbao-Terol et al. (2014) introduce a methodology for evaluating the non-financial performance of government bond funds, with a focus on sustainability. The assessment process utilizes the multi-criteria TOPSIS methodology and an innovative additive version to model relative criteria importance. The resulting scores enable the incorporation of sustainability as an objective in multi-objective portfolio optimization. Besides, Bilbao-Terol et al. (2016) present a tool for sustainable and responsible investors using goal programming with fuzzy hierarchies (GPFH) modeling. GPFH addresses conflicting objectives by utilizing fuzzy relations to model relative importance among criteria. An application to UK-SR mutual funds is showcased.

Examining public attitudes toward sustainable initiatives, Dumbrell et al. (2022) surveyed residents from South Australia and Victoria to assess their preferences for a hydrogen industry. The study found that safety was rated as the most important characteristic, followed by climate change mitigation and affordability, influencing public support for the industry's development in Australia. Socio-demographic and attitudinal factors were found to significantly impact individuals' importance ratings. On the other hand, high-net-worth individuals prioritize real-world impact over value-aligned investments or sustainability for financial risk purposes (Paetzold et al., 2022). They primarily invest in securities aligned with Sustainable Development Goals (SDGs) to mobilize capital for sustainable development, but their strong preference for high financial returns may lead to underfunding of less attractive SDGs. Paetzold and Busch (2014) identify barriers to sustainable investing among private investors. Finally, Riedl and Smeets (2017) investigate investors'

Table 2 Relevant authors by indices and citations

Author	h-Index	g-Index	m-Index	TC	Author	TC	NP	Year
Arenas-Parra M	5	5	0.46	161	Bloom N	708	1	2014
Bilbao-Terol A	5	5	0.46	161	Renneboog L	707	1	2008
Canal-Fernandez V	5	5	0.46	160	Ter Horst J	707	1	2008
Busch T	3	3	0.30	88	Zhang Cd	707	1	2008
Di Falco S	3	3	0.27	98	Mackey A	573	2	2007
Gutsche G	3	4	0.60	56	Mackey Tb	573	2	2007
Smeets P	3	3	0.33	370	Barney Jb	567	1	2007
Adamson D	2	2	0.50	14	Ben-Amar W	379	1	2017
Bauer R	2	2	0.22	92	Chang M	379	1	2017
Belghitar Y	2	2	0.20	63	Mcilkenny P	379	1	2017
Bilbao-Terol C	2	2	0.18	43	Smeets P	370	3	2015
Boiral O	2	3	0.29	192	Millington A	295	2	2004
Botzen Wjw	2	2	0.13	259	Dasgupta P	289	1	2008
Brimble M	2	2	0.12	53	Riedl A	278	1	2017
Brotherton Mc	2	2	0.40	18	Menichetti E	261	1	2012
Brzeszczynski J	2	2	0.40	19	Wustenhagen R	261	1	2012
Chamorro-Mera A	2	2	0.40	8	Botzen Wjw	259	2	2009
Clark E	2	2	0.20	63	Delmas Ma	254	1	2010
Daigneault Aj	2	2	0.33	76	Montes-Sancho Mj	254	1	2010
Dam L	2	2	0.15	68	Aerts Jcjh	242	1	2009
Deshmukh N	2	2	0.20	63	Van Den Bergh Jcjm	242	1	2009
Diouf D	2	2	0.25	199	Flammer C	240	1	2021
Doskeland T	2	2	0.25	44	Brammer S	231	1	2004
Edenhofer O	2	2	0.29	15	Cox P	231	1	2004
Fankhauser S	2	2	0.11	203	Fankhauser S	203	2	2005
Fu Xw	2	2	0.50	17	Diouf D	199	2	2016
Gond Jp	2	2	0.17	159	Boiral O	192	3	2017
Hackel B	2	2	0.29	40	Hoepner Agf	169	2	2011
Hassan Mk	2	2	0.33	20	Tol Rsj	166	1	2005
Hepburn C	2	2	0.11	113	Pastor L	164	1	2021
Hoepner Agf	2	2	0.15	169	Stambaugh Rf	164	1	2021
Howarth Rb	2	2	0.10	71	Taylor La	164	1	2021
Hudson R	2	2	0.11	51	Arenas-Parra M	161	5	2013
Komendantova N	2	2	0.17	136	Bilbao-Terol A	161	5	2013
Mackey A	2	2	0.12	573	Canal-Fernandez V	160	5	2013
Mackey Tb	2	2	0.12	573	Gond Jp	159	2	2012
Mattauch L	2	2	0.33	9	Slager R	158	2	2012
Mendez-Rodriguez P	2	2	0.20	40	Nilsson J	157	2	2008
Millington A	2	2	0.10	295	Deschenes O	153	1	2017
Nilsson J	2	2	0.13	157	Greenstone M	153	1	2017
Paetzold F	2	2	0.20	46	Shapiro Js	153	1	2017

(continued)

Table 2 (continued)

Author	h-Index	g-Index	m-Index	TC	Author	TC	NP	Year
Palacios-Gonzalez Mm	2	2	0.40	8	Moon J	148	1	2012
Patt A	2	2	0.17	136	Hochstadter Ak	146	1	2015
Pedersen Ljt	2	2	0.25	44	Scheck B	146	1	2015
Perez-Gladish B	2	2	0.17	79	Rammal Hg	144	1	2011
Scholtens B	2	2	0.22	65	Rezec M	144	1	2011

motivations for choosing socially responsible mutual funds. They found that investors choose these funds due to social preferences and signaling rather than purely financial motives. The study reveals that investors expect lower returns and higher management fees, reflecting their willingness to prioritize social values over financial performance.

Research groups headed by Gutsche (Gutsche et al., 2021, 2023; Gutsche & Ziegler, 2019) reviewed individual sustainable investment behavior. Positive correlations were found between socially responsible investments (SRIs) and warm glow feelings, social norms in environmental organizations, and peer behaviors. In terms of countries, Japan's sustainable investment was in its early stage with information deficits, while in Germany, strong preferences for sustainable funds were driven by pecuniary and non-pecuniary factors, particularly warm glow feelings. These findings carry significant implications for promoting sustainable development through individual investors.

Scholarly explorations into climate change adaptation, juxtaposed with agricultural investment decisions, have been undertaken by a cadre of distinguished researchers. In their seminal work, Di Falco and Bulte (2013) elucidate that within the confines of rural Ethiopia, mandatory risk-sharing notably diminishes the impetus for weather shock self-shielding. Further, Di Falco and Sharma (2018) unveil that profound pro-environmental convictions directly augment adaptive propensities, with environmentally conscious incentives amplifying adaptive investments in the picturesque Fiji Islands. Complementing this, Di Falco et al. (2019) manifest how income fluctuations, steered by rainfall discrepancies, modulate farmers' subjective discount rates, subsequently influencing agricultural forays in rural Ethiopia. Collectively, these erudite studies proffer salient insights into the intricacies of climate change adaptation interwoven with agricultural decision-making paradigms. In a parallel vein, Rodriguez Lopez et al. (2017) determine that regulation-induced ambivalence molds corporate endeavors toward reducing environmental impact, even though generic regulatory uncertainties yield no palpable consequence. Mallawaarachchi et al. (2020), on the other hand, delve into the byzantine pro-farmer water stratagems of Australia's Murray-Darling Basin, advocating fervently for the imperative of water commerce and reformative strategies that champion efficacious adaptation through synergistic collaboration.

In the discussion of corporate sustainability and ethical investment, various scholars contribute multidimensional perspectives. One line of research, including work by Ben-Amar et al. (2017) and Cox et al. (2004), suggests that board gender diversity and long-term institutional investment are influential factors in Corporate Social Performance and sustainability disclosure. This reflects the broader themes of stakeholder management and governance structure.

Another aspect worth noting is the role of uncertainty. Bloom (2014) and Botzen et al. (2009) both address this, albeit from different angles. The former examines economic uncertainty across business cycles, while the latter considers how homeowners deal with climate-related uncertainties through insurance.

Policy considerations also come into play, with contributions from Dasgupta (2008) and Fankhauser (2017). While Dasgupta challenges the conventional approaches to social discount rates by incorporating ethical considerations, Fankhauser delves into the economic challenges of climate adaptation. Both suggest that policy plays a key role, but they approach the issue from unique methodological standpoints.

On the investment side, Flammer (2021) examines the impact of corporate green bonds on investor perception, a theme echoed in the work of Hoepner and Schopohl (2018), who explore ethical dimensions of various investment styles, including sector-based exclusions and Islamic mutual funds (Hoepner et al., 2011).

The intricate economic reverberations of climate change have been astutely dissected by pioneering intellects such as Fankhauser and Tol (2005). They eloquently delineate the multifaceted ramifications of climate-induced variations upon capital proliferation and savings inclinations, extending the scholarly discourse from immediate societal well-being to protracted economic horizons. In sum, such perspicacious inquiries cumulatively posit that governance structures, nuanced investor predispositions, policy contemplations, and the multifarious economic elements are paramount in decoding corporate oscillations toward sustainability and the looming climate conundrum.

Venturing into the sophisticated domain of Corporate Social Responsibility (CSR), a myriad of scholarly endeavors adopts variegated methodologies to fathom investor behaviors and their expansive ramifications. The erudition of Mackey et al. (2022) invokes the mathematical elegance of game theory to decipher the modalities by which investors can galvanize socially responsible comportment within corporate entities. This investigative thrust finds harmonious resonance with the earlier musings of Fankhauser and Tol (2005), who elucidated the kinetic interplay of investor behavior vis-à-vis climate-centric perils. Mackey and his consortium's keen insights into the dialectics of reward versus retribution as fulcrums for CSR stand as a noteworthy scholarly beacon.

Pursuing this intellectual trajectory, an incisive exploration by Mackey et al. (2007) grapples with the nebulous intersections of market valuations and CSR. Their scholarship finds a kindred spirit in the investigations of Wüstenhagen and Menichetti (2012), who illuminate the strategic topographies of renewable energy capital commitments. Such oeuvres accentuate the convoluted lattice of investment deliberations, tethered invariably to policy frameworks, market voracities, and the moral compass of social accountability. Mol et al. (2020), adopting a tangential perspective, delve into the psychological substrates motivating self-insurance against nature's caprices. Their discernments embellish our comprehension of fiscal stratagems amidst climatic vicissitudes, augmenting the dialogue initiated by Fankhauser and Tol (2005). Along analogous lines, Pástor et al. (2021) embark on an odyssey into the equilibria of sustainable investing, elucidating that verdant assets, albeit perhaps offering modest returns, serve as bulwarks against climate adversities. This thematic confluence resonates harmoniously with the critical appraisals of Renneboog et al. (2008) and is subsequently echoed in the reflections of Riedl and Smeets (2017)—all collectively interrogating the delicate equipoise between pecuniary augmentation and the loftier aspirations of societal beneficence in investment decision-making.

Finally, research by Delmas and Montes-Sancho (2010) scrutinizes the effectiveness of voluntary environmental agreements, raising important questions about the real impact of corporate commitments to sustainability. Wong and Millington (2014) then bring these discussions full circle, examining corporate social disclosures from a stakeholder viewpoint. Their findings shed light on both corporate transparency and the behavior of socially responsible investors.

Most of the mentioned authors made their contributions through quantitative methods, although there are also theoretical proposals (Bloom, 2014; Dasgupta, 2008; A. Mackey et al., 2007; Pástor et al., 2021) and reviews (Fankhauser, 2017; Renneboog et al., 2008). Among the different techniques, the use of surveys is very common (Bauer et al., 2021; Bauer & Smeets, 2015; Ben-Amar et al., 2017; Botzen et al., 2009; Di Falco & Sharma, 2018; Gutsche et al., 2021; Riedl & Smeets, 2017; Rodriguez Lopez et al., 2017; Wong & Millington, 2014), which might be easily understood since they address perceptions and responses. Regression models widely serve to test hypotheses (Bauer & Smeets, 2015; Cox et al., 2004; Di Falco et al., 2019; Gutsche et al., 2021, 2023; Riedl & Smeets, 2017) and probit models (Ben-Amar et al., 2017; Di Falco & Bulte, 2013; Fankhauser, 2017; Gutsche et al., 2019, 2021; Riedl & Smeets, 2017; Rodriguez Lopez et al., 2017) and logit models (Bauer et al., 2021; Gutsche et al., 2021; Gutsche & Ziegler, 2019; Wong & Millington, 2014) often appear among them. The need of testing models and finding causal relationships lead also to the use of framed experiments and path analyses (Di Falco & Sharma, 2018; Riedl & Smeets, 2017), difference-in-difference techniques (Flammer, 2021), or EVE-CVaR (Bilbao-Terol et al., 2013). Structured interviews are employed in qualitative studies (Paetzold & Busch, 2014) and even new methods are proposed (Bilbao-Terol et al., 2014).

2.5 Relevant Topics and Intellectual Framework

Regarding publications covered, we may observe significant changes between the periods of observation. The number of themes doubles and, consequently, it opens more nuances to the study of the topic. This situation correlates with the increase in the number of keywords. It is also noticeable that a small percentage of keywords remains in the second period, implying not only a growth in contributions and related topics, but also a relatively significant change in the focus of research in the second one (see Fig. 4).

As we mentioned, after realizing the differences between the two periods, we split the analysis of the topics and their intellectual framework into two different periods. The first covers publications until 2015 and the second is centered on those elaborated afterward.

2.5.1 Until 2015

In this period, there were just four themes in contributions. SRI and behavior are the motor themes, whereas those on climate change and WTP seem to progress toward being basic and niche themes, respectively (see Fig. 5 Strategic diagram until 2015.).

Analyzing the thematic networks, SRI theme is associated with the investments, their performance, and CSR, and linked to mutual funds. This gives importance to the firm's conduct and the financial vehicles in performing these kinds of investments. Behavior theme is associated with the investors' ethics and responsibility, which is clearly linked to the topic of this chapter. And climate change is associated with its perception, the uncertainty and its risk, and renewable energy (see Fig. 6).

Looking into the intellectual framework of the period, there are four main pillars based on the main co-occurrences of references (Fig. 7). First, the body of work on Socially Responsible Investing (SRI) covers diverse aspects, from its potential mainstream acceptance (Dunfee, 2003) to the psychological factors influencing investor choices (Glac, 2009; Kahneman & Tversky, 1979). Studies also examine

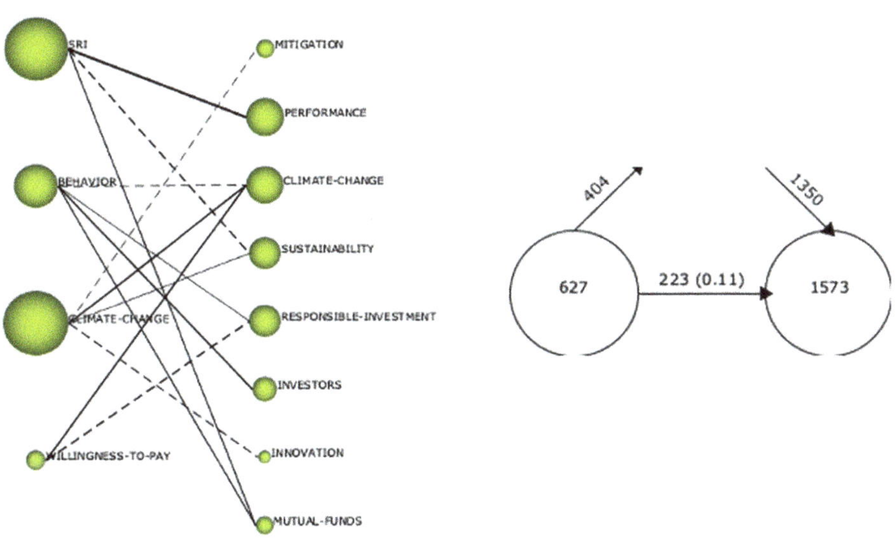

Fig. 4 Evolution of themes and stability between periods. Source: own elaboration with SCImat. Left picture shows the evolution of themes over periods. Right picture shows the number of keywords considered in each period, the number of them that remain in both periods, and the new ones

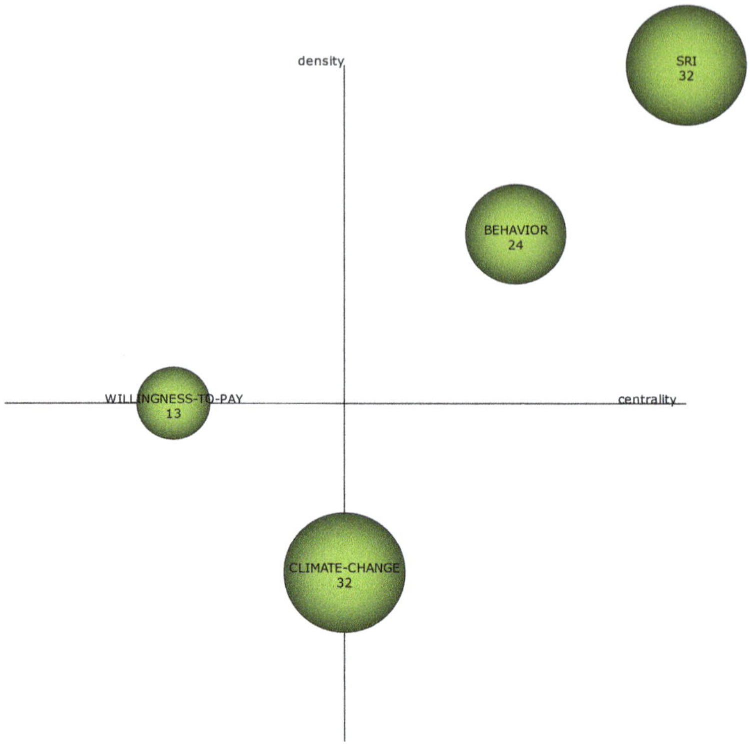

Fig. 5 Strategic diagram until 2015. Source: Own elaboration with SCImat. Theme nodes are placed according to their co-occurrence's centrality and density. Size of the nodes signifies the relative importance according to the h-index of each one

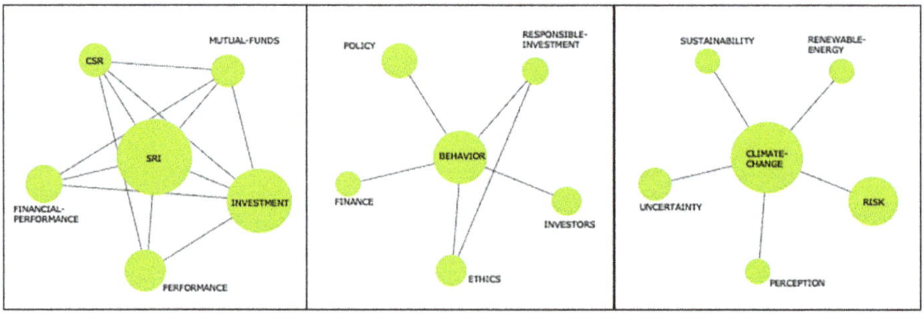

Fig. 6 Thematic networks until 2015. Source: Own elaboration with SCImat

the motivations and characteristics of ethical versus ordinary investors (Lewis, 2001; Rosen et al., 1991), as well as the complexity and diversity within the SRI landscape (Sandberg et al., 2009). Other research discusses the ethical foundations of SRI and its evolving relationship with corporate social responsibility (Sparkes,

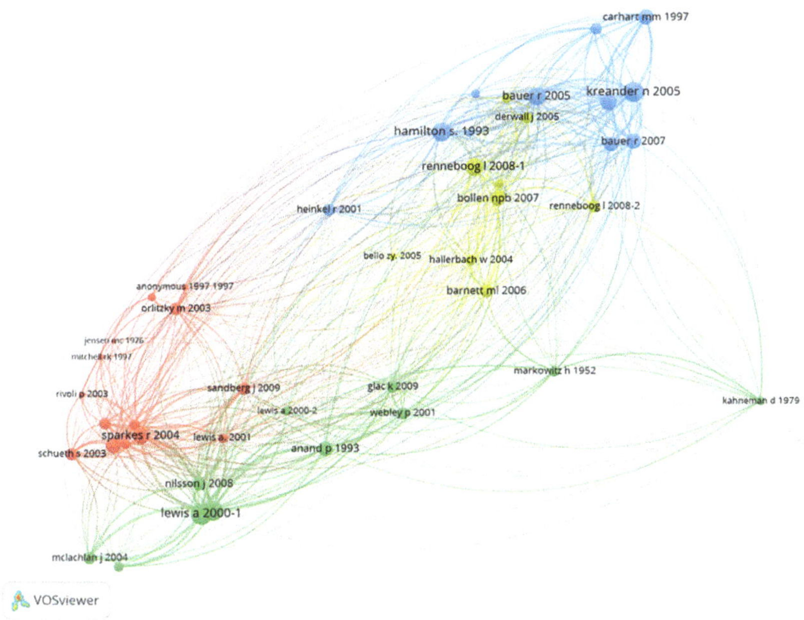

Fig. 7 Intellectual framework until 2015. Co-citation map. Source: Own elaboration with bibliometrix and VOSviewer

2001; Sparkes & Cowton, 2004). Finally, Markowitz (1952) utility function theory and Webley et al. (2001) experimental approach contribute to understanding the risk and commitment aspects of SRI. Together, these studies underscore SRI's growing importance and complexity.

The scholarly landscape on ethical funds and business ecosystems is both nuanced and interconnected. Barnett et al. (2006) set the stage by examining the concept of corporate reputation and the role of "keystone" companies within business ecosystems, a consideration that becomes particularly pertinent when applied to ethical investing. Performance-wise, studies by Bauer et al. (2005, 2007), Kreander et al. (2005), and Mallin et al. (1995) corroborate the idea that ethical and traditional funds show comparable returns, a notion further reinforced by Bollen (2007) and Carhart (1997), who emphasize the stability and persistence of ethical funds. However, Hamilton et al. (1993) and Renneboog et al. (2008) offer a more complex picture, indicating mixed performance outcomes that warrant further analysis. Beyond performance metrics, Hallerbach et al. (2004) contribute by offering a framework for managing ethical portfolios, while Heinkel et al. (2001) posit that ethical investing itself can influence corporate behavior. These discussions should also be contextualized within Merton's (1987) insights into market anomalies due to informational inefficiencies, which could impact ethical funds.

In examining corporate social responsibility and stakeholder management, Margolis and Walsh (2003) scrutinize the tension between economic objectives and broader societal roles for businesses, suggesting that embracing this tension could offer new avenues for research. Mitchell et al. (1997) contribute to stakeholder theory by creating a typology based on the attributes of power, legitimacy, and urgency, aimed at helping managers identify key stakeholders. On the other hand, Orlitzky et al. (2003) provide a meta-analysis affirming that corporate social and environmental performance positively correlates with financial performance, challenging the notion that these elements are too fractured for general conclusions. Together, these studies explore the intricate relationships between corporations, stakeholders, and society, while questioning traditional economic paradigms.

2.5.2 Since 2016

Contributions of the sample provide new themes evolving to new related topics and detailing previously defined ones (Fig. 8). Thus, we found eight fundamental themes. Performance, sustainability, and climate change are motor themes in this period. In turn, responsible investment and investors themselves are a basic theme.

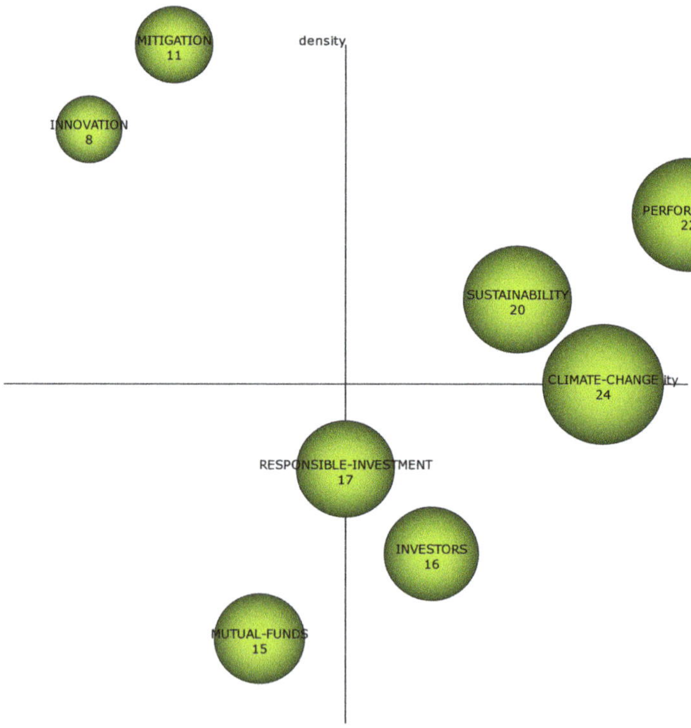

Fig. 8 Strategic diagram 2016–2022. Source: Own elaboration with SCImat. Theme nodes are placed according to their co-occurrence's centrality and density. Size of the nodes signifies the relative importance according to the h-index of each one

Innovation and mitigation appear as niche themes and mutual funds is a declining theme (Fig. 9).

The topics of the period are grounded in three identified pillars that conform to the intellectual framework (Fig. 10). On one hand, Bollen (2007) study on the lower volatility of cash flows in socially responsible funds laid the groundwork for understanding investor loyalty, which complements the findings of Dyck et al. (2019) and Hartzmark and Sussman (2019), who respectively linked institutional ownership to better environmental and social firm performance and revealed how sustainability categorization steers fund flows. Meanwhile, Edmans (2011) research disrupted traditional asset valuation by demonstrating that firms with happy employees have higher stock returns, a sentiment echoed by Krueger et al. (2020) who noted the growing priority institutional investors place on climate risks. Fama and French (2007, 2015) in their seminal work expanded asset pricing models to encompass not just financial metrics but also investor tastes and a broader range of factors, dovetailing with Pástor et al. (2021) model that shows how ESG investing alters asset prices and corporate behavior. Heinkel et al. (2001) and Hong and Kacperczyk (2009) offered contrasting viewpoints; the former established that ethical investing can punish polluters by elevating their cost of capital, whereas the latter highlighted the high returns of "sin stocks" often avoided for ethical reasons. Lins et al. (2017) emphasized the protective role of social capital in times of crisis, which parallels the conclusions of Kulich et al. (2011) and Riedl and Smeets (2017) that non-financial attributes like ethics and social preferences are potent drivers in socially responsible investments.

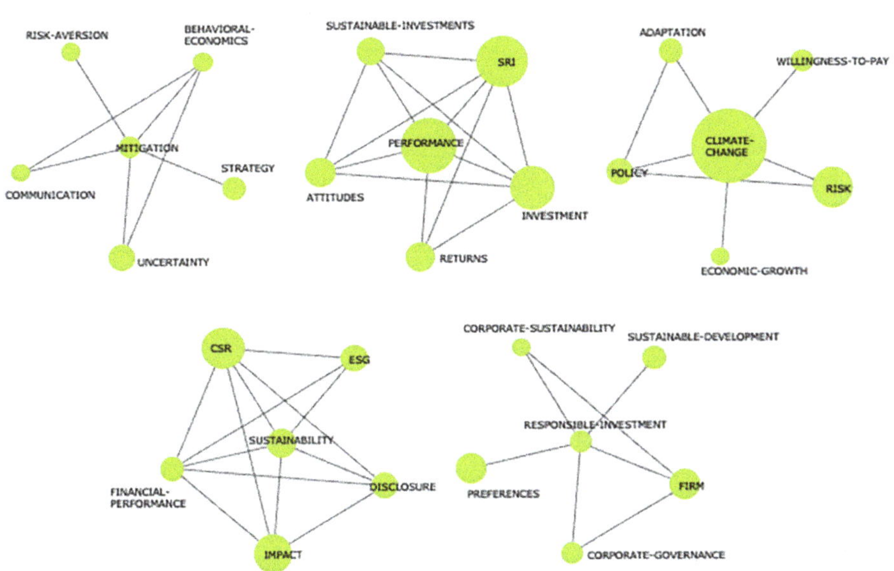

Fig. 9 Thematic networks 2016–2022. Source: Own elaboration with SCImat

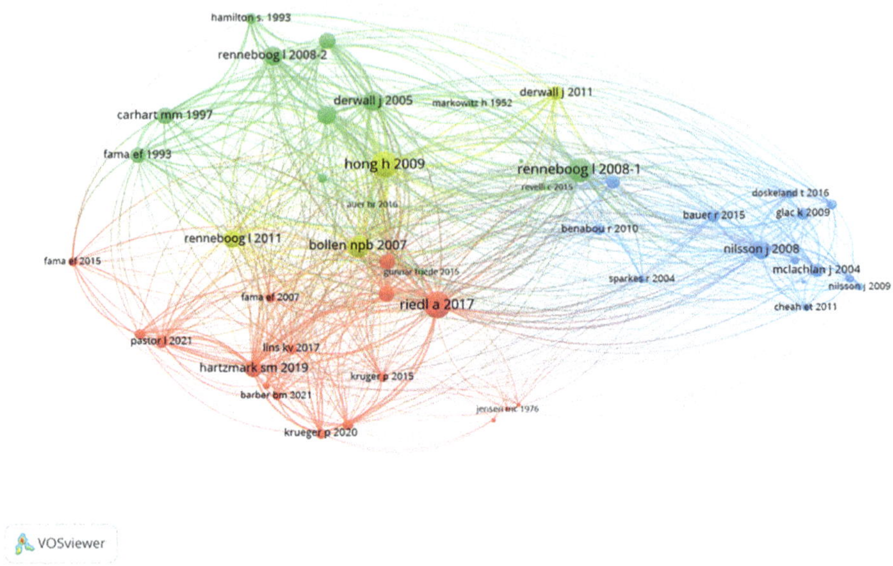

Fig. 10 Intellectual framework since 2016. Co-citation map. Source: Own elaboration with bibliometrix and VOSviewer

In the second pillar of this period, the literature on SRI is combined with multiple facets, from investor behavior to institutional dynamics. Barreda-Tarrazona et al. (2011) and Bauer and Smeets (2015) focus on the role of social preferences and identification in SRI decisions. Demographics play a significant role in SRI attitudes as per Cheah et al. (2011), while Derwall et al. (2011) explore the divide between values-driven and profit-seeking investors. Benabou and Tirole (2010) provide a broader context of societal demands for social responsibility. Glac (2009) and Nilsson (2008) delve into the cognitive aspects and trade-offs involved in SRI. Hong and Kostovetsky (2012) highlight political leanings as an influencing factor. McLachlan and Gardner (2004), Renneboog et al. (2008), and (Sparkes & Cowton, 2004) examine investor behavior, performance, and the maturation of SRI. Taken together, these studies suggest that SRI is a multifaceted field influenced by individual, demographic, and societal factors.

The performance of SRI and ethical mutual funds is a debated topic, and it constitutes another pillar of the framework. Bauer et al. (2005) and Carhart (1997) find no significant difference in risk-adjusted returns compared to conventional funds. Contradictorily, Kempf and Osthoff (2007) report that SRI can yield abnormal returns. Derwall et al. (2005) assert that eco-efficient portfolios outperform, while Hamilton et al. (1993) find SRI funds' performance comparable to conventional ones. Nofsinger and Varma (2014) note that SRI funds excel during crises but lag otherwise. Renneboog et al. (2008) find underperformance in SRI funds globally. These varying findings, rooted in Markowitz's (1952) utility theory, indicate that ethical investing outcomes are nuanced and context dependent.

As a result, empirical evidence shown in the contributions suggests that investor demand for socially responsible investment options is prompting corporations to adopt more sustainable environmental practices. Specifically, recent studies show that firms with better environmental records attract more investment from funds with a stated preference for "green" investments. This has led to a virtuous cycle where companies not only seek to improve their environmental impact to comply with regulatory standards but also to appeal to a growing segment of the investment community that prioritizes sustainability. Furthermore, exploring the role of environmental, social, and governance (ESG) metrics as a common language, align investor preferences with corporate behavior. These metrics provide a framework through which investors can assess corporate responses to climate change, leading to more informed investment decisions that, in turn, incentivize companies to innovate and implement environmentally friendly policies. Those mechanisms, supported by both quantitative data and qualitative analysis, provide a comprehensive understanding of how social investment preferences are currently shaping corporate strategies in the realm of climate change.

3 Conclusions

The research contributions on investor attitudes and preferences toward climate change and sustainable investments experienced a sharp increase over the last decades and built a significant knowledge corpus. We identified two periods in which seminal works preceded others devoted the related themes. This growth was significantly helped by specialized journals and the cross-impact with journals on finance, economics, and management. We think that they play a key role in that it would be necessary to achieve a maturation stage, which usually follows this pattern of development.

The main themes behind climate change perception evolved over the above-mentioned periods. In the first one, SRI and behavior have a main role, whereas in the second, sustainability and performance become motor themes, and the role of investors and responsible investment developed as basic themes. Mitigation of risk in a context of uncertainty is linked to the strategy, and sustainability is usually linked to its disclosure and CSR. Responsible investment ties preferences to institutionalized concepts like sustainable development, corporate governance, and corporate sustainability.

Since the climate change perception is initially based on SRI, the most relevant authors contributed to the topic addressing a variety of themes often related to the disciplines of corporate governance, management, and finance. They include the role of board diversity and long-term investment in corporate social performance, the impact of economic and climate-related uncertainties, and the importance of policy in shaping sustainable practices. Additionally, various investment styles are explored, ranging from green bonds to ethical and sector-based choices. Studies also investigate how investor behavior can incentivize corporate social responsibility and delve into the complex factors affecting investment decisions, such as policy and market demand. Behavioral motivations for self-insurance and the balance between financial gains and social benefits in sustainable investing are also discussed. In sum, they highlight how governance, investor behavior, policy, and economic factors are crucial in understanding corporate financial sustainability and the perception and preferences toward climate change-related issues.

Regarding the intellectual framework behind the contributions on the topic, the body of research is diverse, nuanced, and centered on Socially Responsible Investing (SRI). It covers different grounding pillars, from investor behavior and motivations to performance metrics and corporate influence. References indicate that SRI is gaining mainstream acceptance and is influenced by a range of psychological and demographic factors. On the performance side, the consensus is less clear. While some of the provided research documents suggest that ethical funds perform comparably to traditional funds in terms of returns, others point to mixed or even underperforming results. Importantly, these outcomes are influenced by various market conditions and investor characteristics. Ethical investing not only has the potential to shape corporate behavior but also responds to broader societal demands for social responsibility. Additionally, research in stakeholder management and corporate social responsibility expands the understanding of SRI by questioning traditional economic paradigms and emphasizing the correlation between social/environmental and financial performance. The field is also impacted by evolving asset pricing models that consider not just financial metrics but investor preferences and ethical considerations. In summary, SRI is configured as a multifaceted and complex field that is shaped by individual, institutional, and societal factors, with ongoing debates about its financial viability and broader impact.

The existing literature on investor attitudes and preferences toward climate change and sustainable investments reflects a dynamic evolution of themes and methodologies, as well as a growing sophistication in how these topics are explored and understood. However, a critical gap remains in the integration and application of these insights to actual investment practices and policymaking. There is a need for translational research that bridges the divide between academic findings and practical implementation. Current contributions often stop at the theoretical level or in controlled empirical settings, which may not fully capture the complexities of real-world markets and investor behaviors. The challenge lies in translating the nuanced understanding of investor psychology and market dynamics into actionable strategies that can be adopted by practitioners and policymakers. This translation requires interdisciplinary approaches that blend finance, economics, psychology, and sustainability studies to create robust frameworks that can withstand the test of practical application. Furthermore, it necessitates the development of new tools and metrics that can effectively measure the impact of SRI and climate change considerations on investment returns and risk profiles in a way that resonates with a broad range of investors, from individuals to large institutions. Only through such efforts can the field move beyond the maturation stage and make a tangible impact on fostering sustainable investment practices that align with the urgent need to address climate change.

3.1 Lines of Future Research

In the evolving landscape of sustainable investment, new research should redefine the interplay between finance and climate change. The development of standardized metrics for impact measurement is pivotal, enabling investors to truly understand the real-world effects of their capital allocations. This pursuit is complemented by an exploration of the psychological underpinnings of investor behavior, where the focus is on aligning communications and policies with investors' intrinsic motivations and biases. Moreover, the integration of emerging technologies such as AI and blockchain is viewed as a game changer, offering transparency and revolutionizing environmental impact analysis and investment resilience in the face of climate-related risks. An in-depth understanding of the financial implications of extreme climate events is also critical, as it can reveal how market behaviors, asset valuations, and investor sentiment respond to such phenomena. By synthesizing these elements—technological advancements, behavioral insights, climate change realities, and policy synergy—researchers can provide the investment community with sophisticated strategies that balance economic soundness with environmental sensitivity, paving the way for a regenerative economy and the next era of responsible investing.

Acknowledgments The authors acknowledge "Impact Investments" Research Group at Universidad Villanueva for providing the framework in the elaboration of the chapter.

Carlos Estevez-Mendoza acknowledges the support provided by the PID2021-126516NB-I00 project funded by MCIN/AEI/10.13039/501100011033/FEDER, UE.

References

Alexander, J. C., & Mabry, R. H. (1994). Relative significance of journals, authors, and articles cited in financial research. *The Journal of Finance, 49*(2), 697–712. https://doi.org/10.2307/2329169

Aria, M., & Cuccurullo, C. (2017). Bibliometrix: An R-tool for comprehensive science mapping analysis. *Journal of Informetrics, 11*(4), 959–975.

Bailón-Moreno, R., Bailón-Moreno, R., Jurado-Alameda, E., Ruiz-Baños, R., & Courtial, J. P. (2005). Analysis of the field of physical chemistry of surfactants with the Unified Scientometric Model. Fit of relational and activity indicators. In *Budapest Scientometrics, and Springer* (Vol. 63, Issue 2).

Barnett, M. L., Jermier, J. M., & Lafferty, B. A. (2006). Corporate reputation: The definitional landscape. *Corporate Reputation Review, 9*(1), 26–38. https://doi.org/10.1057/palgrave.crr.1550012

Barreda-Tarrazona, I., Matallín-Sáez, J. C., & Balaguer-Franch, M. R. (2011). Measuring investors' socially responsible preferences in mutual funds. *Journal of Business Ethics, 103*(2), 305–330. https://doi.org/10.1007/s10551-011-0868-z

Bauer, R., Derwall, J., & Otten, R. (2007). The ethical mutual fund performance debate: New evidence from Canada. *Journal of Business Ethics, 70*(2), 111–124. https://doi.org/10.1007/s10551-006-9099-0

Bauer, R., Koedijk, K., & Otten, R. (2005). International evidence on ethical mutual fund performance and investment style. *Journal of Banking and Finance, 29*(7), 1751–1767. https://doi.org/10.1016/j.jbankfin.2004.06.035

Bauer, R., Ruof, T., & Smeets, P. (2021). Get Real! Individuals prefer more sustainable investments. *Review of Financial Studies, 34*(8), 3976–4043. https://doi.org/10.1093/rfs/hhab037

Bauer, R., & Smeets, P. (2015). Social identification and investment decisions. *Journal of Economic Behavior and Organization, 117*, 121–134. https://doi.org/10.1016/j.jebo.2015.06.006

Benabou, R., & Tirole, J. (2010). Individual and corporate social responsibility. *Economica, 77*(305), 1–19. https://doi.org/10.1111/j.1468-0335.2009.00843.x

Ben-Amar, W., Chang, M., & McIlkenny, P. (2017). Board gender diversity and corporate response to sustainability initiatives: Evidence from the carbon disclosure project. *Journal of Business Ethics, 142*(2), 369–383. https://doi.org/10.1007/s10551-015-2759-1

Bilbao-Terol, A., Arenas-Parra, M., Cañal-Fernández, V., & Antomil-Ibias, J. (2014). Using TOPSIS for assessing the sustainability of government bond funds. *Omega (United Kingdom), 49*, 1–17. https://doi.org/10.1016/j.omega.2014.04.005

Bilbao-Terol, A., Arenas-Parra, M., Cañal-Fernández, V., & Bilbao-Terol, C. (2013). Selection of socially responsible portfolios using hedonic prices. *Journal of Business Ethics, 115*(3), 515–529. https://doi.org/10.1007/s10551-012-1411-6

Bilbao-Terol, A., Arenas-Parra, M., Cañal-Fernández, V., & Jiménez, M. (2016). A sequential goal programming model with fuzzy hierarchies to Sustainable and responsible portfolio selection problem. *Journal of the Operational Research Society, 67*(10), 1259–1273. https://doi.org/10.1057/jors.2016.33

Bloom, N. (2014). Fluctuations in uncertainty. *Journal of Economic Perspectives, 28*(2), 153–176. https://doi.org/10.1257/jep.28.2.153

Bollen, N. P. B. (2007). Mutual fund attributes and investor behavior. *Journal of Financial and Quantitative Analysis, 42*(3), 683–708. https://doi.org/10.1017/s0022109000004142

Botzen, W. J. W., Aerts, J. C. J. H., & van den Bergh, J. C. J. M. (2009). Willingness of homeowners to mitigate climate risk through insurance. *Ecological Economics, 68*(8–9), 2265–2277. https://doi.org/10.1016/j.ecolecon.2009.02.019

Callon, M., Courtial, J. P., & Laville, F. (1991). Co-word analysis as a tool for describing the network of interactions between basic and technological research: The case of polymer chemistry. *Scientometrics, 22*(1), 155–205.

Carhart, M. M. (1997). On persistence in mutual fund performance. *Journal of Finance, 52*(1), 57–82. https://doi.org/10.1111/j.1540-6261.1997.tb03808.x

Casillas, J., & Acedo, F. (2007). Evolution of the intellectual structure of family business literature: A bibliometric study of FBR. *Family Business Review, 20*(2), 141–162. http://0-search.proquest.com.cisne.sim.ucm.es/docview/211114720?accountid=14514

Cheah, E. T., Jamali, D., Johnson, J. E. V., & Sung, M. C. (2011). Drivers of corporate social responsibility attitudes: The demography of socially responsible investors. *British Journal of Management, 22*(2), 305–323. https://doi.org/10.1111/j.1467-8551.2011.00744.x

Cobo, M. J., López-Herrera, A. G., Herrera-Viedma, E., & Herrera, F. (2011). An approach for detecting, quantifying, and visualizing the evolution of a research field: A practical application to the Fuzzy Sets Theory field. *Journal of Informetrics, 5*(1), 146–166. https://doi.org/10.1016/j.joi.2010.10.002

Cobo, M. J., Lõpez-Herrera, A. G., Herrera-Viedma, E., & Herrera, F. (2012). SciMAT: A new science mapping analysis software tool. *Journal of the American Society for Information Science and Technology, 63*(8), 1609–1630. https://doi.org/10.1002/asi.22688

Cox, P., Brammer, S., & Millington, A. (2004). An empirical examination of institutional investor preferences for corporate social performance. *Journal of Business Ethics, 52*, 27–43.

Culnan, M. J. (1987). Mapping the intellectual structure of MIS, 1980–1985: A co-citation analysis. *MIS Quarterly, 11*(3), 341–353. https://doi.org/10.2307/248680

Culnan, M. J., O'Reilly, C. A., III, & Chatman, J. A. (1990). Intellectual structure of research in organizational behavior, 1972–1984: A cocitation analysis. *Journal of the American Society for Information Science, 41*(6), 453–458. https://doi.org/10.1002/(SICI)1097-4571(199009)41:6<453::AID-ASI13>3.0.CO;2-E

Danvila-del-Valle, I., Estévez-Mendoza, C., & Lara, F. J. (2019). Human resources training: A bibliometric analysis. *Journal of Business Research, 101*, 627–636. https://doi.org/10.1016/j.jbusres.2019.02.026

Dasgupta, P. (2008). Discounting climate change. *Journal of Risk and Uncertainty, 37*(2–3), 141–169. https://doi.org/10.1007/s11166-008-9049-6

De Bakker, F. G. A., Groenewegen, P., & Den Hond, F. (2005). A bibliometric analysis of 30 years of research and theory on corporate social responsibility and corporate social performance. *Business & Society, 44*(3), 283–317.

Delmas, M. A., & Montes-Sancho, M. J. (2010). Voluntary agreements to improve environmental quality: Symbolic and substantive cooperation. *Strategic Management Journal, 31*(6), 575–601. https://doi.org/10.1002/smj.826

Derwall, J., Guenster, N., Bauer, R., & Koedijk, K. (2005). The eco-efficiency premium puzzle. Source: Financial Analysts Journal, 61(2), 51–63.

Derwall, J., Koedijk, K., & Ter Horst, J. (2011). A tale of values-driven and profit-seeking social investors. *Journal of Banking & Finance, 35*(8), 2137–2147. https://doi.org/10.1016/j.jbankfin.2011.01.009

Di Falco, S., Berck, P., Bezabih, M., & Köhlin, G. (2019). Rain and impatience: Evidence from rural Ethiopia. *Journal of Economic Behavior and Organization, 160*, 40–51. https://doi.org/10.1016/j.jebo.2019.02.016

Di Falco, S., & Bulte, E. (2013). The impact of kinship networks on the adoption of risk-mitigating strategies in Ethiopia. *World Development, 43*, 100–110. https://doi.org/10.1016/j.worlddev.2012.10.011

Di Falco, S., & Sharma, S. (2018). Investing in climate change adaptation: Motivations and green incentives in the Fiji Islands. *Ecological Economics, 154*, 394–408. https://doi.org/10.1016/j.ecolecon.2018.08.015

Dumbrell, N. P., Wheeler, S. A., Zuo, A., & Adamson, D. (2022). Public willingness to make trade-offs in the development of a hydrogen industry in Australia. *Energy Policy, 165*. https://doi.org/10.1016/j.enpol.2022.112987

Dunfee, T. W. (2003). Social investing: Mainstream or backwater? *Journal of Business Ethics, 43*, 247–252.

Durisin, B., & Puzone, F. (2009). Maturation of corporate governance research, 1993–2007: An assessment. *Corporate Governance, 17*(3), 266–291. https://doi.org/10.1111/j.1467-8683.2009.00739.x

Dyck, A., Lins, K. V., Roth, L., & Wagner, H. F. (2019). Do institutional investors drive corporate social responsibility? International evidence. *Journal of Financial Economics, 131*(3), 693–714. https://doi.org/10.1016/j.jfineco.2018.08.013

Edmans, A. (2011). Does the stock market fully value intangibles? Employee satisfaction and equity prices. *Journal of Financial Economics, 101*(3), 621–640. https://doi.org/10.1016/j.jfineco.2011.03.021

Estevez-Mendoza, C. (2022). Is it all said on diversity?: A bibliometric study of research literature on diversity in management. In *Global perspectives on maintaining gender, age, and religious diversity in the workplace* (pp. 1–29). IGI Global.

Estevez-Mendoza, C. (2023). Gobierno Corporativo e Inversión Socialmente Responsable: Un análisis bibliométrico. In J. Infante, M. Del Rio Caballero, & M. Malmierca-Ordoqui (Eds.), *Inversiones socialmente responsables*. McGraw Hill.

Etemad, H., & Lee, Y. (2003). The knowledge network of international entrepreneurship: Theory and evidence. *Small Business Economics, 20*(1), 5–23. https://doi.org/10.1023/A:1020240303332

Fagerberg, J., Fosaas, M., & Sapprasert, K. (2012). Innovation: Exploring the knowledge base. *Research Policy, 41*(7), 1132–1153. https://doi.org/10.1016/j.respol.2012.03.008

Fama, E. F., & French, K. R. (2007). Disagreement tastes and asset prices. *Journal of Financial Economics, 83*(3), 667–689. https://doi.org/10.1016/j.jfineco.2006.01.003

Fama, E. F., & French, K. R. (2015). A five-factor asset pricing model. *Journal of Financial Economics, 116*(1), 1–22. https://doi.org/10.1016/j.jfineco.2014.10.010

Fankhauser, S. (2017). Adaptation to climate change. In: *Annual review of resource economics* (Vol. 9, pp. 209–230). Annual Reviews Inc. https://doi.org/10.1146/annurev-resource-100516-033554.

Fankhauser, S., & Tol, R. S. J. (2005). On climate change and economic growth. *Resource and Energy Economics, 27*(1), 1–17. https://doi.org/10.1016/j.reseneeco.2004.03.003

Fernandez-Alles, M., & Ramos-Rodríguez, A. (2009). Intellectual structure of human resources management research: A bibliometric analysis of the journal *Human Resource Management*, 1985–2005. *Journal of the American Society for Information Science and Technology, 60*(1), 161–175. https://doi.org/10.1002/asi.20947

Flammer, C. (2021). Corporate green bonds. *Journal of Financial Economics, 142*(2), 499–516. https://doi.org/10.1016/j.jfineco.2021.01.010

Glac, K. (2009). Understanding socially responsible investing: The effect of decision frames and trade-off options. *Journal of Business Ethics, 87*(SUPPL. 1), 41–55. https://doi.org/10.1007/s10551-008-9800-6

Groos, O. V., & Pritchard, A. (1969). Documentation notes. *Journal of Documentation, 25*(4), 344–349. https://doi.org/10.1108/eb026482

Gutsche, G., León, A. K., & Ziegler, A. (2019). On the relevance of contextual factors for socially responsible investments: An econometric analysis. *Oxford Economic Papers, 71*(3), 756–776. https://doi.org/10.1093/oep/gpy051

Gutsche, G., Nakai, M., & Arimura, T. H. (2021). Revisiting the determinants of individual sustainable investment—The case of Japan. *Journal of Behavioral and Experimental Finance, 30*. https://doi.org/10.1016/j.jbef.2021.100497

Gutsche, G., Wetzel, H., & Ziegler, A. (2023). Determinants of individual sustainable investment behavior – A framed field experiment. *Journal of Economic Behavior and Organization, 209*, 491–508. https://doi.org/10.1016/j.jebo.2023.03.016

Gutsche, G., & Ziegler, A. (2019). Which private investors are willing to pay for sustainable investments? Empirical evidence from stated choice experiments. *Journal of Banking and Finance, 102*, 193–214. https://doi.org/10.1016/j.jbankfin.2019.03.007

Hallerbach, W., Ning, H., Soppe, A., & Spronk, J. (2004). A framework for managing a portfolio of socially responsible investments. *European Journal of Operational Research, 153*(2), 517–529. https://doi.org/10.1016/S0377-2217(03)00172-3

Hamilton, S., Jo, H., & Statman, M. (1993). Doing well while doing good? The investment performance of socially responsible mutual funds. *Financial Analysts Journal*, November–December, 62–66.

Hartzmark, S. M., & Sussman, A. B. (2019). Do investors value sustainability? A natural experiment examining ranking and fund flows. *Journal of Finance, 74*(6), 2789–2837. https://doi.org/10.1111/jofi.12841

Heinkel, R., Kraus, A., & Zechner, J. (2001). The effect of green investment on corporate behavior. In: *Source: The Journal of Financial and Quantitative Analysis* (Vol. 36, Issue 4).

Hoepner, A. G. F., Rammal, H. G., & Rezec, M. (2011). *Islamic mutual funds' financial performance and international investment style: Evidence from 20 countries*. http://ssrn.com/abstract=1475037

Hoepner, A. G. F., & Schopohl, L. (2018). On the price of morals in markets: An empirical study of the Swedish AP-Funds and the Norwegian Government Pension Fund. *Journal of Business Ethics, 151*(3), 665–692. https://doi.org/10.1007/s10551-016-3261-0

Hong, H., & Kacperczyk, M. (2009). The price of sin: The effects of social norms on markets. *Journal of Financial Economics, 93*(1), 15–36. https://doi.org/10.1016/j.jfineco.2008.09.001

Hong, H., & Kostovetsky, L. (2012). Red and blue investing: Values and finance. *Journal of Financial Economics, 103*(1), 1–19. https://doi.org/10.1016/j.jfineco.2011.01.006

Kahneman, D., & Tversky, A. (1979). Prospect theory: An analysis of decision under risk. *Econometrica, 47*(2), 263–292.

Kempf, A., & Osthoff, P. (2007). The effect of socially responsible investing on portfolio performance. *European Financial Management, 13*(5), 908–922. https://doi.org/10.1111/j.1468-036X.2007.00402.x

Kreander, N., Gray, R. H., Power, D. M., & Sinclair, C. D. (2005). Evaluating the performance of ethical and non-ethical funds: A matched pair analysis. *Journal of Business Finance and Accounting, 32*(7–8), 1465–1493. https://doi.org/10.1111/j.0306-686X.2005.00636.x

Krueger, P., Sautner, Z., & Starks, L. T. (2020). The importance of climate risks for institutional investors abstract. *The Review of Financial Studies, 33*(3), 1067–1111. https://doi.org/10.1093/rfs/hhz137

Kulich, C., Trojanowski, G., Ryan, M. K., Alexander Haslam, S., & Renneboog, L. D. R. (2011). Who gets the carrot and who gets the stick? Evidence of gender disparities in executive remuneration. *Strategic Management Journal, 32*(3), 301–321. https://doi.org/10.1002/smj.878

Landström, H., Harirchi, G., & Åström, F. (2012). Entrepreneurship: Exploring the knowledge base. *Research Policy, 41*(7), 1154–1181. https://doi.org/10.1016/j.respol.2012.03.009

Lewis, A. (2001). A focus group study of the motivation to invest: "Ethical/green" and "ordinary" investors compared. *The Journal of Socio-Economics, 30*, 331–341.

Lins, K. V., Servaes, H., & Tamayo, A. (2017). Social capital, trust, and firm performance: The value of corporate social responsibility during the financial crisis. *Journal of Finance, 72*(4), 1785–1824. https://doi.org/10.1111/jofi.12505

Lόpez-Herrera, A. G., Cobo, M. J., Herrera-Viedma, E., Herrera, F., Bailόn-Moreno, R., & Jiménez-Contreras, E. (2009). Visualization and evolution of the scientific structure of fuzzy sets research in Spain. *Information Research, 14*(4) http://informationr.net/ir/14-4/paper421.html

Mackey, A., Mackey, T. B., & Barney, J. B. (2007). Corporate social responsibility and firm performance: investor preferences and corporate strategies. In: *Source: The academy of management review* (Vol. 32, Issue 3). https://about.jstor.org/terms

Mackey, T. B., Mackey, A., Christensen, L. J., & Lepore, J. J. (2022). Inducing corporate social responsibility: should investors reward the responsible or punish the irresponsible? *Journal of Business Ethics, 175*(1), 59–73. https://doi.org/10.1007/s10551-020-04669-0

Mallawaarachchi, T., Auricht, C., Loch, A., Adamson, D., & Quiggin, J. (2020). Water allocation in Australia's Murray–Darling Basin: Managing change under heightened uncertainty. *Economic Analysis and Policy, 66*, 345–369. https://doi.org/10.1016/j.eap.2020.01.001

Mallin, C. A., Saadouni, B., & Briston, R. J. (1995). The financial performance of ethical investment funds. *Journal of Business Finance and Accounting, 22*, 483–483.

Margolis, J. D., & Walsh, J. R. (2003). Misery loves companies: Rethinking social initiatives by business. *Administrative Science Quarterly, 48*, 268–305.

Markowitz, H. (1952). The utility of wealth. In: *Source: Journal of Political Economy* (Vol. 60, Issue 2).

McCain, K. W. (1990). Mapping authors in intellectual space: A technical overview. *Journal of the American Society for Information Science, 41*(6), 433–443. https://doi.org/10.1002/(SICI)1097-4571(199009)41:6<433::AID-ASI11>3.0.CO;2-Q

McLachlan, J., & Gardner, J. (2004). A comparison of socially responsible and conventional investors. *Journal of Business Ethics, 52*, 11–25. https://doi.org/10.1023/B:BUSI.0000033104.28219.92

Merigό, J. M., Yang, J.-B., & Xu, D.-L. (2015). A bibliometric overview of financial studies. In J. Gil-Aluja, A. Terceño-Gόmez, J. C. Ferrer-Comalat, J. M. Merigό-Lindahl, & S. Linares-Mustarόs (Eds.), *Scientific methods for the treatment of uncertainty in social sciences SE –20* (Vol. 377, pp. 245–254). Springer. https://doi.org/10.1007/978-3-319-19704-3_20

Merton, R. (1987). A simple model of capital market equilibrium with incomplete information. *Journal of Finance, 42*(3), 483–510.

Mitchell, R. K., Wood, D. J., & Agle, B. (1997). Toward a theory of stakeholder identification and salience. *Academy of Management Review, 22*(4), 853–886. https://doi.org/10.5465/AMR.1997.9711022105

Mol, J. M., Botzen, W. J. W., & Blasch, J. E. (2020). Behavioral motivations for self-insurance under different disaster risk insurance schemes. *Journal of Economic Behavior and Organization, 180*, 967–991. https://doi.org/10.1016/j.jebo.2018.12.007

Nerur, S. P., Rasheed, A. A., & Natarajan, V. (2008). The intellectual structure of the strategic management field: An author co-citation analysis. *Strategic Management Journal, 29*(3), 319–336. https://doi.org/10.2307/20142019

Nilsson, J. (2008). Investment with a conscience: Examining the impact of pro-social attitudes and perceived financial performance on socially responsible investment behavior. *Journal of Business Ethics, 83*(2), 307–325. https://doi.org/10.1007/s10551-007-9621-z

Nofsinger, J., & Varma, A. (2014). Socially responsible funds and market crises. *Journal of Banking and Finance, 48*, 180–193. https://doi.org/10.1016/j.jbankfin.2013.12.016

Orlitzky, M., Schmidt, F. L., & Rynes, S. L. (2003). Corporate social and financial performance: A meta-analysis. *Organization Studies, 24*(3), 403–441. www.sagepublications.com

Paetzold, F., & Busch, T. (2014). Unleashing the powerful few: Sustainable investing behaviour of wealthy private investors. *Organization and Environment, 27*(4), 347–367. https://doi.org/10.1177/1086026614555991

Paetzold, F., Busch, T., Utz, S., & Kellers, A. (2022). Between impact and returns: Private investors and the sustainable development goals. *Business Strategy and the Environment, 31*(7), 3182–3197. https://doi.org/10.1002/bse.3070

Pástor, Ľ., Stambaugh, R. F., & Taylor, L. A. (2021). Sustainable investing in equilibrium. *Journal of Financial Economics, 142*(2), 550–571. https://doi.org/10.1016/j.jfineco.2020.12.011

Pilkington, A., & Liston-Heyes, C. (1999). Is production and operations management a discipline? A citation/co-citation study. *International Journal of Operations & Production Management, 19*(1), 7–20. https://doi.org/10.1108/01443579910244188

Pilkington, A., & Meredith, J. (2009). The evolution of the intellectual structure of operations management—1980–2006: A citation/co-citation analysis. *Journal of Operations Management, 27*(3), 185–202. https://doi.org/10.1016/j.jom.2008.08.001

Ramos-Rodríguez, A., & Ruíz-Navarro, J. (2004). Changes in the intellectual structure of strategic management research: A bibliometric study of the Strategic Management Journal, 1980–2000. *Strategic Management Journal, 25*(10), 981–1004. https://doi.org/10.1002/smj.397

Renneboog, L., Ter Horst, J., & Zhang, C. (2008). The price of ethics and stakeholder governance: The performance of socially responsible mutual funds. *Journal of Corporate Finance, 14*(3), 302–322. https://doi.org/10.1016/j.jcorpfin.2008.03.009

Riedl, A., & Smeets, P. (2017). Why do investors hold socially responsible mutual funds? *Journal of Finance, 72*(6), 2505–2550. https://doi.org/10.1111/jofi.12547

Rodriguez Lopez, J. M., Sakhel, A., & Busch, T. (2017). Corporate investments and environmental regulation: The role of regulatory uncertainty, regulation-induced uncertainty, and investment history. *European Management Journal, 35*(1), 91–101. https://doi.org/10.1016/j.emj.2016.06.004

Rosen, B. N., Sandler, D. M., & Shani. (1991). Social issues and socially responsible investment behavior: A preliminary empirical investigation. In: *David the Journal of Consumer Affairs* (Vol. 25, Issue 2).

Sandberg, J., Juravle, C., Hedesström, T. M., & Hamilton, I. (2009). The heterogeneity of socially responsible investment. *Journal of Business Ethics, 87*(4), 519–533. https://doi.org/10.1007/s10551-008-9956-0

Schildt, H. A., Zahra, S. A., & Sillanpää, A. (2006). Scholarly communities in entrepreneurship research: A co-citation analysis. *Entrepreneurship Theory and Practice, 30*(3), 399–415.

Sparkes, R. (2001). Ethical investment: Whose ethics, which investment? *Business Ethics: A European Review, 10*(3), 194–205. https://doi.org/10.1111/1467-8608.00233

Sparkes, R., & Cowton, C. J. (2004). The maturing of socially responsible investment: A review of the developing link with corporate social responsibility. *Journal of Business Ethics, 52*, 45–57.

Tahai, A., & Meyer, M. J. (1999). A revealed preference study of management journals' direct influences. *Strategic Management Journal, 20*(3), 279–296.

van Eck, N. J., Waltman, L., Dekker, R., & van den Berg, J. (2010). A comparison of two techniques for bibliometric mapping: Multidimensional scaling and VOS. *Journal of the American Society for Information Science and Technology, 61*(12), 2405–2416. https://doi.org/10.1002/asi.21421

Webley, P., Lewis, A., & Mackenzie, C. (2001). Commitment among ethical investors: An experimental approach. *Journal of Economic Psychology*, *22*, 27–42. www.elsevier.com/locate/joep

Wong, R., & Millington, A. (2014). Corporate social disclosures: A user perspective on assurance. *Accounting, Auditing and Accountability Journal, 27*(5), 863–887. https://doi.org/10.1108/AAAJ-06-2013-1389

Wüstenhagen, R., & Menichetti, E. (2012). Strategic choices for renewable energy investment: Conceptual framework and opportunities for further research. *Energy Policy, 40*(1), 1–10. https://doi.org/10.1016/j.enpol.2011.06.050

Zupic, I., & Cater, T. (2015). Bibliometric methods in management and organization. *Organizational Research Methods, 18*(3), 429–472. https://doi.org/10.1177/1094428114562629

The Role of Financial Institutions in Addressing Climate Change

Sunanda Vincent Jaiwant, Ajitha Haridasan, and Joseph Varghese Kureethara

Abstract Climate change is an impending danger that humanity has been facing for the past few decades. Recurring cyclones, floods, and severe droughts have become common across the globe and cause widespread destruction of resources, infrastructure, and human lives, resulting in large-scale displacement. Businesses and governments are becoming more aware of the significant effects that climate change will have on the finance sector. There are two primary ways that climate change impacts the finance industry. The first is a physical risk that results from harm to assets, facilities, and the environment. The second risk, known as the "transition risk," is brought on by modifications to public and economic sentiment, technological advances, and environmental legislation due to the transition to an economy with a lower carbon footprint. Exposures might differ greatly from one nation to another. Countries with low and average incomes are often more exposed to physical risk. Financial institutions contribute significantly through carbon pricing and other fiscal policies in decreasing emissions and raising funds. To partially offset the cost of natural catastrophes, financial institutions and markets offer financial security using insurance and other risk-sharing arrangements, such as catastrophe bonds. Financial institutions help activate funds for investing in climate mitigation activities and developing resiliency to climate change as responsible responses to financial warnings, such as carbon prices. Financial institutions have begun implementing various green banking measures to reduce the adverse effects. They are attempting to solve the difficulties that are imminent, particularly from the perspective of risk management, and have created environmental policies and launched a variety of green financial instruments. The study describes financial institutions' key role in mitigating climate change and contributing to climate action initiatives. It deliberates upon the improvement limitations and probable prospects of scaling up climate finance by the financial institutions.

S. V. Jaiwant (✉) · A. Haridasan · J. V. Kureethara
CHRIST (Deemed to be University), Bengaluru, Karnataka, India
e-mail: sunanda.vincent@christuniversity.in; ajitha.h@christuniveristy.in; frjoseph@christuniversity.in

N. Naifar (ed.), *Climate Change and Finance*, Sustainable Finance,
https://doi.org/10.1007/978-3-031-56419-2_5

1 Introduction

The escalating issue of climate change increases the significance of making an effort to contribute to the climate transition, and financial markets around the world play an essential function in coordinating investment with net zero. An international response that is strong and powerful is urgently needed to combat the effects and potential risks of the global climate crisis. It is definitely good that the government's commitments to climate change have recently gained traction. The key difficulty still lies in translating heightened political intent on the environmental threat into actions that guarantee a net-zero transition by 2050. Although climate funding is increasing, affluent nations were still USD 20.4 billion behind their target of mobilizing USD 100 billion annually for supporting emerging economies in their green transitions by the end of 2019. International organizations, central banks, finance ministries, and other public agencies have been taking steps during the past few years in order to comprehend the climate transition, including the risks and advantages connected with the road to low-carbon nations (OECD, 2021).

This study addresses the crucial role financial markets must play and the policies required to promote a smooth transition to low-carbon economies. The expansion of investments to enable smooth shifts to low-carbon economies is being hampered by an array of obstacles despite the fact that financial markets are starting to include climate transition risks and possibilities in financial decision-making. Notably, there are still significant challenges for businesses and financial institutions in measuring and managing climate change risks due to a lack of data, monetarily significant measures, and statistical instruments, which emphasizes the need for policy implications. The routes and effects of systematic and chaotic changes, as well as important aspects of physical and transition risks from climate change, are included in these analyses.

2 Banking Policies in Developing and Developed Nations

This is generally agreed that nations that have created unambiguous policies and binding rules to focus public and private funding toward environmentally friendly products provide a favorable atmosphere that encourages local financial institutions to increase their green investments (GIZ, 2019). While regulatory authorities' activities vary in advanced economies, strategies for green banking initiatives differ across advanced and developing nations. For instance, rules-driven the government, like the ones found in France, typically a.

As shown in Table 1, a number of emerging economies have enacted laws that are compulsory for banking institutions to formally execute an environmentally and socially responsible safeguards policy as well as to inform central regulators and financial institutions of any relevant activity (IFC., 2018a, 2018b). In other instances, central banks in emerging economies like Bangladesh and India create explicit loan allocations for industries that are vulnerable to climate change. Multilateral development organizations like the IFC have helped several.

Table 1 Green banking policy interventions around the world

	Country	Name of institution	Type of intervention	Concept	Code of conduct
Developing countries	Bangladesh	Bangladesh Bank	1. Green central bank financing 2. Lending quotas 3. Environmental risk management guidelines	1. Exclusive refinancing windows to encourage green finance initiatives 2. Requires all banks to have at least 5% of their portfolio in green finance 3. Mandates banks to formulate their own environmental and social (E&S) risk management framework and introduced E&S risk assessment tool	Mandatory
	Brazil	1. Banco Central do Brasil 2. Ministry of the Environment, Brazilian Federation of Banks	1. Resolutions 2. Protocol verde	1. Require financial institutions to assess their activities' exposure to E&S risks and formalize an E&S policy for all their activities and issue guidelines on how to implement the policy 2. The commitment of state-owned banks and commercial banks to voluntary green guidelines	1. Mandatory 2. Voluntary
	China	People's Bank of China; China Banking Regulatory Commission	Green credit guidelines	Require banking institutions to report loan balances in 12 green sectors based on international sustainability standards and established a monitoring and evaluating system for green credit	Mandatory
	Colombia	Asobancaria (Association of Banks)	Green protocol and environmental and social risk management guidelines	Require the formalization and implementation of an E&S policy, and require clear E&S performance standards, examples, and tools	Voluntary
	India	Reserve Bank of India	Lending quotas	Require a minimum proportion of bank lending to climate and environment-related sectors	Mandatory
	Indonesia	OJK	Sustainable finance regulations	Impose financial institutions to apply sustainable finance in their business activities	Mandatory
	Lebanon	Banque du Liban	1. Green prudential regulation policy 2. Differential capital requirements for green projects	1. Incorporation of climate, environmental, and sustainability considerations into regulation 2. Incentivize banks to increase their loan portfolio in renewable energy and energy efficiency by exempting them from part of the required reserves to finance RE and EE projects at low cost	Mandatory
	Mongolia	Mongolian Bankers Association; Bank of Mongolia	Sustainable finance principles	Help banks integrate E&S considerations into lending decisions and product design	Voluntary
	Nepal	Nepal Rastra Bank	Guideline on environmental and social risk management for financial institutions	Help financial institutions evaluate the environmental and social risks that could arise from transactions	Mandatory
	Nigeria	Central Bank of Nigeria	Nigerian sustainable banking principles	Require the formalization of an ESG policy, providing financial institutions with three additional sector guidelines for the most sensitive sectors (oil & gas, power, and agriculture)	Mandatory
	Pakistan	State Bank of Pakistan	Green Banking Guidelines	Provide guidelines to banks and development financial institutions for environmental risk management, green business facilitation, and own impact reduction	Voluntary
	Vietnam	State Bank of Vietnam	2015 Directive; 2016 Circular; 2018 Scheme	Require credit institutions to formalize their E&S risk management policies and report to the central bank and encourage lending to green projects	Mandatory

(continued)

Table 1 (continued)

	Country	Name of institution	Type of intervention	Concept	Code of conduct
Developed countries	Belgium, Sweden, UK		Disclosure of climate-related financial risks	Encourage information disclosure by firms and investors	Voluntary
	France	Government authorities and regulators such as AMF	Disclosure of climate-related financial risks	Enforce information disclosure by listed firms, banks, and credit providers and investors under the France's Energy Transition Law for Green Growth	Mandatory
	Japan	Bank of Japan	Green central bank financing	Provide concessional loans to banks that lend to environment and energy business	Mandatory
	The Netherlands	DNB	Consideration of ESG factors in asset eligibility criteria	Consider ESG aspects when purchasing assets and accepting collaterals	Voluntary
	Norway	Norges Bank	Consideration of ESG factors in asset eligibility criteria	Consider ESG aspects when purchasing assets and accepting collaterals	Voluntary

Source: (CCCU, 2014); (Volz, 2017); (Campiglio et al., 2018); (TCFD, 2018); (Dikau & Volz, 2021)

Table 2 Examples of green banking products and services

Category	Sub-category	Concept and example
Loan	Corporate	Loans to micro-, small-, medium-, and large-sized enterprises for them to invest in green projects such as renewable energy, energy efficiency, forestry, and climate-smart agriculture
	Personal	Loans to individual clients for them to install small-scale renewable power and more energy efficient and climate-smart equipment, appliances, houses, and vehicles and purchase climate-resilient seeds
	Project finance	Long-term, usually non-recourse and syndicated loans to finance large-scale renewable energy projects and climate-resilient infrastructure projects
Insurance	Auto insurance	Charge lower insurance premiums for eco-friendly actions such as using electric/hybrid vehicles and recycled parts when repairing a damaged vehicle
Securitization	Bonds	Use green bonds including asset-backed securities (ABS) and mortgage-backed securities (MBS) to finance green projects and refinance existing green assets
	Warehousing	Carry out the warehousing of the assets until the target amount is reached
Equity investment	Venture capital	Invest in start-ups and venture firms developing green and climate-smart technologies
	Private equity fund	Invest in a fund dedicated to financing green projects
Brokerage and market making	Brokerage	Buy and sell green bonds and carbon credits on a client's behalf to facilitate and promote green investments
	Market-making	Buy and sell green bonds and carbon credits using a bank's own accounts to help facilitate the market
Technical assistance	Advisory	Offer advisory services with fees or on a pro-bono basis for financial structuring of a project
	Capacity building	Provide capacity-building support and consulting services to borrowers or developers to better access a bank's products

Source: (Park & Kim, 2020)

A majority of wealthy nations have adopted an unrestricted industry-driven strategy, concentrating principally on the disclosure of financial risks associated with climate change as part of TCFD support. As of 2018, the TCFD, which is wholly voluntary, has the endorsement of governments in Belgium, France, Sweden, and the United Kingdom (UK), as well as financial regulators from Australia, Belgium, France, Japan, the Netherlands, Sweden, and the UK. Additionally, France's Energy Transition Law for Green Growth mandated that publicly traded companies, banks, and lenders, besides shareholders, disclose monetary information relating to climate change. Japan is yet another instance of an advanced nation since the Bank of Japan offers discounted loans to banks that finance environmental and energy enterprises. Although required programs that are now in place frequently have insufficient compliance information, it raises questions about which government agencies will be in charge of checking adherence and maintaining oversight (Table 2).

3 Influence of the Low-Carbon Conversion on Asset Value

Substantial reductions in carbon emissions and carbon-intensive activities across almost every industry and businesses in business around the world are necessary for achieving the Paris Agreement targets and making a move to low-carbon economy. In the event that there is the political and private will to accomplish the 2-degree goal, additional drivers of emissions might be required to make up the difference by reducing their carbon15 emissions. The following could happen by means of one of two ways: either a systematic change that ensures the public, private, and financial sectors have the capacity to prepare for, handle, and combat prospective transition risks or in the case that collaborative action is postponed, a disorganized change that might call for abrupt or immediate modifications to policy in order to accomplish the decreases in CO_2 emissions required to meet a 2-degree target toward the close of the century. There are transition hazards connected to both scenarios. However, in the event of a disordered market, such could become substantially more apparent (Fig. 1).

According to the nature of the economy, the chosen carbon price, and the application of monetary policy, the total economic cost of aiding the low-carbon transition can change. The information utilized for calculating the overall predicted losses from efficiency losses and output that will be lost as an outcome of transition operations differ. For instance, Schroders Asset Management draws on a variety of techniques to calculate the transition costs for particular economies in accordance with 30-year returns analysis. The predicted economic cost will be influenced by the carbon price decision and the level of fiscal policy application. For instance, the gap is only 20 or 30 basis points of GDP a year for economies that are more service-oriented, such as many OECD nations, but for economies that have a greater emphasis on commodities, like China and India, the difference is much larger. According

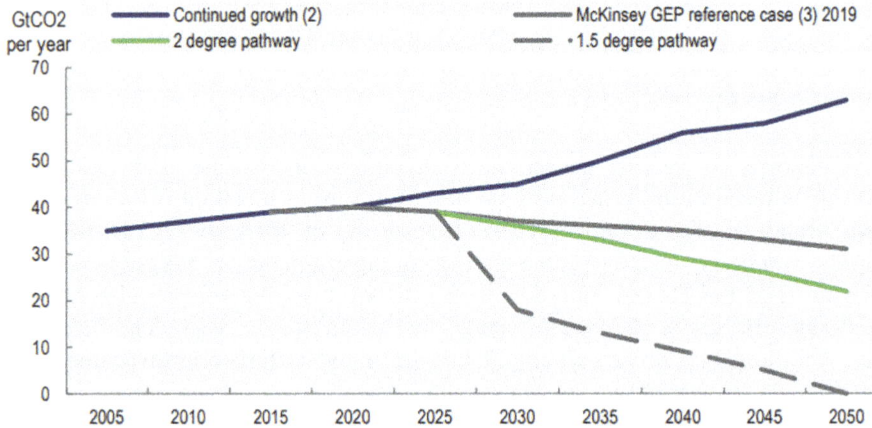

Fig. 1 Achieving the 2-degree target by the end of the century requires rapid declines in CO_2 emissions in the coming years Projected global CO_2 emissions per scenario, metric gigatons of CO_2 ($GtCO_2$) per year. Source: (OECD, 2021)

to the study, certain countries might only incur a small amount in predicted transition costs because they hardly need to reduce carbon emissions. However, such might be doable under milder transition scenarios and with a more effective use of the ensuing profits. On the other side, more carbon-intensive economies may have higher expenses (OECD, 2021).

4 Financing Climate Mitigation

Financing climate mitigation is a critical component of global efforts to address climate change. Climate mitigation refers to actions and strategies aimed at reducing greenhouse gas emissions and limiting the extent of global warming. Financing these mitigation efforts is essential to transition to a low-carbon and more sustainable economy. Governments at the national and local levels often provide public funding for climate mitigation projects. This can include direct investments in renewable energy infrastructure, energy efficiency programs, and public transportation systems. Governments can also allocate resources for research and development of clean technologies. Carbon pricing mechanisms, such as carbon taxes and cap-and-trade systems, provide economic incentives for businesses and individuals to reduce their carbon emissions. Revenues generated from these mechanisms can be reinvested in climate mitigation projects. International organizations and developed countries often provide financial support to developing nations to help them mitigate and adapt to climate change. The Green Climate Fund, for example, was established to help developing countries reduce their emissions and build resilience to the impacts of climate change.

Private sector investment in climate mitigation is crucial. Companies can invest in clean energy technologies, energy-efficient practices, and other environmentally friendly initiatives. This can be driven by a desire to reduce operational costs, meet regulatory requirements, or respond to consumer demands for sustainability. Green bonds are financial instruments specifically designed to fund environmentally friendly projects. They allow organizations, including governments and corporations, to raise capital for climate mitigation initiatives. The proceeds from these bonds must be used for projects that have clear environmental benefits. Sustainable finance focuses on incorporating environmental, social, and governance (ESG) criteria into investment decisions. ESG investors prioritize companies and projects that demonstrate a commitment to climate mitigation and other sustainability goals. This approach encourages the allocation of capital to environmentally responsible initiatives.

Collaboration between the public and private sectors can help finance large-scale climate mitigation projects. Public–private partnerships can leverage both public funds and private sector expertise to drive investment in renewable energy, infrastructure, and other mitigation efforts. Foundations and philanthropic organizations play a significant role in financing climate mitigation projects. They provide grants, donations, and support for research, advocacy, and grassroots initiatives aimed at addressing climate change. Funding research and development of innovative technologies and solutions is crucial for climate mitigation. Public and private investment in clean energy technologies, carbon capture and storage, and sustainable agriculture practices can significantly contribute to reducing emissions. Financial instruments such as insurance, risk-sharing mechanisms, and guarantees can help mitigate the financial risks associated with climate-related projects, making them more attractive to investors. Financing climate mitigation is a complex and multifaceted challenge that requires a combination of public and private sector efforts, international cooperation, and innovative financial tools. It is a crucial element of the global response to climate change and requires a sustained commitment from governments, businesses, investors, and individuals to achieve meaningful results in reducing greenhouse gas emissions (Bachner et al., 2019).

5 Financing Climate Adaptation

Financing climate adaptation is as important as financing climate mitigation, as it involves addressing the impacts of climate change that are already occurring or are expected to occur in the near future. Climate adaptation involves measures to increase resilience to climate-related challenges, such as rising sea levels, extreme weather events, and changing precipitation patterns. Governments at various levels (local, regional, and national) often allocate public funds to support climate adaptation efforts. These funds may be used for building resilient infrastructure, implementing early warning systems, and developing policies that promote adaptive practices. Climate risk insurance mechanisms can help individuals, communities, and businesses recover from climate-related disasters. These insurance schemes are designed to provide financial support in the aftermath of extreme weather events and can be supported by both the public and private sectors. Just as in climate mitigation, international organizations and developed countries provide financial assistance to developing nations for climate adaptation. This support helps vulnerable countries build resilience and adapt to the impacts of climate change. Some cities issue climate adaptation bonds to fund resilience projects and infrastructure improvements. These bonds are backed by the revenue generated from these projects and can be an effective way to finance adaptation at the local level. Companies are increasingly recognizing the need to adapt to climate change to protect their operations and supply chains. Private sector investment can be directed toward making business operations more resilient and supply chains less vulnerable to climate risks. Similar to green bonds for climate mitigation, green bonds or climate resilience bonds are financial instruments specifically designed to fund adaptation projects. The proceeds from these bonds are used to support initiatives that enhance resilience to climate change impacts. ESG (Environmental, Social, and Governance) criteria in investing can extend to climate adaptation. Sustainable finance and ESG investors may favor companies and projects that demonstrate a commitment to resilience and preparedness for climate-related challenges. Collaboration between the public and private sectors is valuable for financing large-scale climate adaptation projects. Such partnerships can combine government funds with private sector expertise and resources to support adaptation initiatives. Foundations and philanthropic organizations provide grants and donations to support adaptation projects, research, and community resilience-building efforts. These contributions can be especially valuable for community-based adaptation initiatives.

Investing in risk assessments and climate adaptation planning is a fundamental step in securing financing for adaptation. Understanding the specific climate risks a region or sector faces is essential for developing effective adaptation strategies. Local communities often play a critical role in climate adaptation, and grassroots initiatives can be supported through community-based funds, grants, and crowdfunding. Financing climate adaptation is essential to reduce the vulnerabilities of communities and ecosystems to the impacts of climate change. This financing is crucial for protecting infrastructure, ensuring food and water security, and preserving livelihoods in the face of a changing climate. It requires a combination of public and private sector contributions, international cooperation, and innovative financial mechanisms to address the urgent and ongoing challenges posed by climate change.

6 Financial Institutions Responding to Climate Change Challenges

Financial institutions are increasingly recognizing the importance of responding to climate change challenges. Climate change poses financial risks and opportunities that can impact the stability and profitability of these institutions. Financial institutions are conducting climate risk assessments to understand how climate change may affect their portfolios. This involves identifying exposure to physical risks (e.g., extreme weather events) and transition risks (e.g., policy changes and market shifts related to climate change). Some institutions are conducting climate stress tests to assess how their assets and investments may perform under different climate scenarios. This helps them identify vulnerabilities and develop mitigation strategies.

Many financial institutions are offering sustainable financial products, including green bonds, green loans, and sustainable investment funds. These products direct capital toward environmentally friendly projects and businesses. Environmental, Social, and Governance (ESG) criteria are increasingly integrated into investment and lending decisions. Financial institutions consider a company's environmental practices and sustainability performance when evaluating its creditworthiness. Many financial institutions are improving their climate-related disclosure, providing more information about their exposure to climate risks and the actions they are taking to address them. This enhances transparency and allows investors to make informed decisions. Some banks and investment firms are committing to achieving net-zero carbon emissions in their own operations and investment portfolios. This involves reducing emissions and offsetting the remaining emissions through investments in carbon removal projects. Financial institutions are engaging with their clients to encourage and support sustainable practices. They may provide advice on energy efficiency, low-carbon technologies, and other climate-related initiatives. Many banks are increasing their funding for renewable energy projects, such as wind and solar farms. This supports the transition to a low-carbon economy.

Financial institutions are investing in research and data analytics to better understand climate-related risks and opportunities. This includes assessing the potential impact of climate policies and market shifts. Some financial institutions are actively advocating for government policies that support climate action, such as carbon pricing, renewable energy incentives, and climate resilience planning. Educating employees, clients, and investors about climate risks and opportunities is an important aspect of financial institutions' responses to climate change. This helps build awareness and understanding. Financial institutions are providing funding for innovative climate solutions, including clean technologies and climate adaptation projects. Many financial institutions are publishing sustainability reports that detail their environmental and social performance. This provides stakeholders with insight into their sustainability efforts. Financial institutions are increasingly collaborating with each other, governments, and environmental organizations to share best practices, coordinate efforts, and mobilize more capital for climate solutions. Responding to climate change challenges is not only about managing risks but also seizing opportunities. Financial institutions play a crucial role in financing the transition to a more sustainable and low-carbon economy. Their actions can influence corporate behavior, shape markets, and contribute to the global effort to mitigate and adapt to climate change.

7 The Financial Sector's Dominating Role in Addressing Climate Change

The financial sector is leading the way in a number of ways on climate change. Financial institutions differ in their individual goals and roles, even if they are all bound by the same fiduciary obligation for the capital of others. Because of this, they are implementing a variety of strategies to manage the possibilities and dangers related to climate change. The report's leadership acts are divided into six categories, which are as follows:

- Investments and financing for energy efficiency and low carbon.
- Reducing emissions through investing and financing.
- Investments and financing for adaptation.
- Transparency and measurement.
- Interaction with businesses.
- Interaction with legislators.

The financial sector is leading the way in a number of ways on climate change. Financial institutions differ in their individual goals and roles, even if they are all bound by the same fiduciary obligation for the capital of others. Because of this,

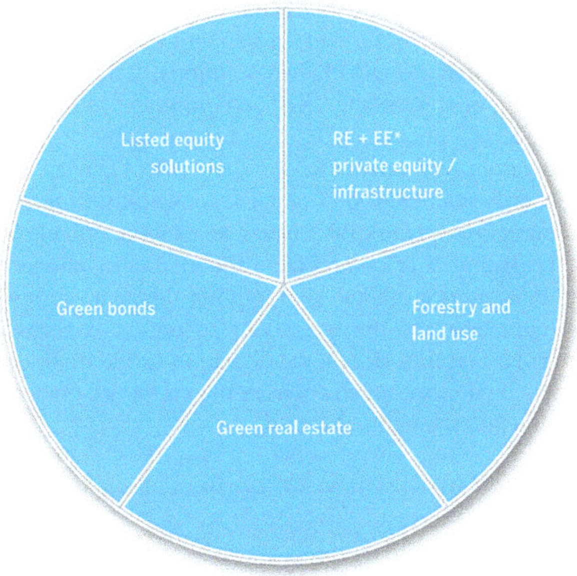

Fig. 2 Low carbon investing and energy efficiency across asset classes. Source: (Guyatt, 2021)

they are implementing a variety of strategies to manage the possibilities and dangers related to climate change.

Some of the areas where financial institutions are exhibiting are shown in Fig. 2.

Leadership in developed and developing nations, divided into groups according to the many asset classes that financial institutions employ when allocating capital.

8 Cases on Financial Institutions in Addressing Climate Change

(Group, 2021) Financial Commitments from Private Financial Institutions (2014–15)

The business sector declared intentions to raise over $200 billion in financial resources for low-carbon and climate-resilient development during the September 2014 UN Climate Summit. Among these is the insurance sector's pledge to boost its green expenditures tenfold by 2020 and to quadruple to $84 billion by the end of 2015. By 2020, three large pension funds promised to devote over $31 billion to "low-carbon" investments, while a group of institutional investors committed to decarbonizing $100 billion. Commercial banks committed to providing $30 billion in fresh climate funding by the end of 2015 through the issuance of green bonds and other creative financing programs during the Summit. The finance industry has made significant announcements about climate finance in the year after the Climate

Summit. This includes a number of significant commitments made by the financial industry, such as Citibank's aim to lend, invest, and facilitate $100 billion over 10 years to environmental and climate change solutions, Barclay's intention to invest £1 billion in the green bond market, AXA's commitment to triple its green investment footprint to reach €3 billion by 2020, and many more.

Prudential Regulatory Authority (PRA) of the Bank of England: Based on a relationship between its fundamental prudential responsibilities and the UK Climate Change Act, the Bank of England's PRA began an examination of climate risk in the insurance industry. It aimed to comprehend the effects of climate change on the stability of the UK insurance market, highlighting the connections between financial policy and regulation and the operations of regulated financial institutions. Through the Financial Stability Board, the Governor of the Bank of England, Marc Carney, has also facilitated a conversation on the effects of climate change on the stability of the financial system.

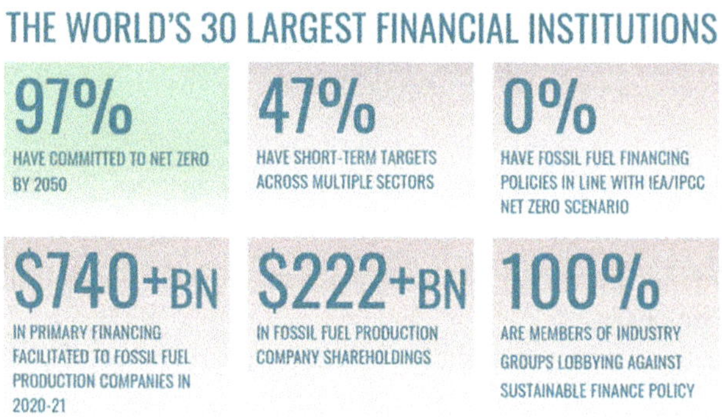

Source: (InfluenceMa**p, 2020**)

Setting reduction targets is a requirement of collaborative involvement for high emitting

Global 500 enterprises. Together, the PRI and CDP have launched a project known as Carbon Action. The organization is interacting with Global 500 businesses that operate in emissions-intensive industries and generate more than 1 million metric tons of emissions annually. The committee met with 22 companies in 2012–2013, and 8 of them have since announced a target or shown progress in establishing one (Guyatt, 2021).

800 shareholder proposals have been filed with US companies that relate to climate change More than 800 shareholder recommendations about climate change have been filed with US listed businesses over the past 10 years by INCR members and other investors working together as part of the Shareholder Initiative on Climate and Sustainability, which is coordinated by Ceres. Consequently, over 350 prominent corporations have pledged to tackle the hazards and prospects associated with climate change. Investors submitted 148 resolutions pertaining to climate change in just 2013–2014; this resulted in 64 agreements for corporate action, demonstrating the increasing momentum behind this activity (Guyatt, 2021).

Investors achieve commitments to source sustainably produced palm oil to reduce tropical deforestation.

Investors and the many businesses that source and utilize palm oil in their products have banded together to bring attention to the problem of unsustainable palm oil production and its effects on tropical forests. Consequently, a minimum of 17 corporations have pledged to procure palm oil that is produced sustainably and does not cause deforestation.

Since 2011, more than 20 firms have received shareholder proposals from INCR members, such as Green Century and the NYS Comptroller, and numerous others have been contacted via letters or conversations. A working group inside the PRI is also focused on this matter. Investors are getting guarantees from businesses that their supply chains will not source palm oil that comes from deforestation (Guyatt, 2021).

9 Green Banking Policy Instruments

Green banking policy instruments are a set of regulatory and market-based tools that governments and financial authorities can use to encourage and facilitate environmentally responsible and sustainable practices within the banking sector. These policies are aimed at aligning financial institutions with broader environmental and sustainability goals. Governments and regulatory bodies can issue guidelines and frameworks that outline the principles and expectations for green banking. These guidelines provide a roadmap for banks to integrate sustainability considerations into their operations and lending practices.

These policies encourage banks to provide preferential credit terms, such as lower interest rates or reduced fees, for loans that support environmentally friendly projects, such as renewable energy, energy efficiency, and sustainable agriculture. Governments and financial regulators can establish standards and criteria for green bonds. Green bonds are financial instruments that are used to fund environmentally beneficial projects. Ensuring the credibility and transparency of green bond standards is crucial for promoting investment in sustainable initiatives. Regulatory authorities can require banks to develop and implement risk assessment and management procedures related to environmental and social risks. This ensures that banks consider the potential environmental and social impacts of their lending activities. Banks may be required to disclose their environmental, social, and governance (ESG) performance through standardized reporting frameworks. This transparency allows stakeholders to assess a bank's sustainability efforts. Governments or regulatory authorities can set targets for banks to allocate a certain percentage of their lending portfolios to green or sustainable projects. Meeting these targets may be tied to regulatory incentives or penalties.

Financial authorities can provide incentives, subsidies, or grants to banks that actively promote and finance green projects. These can include tax incentives, reduced capital requirements, or access to special funding programs. Banks may be encouraged to invest in training and building the capacity of their staff to understand and evaluate sustainability risks and opportunities. Governments can implement carbon pricing mechanisms, such as carbon taxes or cap-and-trade systems, that encourage banks to consider carbon emissions when evaluating investments and lending. Regulatory authorities can encourage banks to develop and offer eco-friendly financial products, such as green mortgages, green savings accounts, and sustainability-linked loans. Encouraging collaboration among financial institutions, regulators, and other stakeholders can facilitate the exchange of best practices and promote sustainable banking. Regulators can require banks to conduct stress tests that evaluate the potential impact of climate-related risks on their portfolios. This helps banks prepare for the financial implications of climate change. Encouraging innovation in sustainable finance, such as the development of new financial products or investment strategies, can accelerate the transition to a green economy. Green banking policy instruments vary from one country to another and depend on local regulatory frameworks and sustainability priorities. Their effectiveness depends on the commitment of regulatory authorities, the willingness of banks to embrace sustainable practices, and the involvement of stakeholders in driving the transition toward a more sustainable financial sector.

10 International Initiatives for Sustainable Infrastructure Financing and Investment

International initiatives for sustainable infrastructure financing and investment are critical for addressing global infrastructure needs in an environmentally responsible and socially inclusive manner. These initiatives bring together countries, financial institutions, and organizations to promote and facilitate investments in infrastructure projects that are economically viable and environmentally sustainable. While not a specific financing initiative, the United Nations Sustainable Development Goals (SDGs) serve as a global framework for sustainable development, including infrastructure. Many international efforts align with these goals to promote investments in infrastructure that contribute to the well-being of people and the planet. Launched by China, the Belt and Road Initiative (BRI) is a massive infrastructure development and investment project that spans multiple countries in Asia, Europe, and Africa. While it has faced criticism regarding environmental and debt sustainability issues, there is an ongoing effort to make BRI projects more environmentally and socially responsible. Hosted by the World Bank, the Global Infrastructure Facility (GIF): provides advice and resources to governments and the private sector for preparing and structuring complex infrastructure projects. It focuses on sustainable and climate-resilient infrastructure development. Sustainable Infrastructure Foundation (SIF) promotes sustainable infrastructure development through the use of the Infrastructure Sustainability (IS) rating scheme. This initiative assesses and certifies infrastructure projects based on their social, environmental, and economic performance. Institutions like the World Bank, Asian Development Bank, and

African Development Bank play a crucial role in financing sustainable infrastructure projects. These Multilateral Development Banks (MDBs) provide financing and technical assistance for projects that align with environmental and social safeguards. Global Infrastructure Basel (GIB) is a non-profit organization that promotes sustainable and resilient infrastructure by developing standards and guidelines for environmental and social risk management in infrastructure projects.

While not specific to infrastructure, Task Force on Climate-related Financial Disclosures (TCFD) provides recommendations on disclosing climate-related financial risks. This has a direct impact on infrastructure projects as investors and financiers increasingly consider climate risks in their decision-making. Global Covenant of Mayors for Climate & Energy is an initiative that brings together thousands of cities committed to addressing climate change. Many of these cities are implementing sustainable infrastructure projects, particularly in areas like transportation, energy efficiency, and renewable energy. Green Climate Fund (GCF) is dedicated to financing projects in developing countries that reduce greenhouse gas emissions and enhance climate resilience. A portion of its funding goes toward sustainable infrastructure projects that align with climate goals. Renewable Energy and Energy Efficiency Partnership (REEEP) supports clean energy projects in developing countries, including sustainable infrastructure projects related to renewable energy and energy efficiency. Public–Private Partnerships (PPPs) are yet another initiative that many countries and international organizations use to promote sustainable infrastructure through PPPs. These partnerships often involve both public and private sector stakeholders and can focus on various sectors, including energy, transportation, and water. Sustainable Infrastructure Investment Principles (SIIP) are the principles, developed by the International Finance Corporation (IFC), that provide guidance for financial institutions on how to integrate sustainability into their infrastructure investments. All these initiatives play a crucial role in mobilizing resources and knowledge for sustainable infrastructure development and ensuring that projects are built with environmental and social considerations in mind. They help address the global infrastructure gap while contributing to broader sustainability objectives, including climate action and social equity.

11 Conclusion

Financial institutions allocate resources to different industries, which plays a crucial effect in influencing carbon emissions. These organizations' investment decisions have the power to either hasten or obstruct the achievement of carbon reduction goals. Financial institutions face increasing financial risks stemming from climate change, including exposure to physical risks (e.g., property damage due to extreme weather events) and transition risks (e.g., policy changes and stranded assets in carbon-intensive industries). The study has explored the critical role of financial institutions in addressing the multifaceted and urgent challenge of climate change. The analysis has revealed key aspects that underscore the significance of financial institutions in the transition to a more sustainable and resilient global economy. First and foremost, financial institutions have a profound influence on climate action, both in terms of fostering positive change and mitigating risks. Their capacity to allocate capital to sustainable projects, encourage responsible corporate behavior, and manage climate-related risks is instrumental in driving the transition toward a low-carbon and climate-resilient future. The complexity of the relationship between financial institutions and climate change is evident. The financial sector is not merely a passive actor but a dynamic influencer of climate policy and practice. It is subject to the evolving landscape of climate agreements, government regulations, and societal expectations. Furthermore, the policy implications of our findings are substantial. Policymakers, regulators, and international organizations must recognize the central role of financial institutions and design policies that incentivize sustainable practices and carbon risk management. The study provides insights that can inform the creation of policies that encourage responsible investing, transparent disclosure of climate risks, and the alignment of financial flows with climate objectives. It is essential to acknowledge the limitations of this research. While the chapter has delved into the broad landscape of financial institutions and climate change, there are areas yet to be explored in detail. Future research can expand on these findings and delve deeper into the evolving dynamics of climate finance.

In a world characterized by increasing climate-related challenges, this research underscores the urgency of addressing climate change through the financial sector. It highlights the need for a comprehensive and collaborative approach. Financial institutions, governments, businesses, and society at large must work together to accelerate the transition to a sustainable, low-carbon economy. The study emphasizes the pivotal role that financial institutions play in addressing climate change. Their commitment to sustainability, responsible investment, and risk management has a direct impact on our ability to mitigate and adapt to climate change. There are indeed numerous challenges and opportunities presented by a rapidly changing climate, financial institutions stand as key allies in the collective effort to secure a more sustainable and resilient future for all.

References

Bachner, G., Mayer, J., & Steininger, K. (2019). Costs or benefits? Assessing the economy-wide effects of the electricity sector's low-carbon transition—The role of capital costs, divergent risk perceptions and premiums. *Energy Strategy Reviews, 26*, 100373. https://doi.org/10.1016/j.esr.2019.100373

Campiglio, E., et al. (2018). Climate change challenges for central banks and financial regulators. *National Climate Change, 8, 462.*

CCCU. (2014). *Climate finance loan schemes: Existing and planned loan schemes in Lebanon.* Climate change coordination unit (CCCU).

Dikau, S., & Volz, U. (2021, June). Central Bank mandates, sustainability objectives and the promotion of green finance. *Ecological Economics, 184*, 107022. https://doi.org/10.1016/j.ecolecon.2021.107022

GIZ. (2019). *The role of National Financial Institutions in the implementation of NDCs*. Retrieved from. https://www.giz.de/en/downloads/giz2019-EN-The-Role-of-National-Financial-Institutions.pdf
Group, W. B. (2021). *Mainstream climate action within financial institutions Emerging practices*. Retrieved from https://www.eib.org/attachments/fi_mainstreaming_epp_overview_en.pdf
IFC. (2018a). *Raising US$ 23 trillion: Greening banks and capital markets for growth. G20 input paper on emerging markets*. IFC.
IFC. (2018b). *Sustainable banking network (SBN) global Progress report*. IFC.
InfluenceMap. (2020). *Finance and climate change*. InfluenceMap.
OECD. (2021). *Financial markets and climate transition opportunities, challenges and policy implications*. OECD. Retrieved from https://www.oecd.org/finance/Financial-Markets-and-ClimateTransition-Opportunities-challenges-and-policy-implications.htm
Park, H., & Kim, J. D. (2020). Transition towards green banking: Role of financial regulators and financial institutions. *Asian Journal of Sustainability and Social Responsibility, 5*.
Spiegel, A., Wiener, D., Schneider-Roos, K., & Diamant, N. (2019). *The missing link: Linking financial stability with environmental stability*. Published by ecos.
TCFD. (2018). TCFD.
Volz, U. (2017). On the role of central banks in enhancing green finance. *UN environment inquiry working paper*.

Green Development Prospects of Fossil Fuel Export-Dependent Countries: A Case Study of the Caspian Sea Region

Liudmila Filipava

Abstract Toward the Global Agenda for Sustainable Development (UN Agenda 2030), the fossil fuel export-dependent states are particularly interested in implementing their energy policies. Despite the necessity for structural transformations, the countries nevertheless pursue enhancing fossil fuel production according to their long-term strategies, reorienting the focus from the green agenda toward stabilizing national economies and achieving comprehensive solutions to geopolitical changes. In the current research, the author conducts an empirical study of 38 fossil fuel-exporting states, analyzing variables in four groups (Green Growth, Climate Change and Green Transition, Fossil Fuel Dependence, and Macroeconomic) to investigate whether fossil fuel export dependence determines the effectiveness of ongoing greening reforms. The author observes a negative correlation between the Global Green Index (GGI) and governmental revenues derived from fossil fuels, thereby substantiating the assertion that a green transition is feasible through diversifying revenue sources and reducing dependence on fossil fuel exports. Increasing the role of renewable energy and green hydrogen will allow the Caspian littoral states to develop export supplies to the EU through the Green Deal plan. Simultaneously, improving the national taxonomy and integrating it into the Global Green Finance (GGF) market will gradually transform the raw material export-oriented model into a resource-saving one, considering the national specificities and priorities of the Caspian Sea Region.

1 Introduction

Caspian littoral states, the world's largest suppliers of hydrocarbon raw material resources, appear as an antipode to achieving the objectives of the *Global Green (GG) Agenda*. Their Strategies for *Sustainable Development* (Agenda 21, 1992) are

L. Filipava (✉)
Department of International Economic Relations, Belarusian State University, Minsk, Belarus

N. Naifar (ed.), *Climate Change and Finance*, Sustainable Finance,
https://doi.org/10.1007/978-3-031-56419-2_6

determined by internal guidelines, as well as geopolitical factors that alter the system of *green priorities* for coordinating national interests in the global arena.

The initial step in the direction of the consistent "greening" of the Caspian region was the ratification of the *Tehran Convention*[1] (2003), which entails the collective accountability of nations toward the Caspian Sea preservation. Subsequent Agreement on Cooperation in the Field of Security (2010) and 5-party intergovernmental Agreements[2] (2014, 2015) have provided a framework for strategic economic planning in the region, leveraging natural resources, and industrial and technological potential.

Today, the countries are implementing long-term development strategies, fulfilling their Nationally Determined Contributions (NDCs). The priority is given to the Net Zero target in accordance with the Paris Agreement (UNFCCC, 2015), since exploitation of fossil resources produces the highest level of Greenhouse Gas emissions (GGEs) (Gütschow et al., 2016). This necessitates the adaptation of the green taxonomy for environmental friendly projects' funding through green financial instruments. To attract investment, the countries provide fiscal incentives, implement an auction system instead of fixed tariffs, etc.

Despite certain success in achieving the "green" objectives (OECD, 2022), the Caspian littoral states are primarily focused on the development of their oil and gas industry to attract foreign investment for immediate economic effects (Shaffer, 2010). States' cooperation is also hindered by political factors, which shift the focus from environmental perspectives to ensuring "long-term and unbreakable security" (VI Caspian Sea Summit, 2022). This determines the region's development toward the stabilization of economic systems in a changing geopolitical environment, demonstrating collaborative problem solving.

Most existing research referred to the specifics of the economic development of fossil fuel export-dependent countries (Anser et al., 2020; Wang & Zhang, 2023). A particular focus is given to their socio-economic challenges, including but not limited to poverty (Barbier, 2020), social inequality (Arze del Granado et al., 2012), ineffectiveness of governmental policy (Zakharova & Medas, 2009), low diversification of country revenues (Karanfil & Omgba, 2023; Shrestha et al., 2021), FDI deficiency (Shear et al., 2023), among others. The Caspian countries, as the largest producers of the oil and gas global market, are traditionally studied in terms of their energy potential (Shaffer, 2010). The present chapter attempts to bridge the gap between climate science and fossil fuel dependence, with the objective of determining whether such dependence has an impact on the implementation of the GG Agenda (UN, 2015) and NDC plan.

The chapter is structured as follows: Section 2 provides an analysis of the Green Growth indicators, such as the SDG-based Indices, Green Growth Index, Global Green Economy Index, and Economic Freedom Index. Section 3 unites climate

---

[1] Framework Convention for the Protection of the Marine Environment of the Caspian Sea (CEP, 2003).

[2] Agreement on the Conservation and Sustainable Use of Aquatic Biological Resources of the Caspian Sea (2014), Agreement on Cooperation in Emergency Prevention and Response in the Caspian Sea (2014), Agreement on Hydrometeorological Cooperation in the Caspian Sea (2015).

economics and international economics by examining the impact of fossil fuel countries' dependence on achieving their green strategies. Section 4 scrutinizes the compliance of the Caspian littoral states with the NDC plan, confirming the conclusion regarding the necessity to diversify their revenues, with particular emphasis on the export of renewable energy and "green" hydrogen to the EU countries in accordance with the Green Deal plan. Finally, Section 5 concludes the chapter by presenting the main findings.

2 Caspian Littoral States in Achieving Green Growth: Analysis of GG Indicators

The conceptual framework for the greening approach was formed within the realm of *Sustainable development* in the late 1980s and was articulated in the "Our Common Future" report by the UN World Commission on Environment and Development (WCED, 1987). As per its assertion, the global economy necessitates the coordination of economic, social, and environmental components (Agenda 21, 1992), analyzing how to boost peoples' well-being while minimizing environmental hazards. In 2008, the United Nations Environment Program (UNEP) initiated the Green Economy Initiative (GEI), which identifies the fundamental principles for sustainable growth*:* (i) the new sources of income and employment; (ii) the reduction in carbon emissions, resource use, waste, and pollution generation; and (iii) the contribution to achieving public objectives of sustainable development, social equity, and poverty reduction (UNEP, 2011).

The *SDG-based Indices* are widely used for evaluating the efficacy of nations' efforts to promote green economic development, and encompass a diverse range of factors, including: (i) the carbon footprint, air pollution, emission of sulfur dioxide and nitrogen; (ii) the accessibility to environmentally friendly fuels and technologies; and (iii) the water supply (UNCTAD, 2019).

The analysis of carbon footprint of the Caspian littoral countries shows that the situation is favorable in Kazakhstan and Turkmenistan, while the other countries demonstrate exceptionally high levels of CO_2 emissions (Table 1). The share of renewable energy in the total volume of primary energy supplies remains remarkably low, hovering around 3% (compared to Norway's 47.6%); in Iran and Turkmenistan, the figures hover around 0% (Table 2). The two largest consumers of fresh water are Turkmenistan and Iran, while consumption in Kazakhstan is below the average, primarily due to the import of water. In general, freshwater consumption in the region is above average, imports are high, and wastewater treatment is inadequate.

In 2011, the Organization for Economic Cooperation and Development (OECD) presented the *Green Growth Index* (GGI) in the "Towards Green Growth: Monitoring Progress" report (OECD, 2011). The indicator is based on an aggregated approach, considering (i) the interdependence between the environment and economy, (ii) the ease of information provided, (iii) the alignment with the OECD

Table 1 SDG-based indices characterizing environmental pollution of the Caspian Sea region, 2022

	SO2 emissions based on production (kg/capita)	Nitrogen emissions based on production (kg/capita)	Export of plastic waste (kg/person)	Electronic waste (kg/capita)	CO2 emissions from fossil fuel combustion and cement production (tCO2/person)	CO2 emissions embodied in imports (tCO2/capita)	CO2 emissions embodied in fossil fuel exports (kg/capita)	CO2 emissions from fuel combustion per total electricity production (MtCO2/TWh) 1.52019–5	Scarce water consumption embodied in imports (m3 of water per capita)	Annual average concentration of particulate matter less than 2.5 microns in diameter (PM2.5) (μg/m^3)
Azerbaijan	5.0	12.0	0.0	8.0	3.7	0.4	11,268	1.5	1028.4	19.2
Iran	11.1	9.5	0.2	9.5	8.9	0.3	4226	2.3	664.9	**39.2**
Kazakhstan	**68.2**	17.9	0.3	9.2	**15.5**	0.9	**16,273**	2.7	**2267.8**	13.6
Russia	13.2	11.6	0.1	**11.3**	10.8	0.4	9224	1.7	1384.0	15.6
Turkmenistan	5.7	**35.7**	N/A	6.5	12.5	0.7	N/A	**4.0**	1513.1	19.9

Source: Sustainable Development Report 2022

Table 2 SDG-based Indices of the Caspian Sea region, 2022

	Freshwater withdrawals (in % of available freshwater resources)	Anthropogenic wastewater receiving treatment (%)	Access to clean fuels and technologies (% of population)	Renewable energy sources in total primary energy supply (%)	Access to an improved water source, piped water (% of urban population)	Unemployment rate (% of total labor force)	Research & development expenditure (% of GDP)	Average area of protected marine sites important for biodiversity (%)	Fish caught from overexploited or disturbed stocks (% of total catch)	Average area protected in terrestrial areas important for biodiversity (%)	Adjusted GDP growth rate (%)	Government expenditure on health and education (% of GDP)
Azerbaijan	53.7	3.8	96.7	1.4	100	6.6	0.2	N/A	N/A	36.6	–3.2	4.0
Iran	81.3	3.7	96.5	0.7	99.5	**11.1**	0.8	68	14.4	43.6	–6.9	7.0
Kazakhstan	32.7	**28.6**	97.6	1.5	94.0	4.9	0.1	N/A	N/A	13.1	–0.7	4.5
Russia	4.0	18.5	90.1	2.6	96.9	4.7	1.0	23.6	35.6	25.1	1.6	8.1
Turkmenistan	**144**	9.8	99.9	0.0	77.4	5.1	N/A	N/A	N/A	14.0	–0.8	4.3
Average	63	12.9	96	1.24	93.5	6.5	0.4	N/A	N/A	26	–2	5.6

Source: Sustainable Development Report 2022

methodology, and (iv) the adaptability to the global context. The suggested metrics evaluate the effectiveness of resources utilized (Russia leads with a score of 50.55), the protection of natural resources (Azerbaijan and Iran score 65.96 and 56.19, respectively), and the enhancement of ecological sustainability (Russia leads with a score of 37.17) (Table 3). Nevertheless, the Caspian Sea region's average values remain relatively low.

Subsequently, OECD in collaboration with the Global Green Growth Initiative (GGGI), UNEP, and the WB, compiled the report "Moving towards a Common Approach on Green Growth Indicators" (OECD, 2013) as a component of the Green Growth Knowledge Platform (GGKP). The document incorporated modifications to the existing approach, including an assessment of carbon and resource effectiveness, multifactor productivity, the impact of air pollution on the populace, and other effects. In 2019, the GGGI has proposed a methodology for analyzing developing nations efforts to achieve sustainable growth, based on evaluating the *Global Green Economy Index (GGEI)*, which measures (i) the alignment of national growth plans with the fundamental principles of environmental responsibility (diagnostic indicators); (ii) the generation of alternative green growth scenarios and the identification of dependencies between environmental concerns (planning indicators); and (iii) the facilitation of progress tracking and evaluation of the outcomes of green growth planning (GGGI, 2019).

GGEI ranks Kazakhstan as the top Caspian littoral country, followed by Azerbaijan and Russia (Table 4). Turkmenistan demonstrates the lowest index values, while Iran is not represented in the ranking.

Table 3 OECD green growth index, 2020, 2022

	Criteria				Green Growth Index		
	Efficient use of resources	Protection of natural resources	Opportunities for economy greening	Social inclusivity	Points	Level	Place
Japan (Asian region leader)	55.74 60.68	71.10 71.04	44.88 32.41	82.16 80.48	61.83 65.03	High High	1 1
Azerbaijan	44.94 43.97	65.96 64.89	21.17 19.84	64.01 59.54	44.77 49.98↑	Moderate Moderate	13 21
Kazakhstan	45.95 51.21↑	43.21 47.73	8.58 17.60↑	81.40 78.30	34.32 51.19↑	Low Moderate	26 20
Iran	48.96 21.51↓	56.19 57.30	N/A 16.50	67.88/ 60.20	N/A 38.27	N/A Low	N/A 35
Turkmenistan	N/A 16.55	45.49 45.20	N/A	N/A 71.73	N/A	N/A	N/A
Sweden/ Austria (European region leaders)	87.78 78.97	78.14 80.28	59.53 38.99	94.06 93.45	78.72 77.78	High High	1 1
Russia	50.55 53.86	55.84 55.93	37.17 19.93↓	77.88 77.26	53.46 54.96	Moderate Moderate	29 37

Source: Green Growth Index 2020/2022 GGGI Report (GGGI, 2020; GGGI, 2022)

Table 4 Global green economy index, 2022

State	Aggregated result		State	Success of measures taken	
	Percentile	Place		Percentile	Place
Sweden (leader)	0.799	1	Jordan (leader)	0.780	1
Kazakhstan	0.491	**84**	Kazakhstan	**0.634**	**15**
Azerbaijan	0.448	114	Azerbaijan	0.497	96
Russia	0.414	136	Turkmenistan	0.480	109
Turkmenistan	0.318	158	Russia	0.461	116
Iran	N/A	N/A	Iran	N/A	N/A
Oman (last)	0.262	160	Mozambique (last)	0.234	160

Source: Global Green Economy Index 2022 Report (Dual Citizen, 2022)

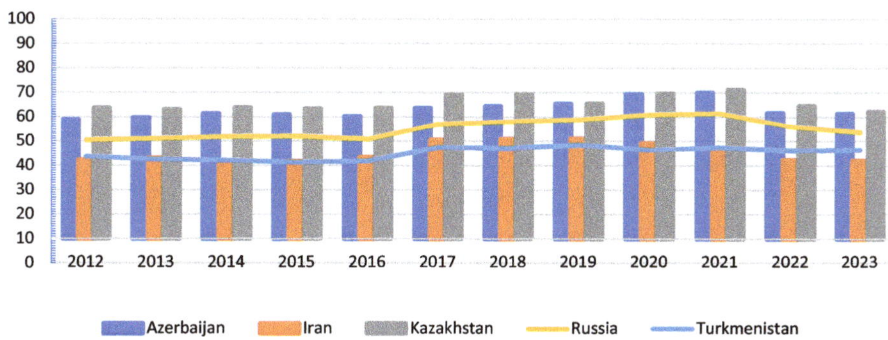

Fig. 1 Economic Freedom Index, %. Source: Compiled by the author based on the 2023 Index of Economic Freedom Report

Finally, the *Economic Freedom Index* (EFI) measures economic openness reflecting the success of ongoing reforms toward liberalization and consistency in sustainable development taken. EFI demonstrates that, over a 10-year period, the Caspian littoral states have not made significant progress in opening up, whereas since 2021, there has been a steady decline in economic openness (Fig. 1)

3 Fossil Fuel Export-Dependent Countries in Implementing a Global Green Agenda: An Empirical Analysis

In the preceding chapter, the examination of indicators revealed that not all nations have made significant progress in sustainable growth. Therefore, a logical question arises: *Does the economic dependence on fossil fuel exports affect the implementation of the GG Agenda?*

To answer this question, the author examined the data of 36 countries—fossil fuels exporters, including 11 high-income, 11 upper-middle income, 8 lower-middle income, and 6 low-income economies (Appendix A). The brief examination of the country data has led to the conclusion that the export orientation toward fuel fossils is solely a characteristic of emerging economies. The percentage of resource (fossil fuel) revenue in total government revenues makes Iraq and Libia the leaders (>90%), followed by South Sudan and Brunei (70–90%), Kuwait and Saudi Arabia (60–70%), the Democratic Republic of Congo, Angola, and Azerbaijan (50–60%). Other states exhibit results that are below 50%.

It may be anticipated that low-income economies exhibit the highest level of ND-GAIN Vulnerability Score, which measures sensitivity and capacity to adapt to the adverse effects of climate change. Simultaneously, certain low-income African states, including the Democratic Republic of Congo, Mozambique, and Chad, demonstrate the highest percentages of Renewable Energy Consumption (REC) of total final energy consumption, namely 96.16%, 80.91%, and 73.74%, respectively. This indicates the efficacy of ongoing international programs (SWERA, AFREA, AREED) aimed at promoting renewable energy production on the African continent.

Four groups of variables were considered for further empirical studies (Table 5):

- *Green Growth* Variables: SDG Index, Green Growth Index
- *Climate Change* and *Green Transition* Variables: CO_2 Total Emissions, ND-GAIN Vulnerability Score, Renewable Energy Consumption
- *Fossil Fuel Dependence* Indicators: Fuel Exports, Resource Revenue, Dependency Index
- *Macroeconomic* Variables: GDP Growth, GDP per Capita, Industry value added

Following an analysis of empirical literature (Wang & Zhang, 2023; Jensen, 2023), the author forms a correlation matrix (Table 6). The resulting values indicated that economically robust countries, according to GDP_CAP data, characterized by elevated financial capacity and technological resilience, are expected to demonstrate significant success in achieving the Sustainable Development Goals (SDGs) (positive correlation 0.428 GDP_CAP/SDG_I). These findings are consistent with existing empirical studies, such as Selomane et al. (2015) and Wu et al. (2023). However, the GDP_CAP/GGI correlation is not so clear-cut (Afshan et al., 2023). Nevertheless, nations with lower GDP per capita values are more susceptible to climate vulnerability, which can be attributed to the ineffectiveness of fiscal policy (Zakharova & Medas, 2009), instability of financial markets (Naifar, 2023), FDI deficiency (Shear et al., 2023), and other factors.

With some degree of certainty, it can be contended that nations with higher levels of development are more likely to reduce their fuel exports (as exemplified by a negative correlation of −0.322 GDP_GR/FUEL_EX) and orient their economies toward sustainable development (a positive correlation of 0.328 GDP_GR/SDG_I). This is also augmented by regional development strategies aimed at ensuring energy independence (EU), enhancing the volume of exported oil by increasing internal solar energy consumption (OPEC), safeguarding fossil fuel reserves for future generations (Middle Eastern countries), or guaranteeing access to electricity for rural and remote areas (low-income economies) (UN, 2021).

Table 5 Variables' description

Green growth variables		
SDG_IN	SDG Index	Countries are ranked by their overall score, which measures the total progress toward achieving all 17 SDGs. The score can be interpreted as a percentage of SDG achievement. A score of 100 indicates that all SDGs have been achieved.
GGI	Green Growth Index	Measures country performance in achieving sustainability targets including SDGs, Paris Climate Agreement, and Aichi Biodiversity Targets for four green growth dimensions: efficient and sustainable resource use, natural capital protection, green economic opportunities, and social inclusion.
Climate change and green transition variables		
CO2_EM	CO_2 Total Emissions	In mil tons. Contains sector-level greenhouse gas (GHG) emissions data, including emissions of the six major GHGs from most major sources and sinks.
NDG_VUL	ND-GAIN Vulnerability Score	ND-GAIN measures overall vulnerability by considering six life-supporting sectors food, water, health, ecosystem services, human habitat, and infrastructure.
REN_CONS	Renewable Energy Consumption	Percent of total final energy consumption
Fossil fuel dependence indicators		
FUEL_EX	Fuel Exports	Percent of total exports, annual average
FUEL_REV	Resource Revenue	Percent of government revenue
DI	Dependency Index	Average of the fuel share of exports and fossil fuel share of revenue (latter as the resource revenue share of total revenue adjusted by the ratio of fossil fuel rents to total non-renewable rents). For countries with missing revenue estimates, the index is just the fuel share.
Macroeconomic variables		
GDP_GR	GDP Growth	Annual, %
GDP_CAP	GDP per Capita	In USD
INDUST	Industry (inc. construction), value added	Percent of GDP

The above-mentioned thesis is also supported by the positive correlation (0.304) between the SDG Index and economic productivity (INDUST). It is anticipated that the INDUST relies on the consumption of fossil fuels, as substantiated by Anser et al. (2020) study. However, it is also rebutted that countries dependent on fossil fuel exports have significant effects on CO_2 emissions (correlation −0.06).

Lastly, there is no apparent dependence between fossil fuel exports and the green agenda implementation. Nonetheless, there is a negative correlation (−0.434)

Table 6 Correlation matrix. Source: Composed by the author

	Green Growth Variables		Climate Change & Green Transition Variables			Fossil Fuel Dependence Indicators			Macroeconomic Variables		
	SDG_I	*GGI*	*CO2_EM*	*NDG_VUL*	*REN_CONS*	*FUEL_EX*	*FUEL_REV*	*DI*	*GDP_GR*	*GDP_CAP*	*INDUST*
SDG_I	1										
GGI	0.493188	1									
CO2_EM	0.214936	0.06468	1								
NDG_VUL	-0.89316	-0.50954	-0.17591	1							
REN_CONS	-0.55307	0.251899	-0.10112	0.556606	1						
FUEL_EX	-0.10518	-0.15567	-0.06357	-0.11733	-0.08543	1					
FUEL_REV	-0.14682	-0.4348	-0.03365	-0.05538	-0.2908	0.699107	1				
DI	-0.10119	-0.31633	-0.06747	-0.10149	-0.2213	0.918409	0.920848	1			
GDP_GR	0.328926	-0.07514	0.030238	0.00739	-0.01262	-0.32228	-0.12967	-0.2833	1		
GDP_CAP	0.428214	0.244329	-0.08578	-0.55165	-0.18917	0.149858	-0.00177	0.088752	0.158996	1	
INDUST	0.304393	-0.04386	-0.08445	-0.34786	-0.25487	0.596266	0.627015	0.66015	0.10946	0.410558	1

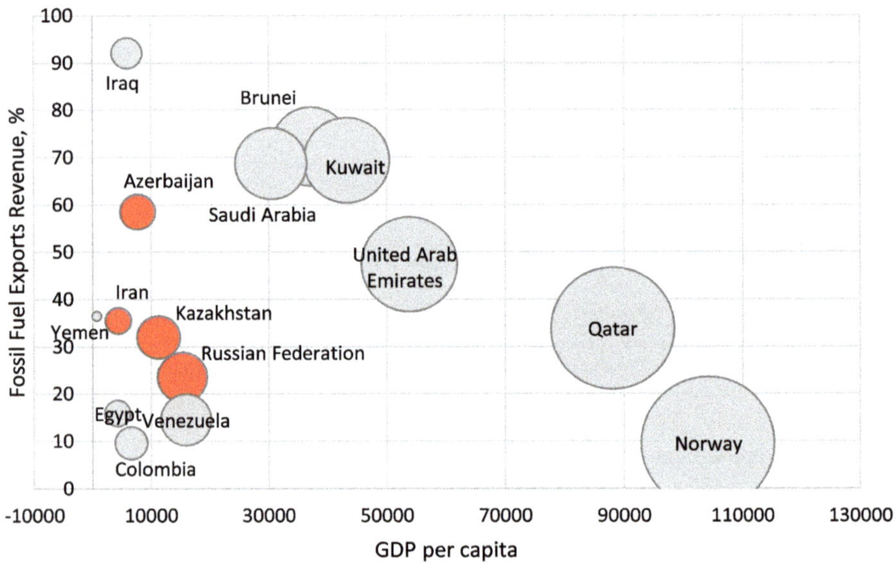

Fig. 2 GDP per capita, USD/Fossil Fuel Exports Revenue, % of Total Governmental Revenues. Source: compiled by the author based on the WB database and International Energy Agency (IEA) data. *Note:* The size of bubbles measures GGI

between GGI and FUEL_REV. Thus, more developed economies (Norway, Qatar, UAE), geared toward eco-friendly development, exhibit a high degree of diversification of government revenues, not limited to fossil fuel exports (Fig. 2). These findings are consistent with Karanfil and Omgba (2023), Solano-Rodríguez et al. (2021), and Shrestha et al. (2021) studies.

Thus, fossil fuel export dependence does not limit countries' capacity for green transformation. However, their "green" achievements are determined by institutional, financial, and technological resiliency, as well as local and regional priorities.

4 Caspian Littoral States in Complying with the Nationally Determined Contribution (NDC) Plan

4.1 Nero Zero Target

According to the National Development Plans (NDPs) of the Caspian states (UNFCCC, 2022), reducing greenhouse gas emissions (GGEs) appears to be the most significant challenge, necessitating transformation within the Global Greenhouse Hub (GGH).

Kazakhstan first declared its intention to achieve "carbon neutrality" (net zero CO_2 emissions) by 2060 during the Climate Ambition Summit held in December 2020. In September 2021, the government of Kazakhstan issued a Doctrine (strategy) for achieving carbon neutrality until 2060. The target covers the most important greenhouse gases (GHGs) and provides some assumptions on intended removals by land- and technology-based solutions. As per the recently implemented Environmental Code, it is anticipated that the 50 largest companies, which account for 80% of emissions in Kazakhstan, will replace their outdated technologies with the most advanced technologies (BATs) by the year 2025.

In 2021, *Russia* enacted a law aimed at regulating the emission of greenhouse gases. Unlike the initial version, it does not mandate emission quotas or impose penalties on large emitters of GHGs. Instead, it simply requires companies to report their emissions. The Russian Energy Strategy to 2035 focuses exclusively on promoting fossil fuel extraction, consumption, and exports to the rest of the world. However, such a strong focus on increasing reliance on fossil fuel revenues poses a significant economic risk in a future compatible with the Paris Agreement's 1.5 °C temperature limit.

Azerbaijan introduced its NDGs for preventing global climate change in October 2015, with the aim of reducing GGEs to 35% by 2030.

Iran has pledged to mitigate climate change and advance clean energy (Mahoozi, 2021). Nonetheless, as a consequence of the deteriorating relations with Western nations and the imposition of sanctions, experts have observed negative trends and expressed apprehension regarding the prospects for Iran's further green transition.

Turkmenistan is not represented in the NDC tracker, which indicates a significant lag in the GG Agenda implementation. However, the diversification of hydrocarbon export flows toward China holds the potential to result in a more favorable climate policy. In the year 2022, Turkmenistan formally accepted membership in the Global Gender Gap Initiative (GGGI) and commenced methodological research on the NDCs.

According to the Climate Action Tracker (CAT), Caspian states exhibit either "insufficiency" or "highly insufficiency" in their commitment toward the Paris Agreement (Table 7). This implies that the countries' climate policies are inconsistent and require significant enhancements. The absence of significant contributions from nations to global climate objectives, as well as the absence of sufficient

Table 7 Country's Climate Policies and Commitments toward Paris Agreement, 2022

State	Policies and action against fair share	NDC target against modelled domestic pathways	NDC target against fair share	Climate finance	Overall rating
Kazakhstan	Insufficient < 3 °C world	Almost sufficient < 2 °C world	Insufficient < 3 °C world	Not assessed	**Insufficient**
Russian Federation	Highly insufficient < 4 °C world	Highly insufficient < 4 °C world	Critically insufficient 4 °C+ world	Critically insufficient	**Critically insufficient**
Iran	Critically insufficient 4 °C+ world	Critically insufficient 4 °C+ world	Critically insufficient 4 °C+ world	Not assessed	**Critically insufficient**
Azerbaijan	N/A	N/A	N/A	N/A	N/A
Turkmenistan	N/A	N/A	N/A	N/A	N/A

Source: Climate Action Tracker

domestic climate policy measures, resulted in an overall rating as "critically insufficiency." Azerbaijan and Turkmenistan are not included in this rating.

With regulatory modifications, achieving the Nero Zero objective will necessitate the advancement of green taxonomy (UNECE, 2022). Several Caspian Sea nations have already experienced a certain success. Kazakhstan authorized the adoption of a taxonomy[3] for environmentally friendly projects that could be financed through green instruments. According to the ECE report (2022), Baiterek National Managing Holding JSC possesses significant expertise in green banking. It is currently exploring a green taxonomy for new financial products (e.g., loans for energy-efficient housing, green infrastructure lending, and loans for sustainable development for companies and SMEs) to improve their financial performance. In 2022, the AIFC Green Finance Center was launched in order to implement the issuance of green bonds on the KASE and AIX exchanges.

In 2021, the Russian government, in cooperation with VEB, developed a green taxonomy,[4] aiming to reduce carbon dioxide emissions, increase recycling of materials, and protect and restore biodiversity and natural sites. The Carbon Border Adjustment Mechanism (CBAM) is expected to be implemented by 2026, with the objective of preventing carbon leakage, promoting greening production, and achieving EU climate change mitigation targets (PCGCC, 2009) in accordance with the WTO regulations (2022).

[3] Order of the Government of the Republic of Kazakhstan of December 31, 2021, No. 996 about Approval of Classification (Taxonomy) of the "Green" Projects which are Subject to Financing through "Green" Bonds and the "Green" Credits.

[4] Resolution no. 1587 of September 21, 2021, on the Approval of the Criteria for Sustainable (Including Green) Development Projects in the Russian Federation and the Verification System Requirements

The financial markets of Iran and Turkmenistan are not yet sufficiently developed to support an ESG Agenda. However, governments are considering the possibility of utilizing green financing instruments in the future.

4.2 Energy Sector

The economies of the Caspian nations are heavily dependent on fossil fuels. The size of fossil fuel exports in 2022 significantly exceeded the world average (Azerbaijan 90.6%, Turkmenistan 89.8%, Iran 66.3%, Kazakhstan 62.8%, and Russian Federation 53% in total exports), as well as the percentage of government revenues from oil exports (the leaders were Turkmenistan 62.1% and Azerbaijan 58.5%) (Appendix A).

The dependence of the Caspian states on energy resources is transforming into a real concern, aggravated by the high volatility of the segment and geopolitical factors, the most significant of which being the Russian invasion of Ukraine. The sanctions imposed by numerous nations resulted in a decrease of 18% in *Russias* exports compared to the record level achieved at the commencement of the invasion (February–March 2022) primarily attributed to the reductions in the exports of pipeline gas, oil products, and coal (CREA, 2022). By the end of 2022, the EU had imposed a partial embargo on the import of Russian oil and petroleum products and stopped the import of Russian coal.

Sanctions had a significant impact not only on the Russian economy but also on the economies of neighboring countries. Most of Kazakhstan's oil exports pass through Russia to Ukraine, from which they are fed into the European pipeline network. As a consequence of the conflict, a number of export routes were halted, and prices surged due to the elevated war risks (O'Byrne, 2022).

Iran, being under the influence of international sanctions, is facing a shortage of fossil gas due to high domestic demand and a lack of funding for industrial modernization (US EIA, 2022).

Despite the necessity for structural modifications, the Caspian nations persist in implementing their long-term strategies to enhance oil production. The Russian energy plan for the next few decades includes continuing to supply natural gas to the global marketplace, with a gradual increase in the proportion of energy derived from sustainable sources. Kazakhstan intends to enhance its oil production by 15.8% by 2025 (IEA, 2022) and plans to expand its energy sector by coaled methane exploration. According to Iran's development plan for 2023–2027, there will be no reduction in electricity production from fossil fuels.

Such an approach appears to be an obstacle to achieving the Agenda 21 (UN, 1992) objectives and negates all states' attempts to reduce emissions.

Despite the substantial potential of the Caspian nations in the use of renewable energy sources (Sachs et al., 2022), their contribution to the total electricity production is currently insignificant, with the exception of Azerbaijan, which accounts for 18% (Table 8). However, most countries intend to increase these numbers according

Table 8 Renewable energy share in the energy balance, %

State	2020	by 2030
Azerbaijan	**18**	**30**
Iran	1	16
Kazakhstan	3	15
Russia	**0.5**	**2**
Turkmenistan	0	N/A

Source: Sustainable Development Report 2022

Table 9 Renewable energy in the energy balance in TJ, 2020

State	Solar energy	Wind energy	Hydropower	Solid biofuels	Renewable waste	Geothermal energy
Azerbaijan	169	346	3850	1383	3018	N/A
Iran	2102	2113	**79,225**	341	N/A	N/A
Kazakhstan	6442	3876	**34,777**	5427	N/A	N/A
Russia	4497	5044	768,605	277,059	N/A	4255
Turkmenistan	N/A	N/A	N/A	N/A	N/A	N/A

Source: IRENA 2020 Report

to their long-term strategies. Azerbaijan anticipates enhancing the proportion of REC in the country's energy balance to 30% by 2030. Iran and Kazakhstan stand for less ambitious plans—15% and 16% respectively. There is considerable uncertainty regarding Russia and Turkmenistan.

Among other renewable energy sources, the greatest potential demonstrates hydropower (Table 9). However, the predicted increase in water scarcity due to global warming is driving interest in wind and solar energy. In collaboration with international financial organizations, within the long-term programs, such as the Kazakhstan National Wind Power Development Program (UNDP, 2011), the Offshore Wind Roadmap for Azerbaijan (WB, 2022b), the practice of erecting wind turbines is rapidly spreading across nations.

Recently, the public funding for renewable energy in the Caspian countries has experienced a significant increase, however, the majority of investment is sourced from international financial organizations (IRENA, 2019). In order to attract private investment, Kazakhstan and Azerbaijan implemented an auction system instead of the previously used fixed tariffs. Foreign investment in renewable energy in Iran and the Russian Federation has been stalled as a result of economic sanctions.

4.3 Transport

Kazakhstan, the largest emitter of GHGs in Central Asia, is proactively implementing measures to mitigate the detrimental environmental repercussions entailed by the growth of the transportation sector. Municipal authorities are engaged in enhancing the technologies utilized for public transport management and enhancing air quality. Two automobile manufacturers, namely AZIA AVTO and Saryarka AvtoProm, are launching domestic electric vehicles, while the Ministry of Energy is promoting the construction of charging stations (WB, 2018). Kazakhstan also engages in interstate initiatives aimed at enhancing the environmental safety of railway transportation. The JSC National company "Kazakhstan Temir Zholy" has signed an agreement on the development of cooperation in environmental protection with the CIS countries and members of the International Union of Railways (IUR).

In the year 2021, *Russia* approved the Transport Strategy,[5] which encompasses measures to develop electric vehicles, low-carbon infrastructure, and alternative fuels, with the objective of reducing transport emissions to 1.2% by 2030. As a result of the measures implemented, which included the provision of subsidies covering up to 25% of the expenses of domestically produced electric vehicles, their sales surpassed a triple by the end of 2021 (Bouckley, 2022). Such demand could have allowed it to achieve the target of 10% EV in total vehicle production by 2030. However, the economic sanctions imposed on Russia have restricted the imports of crucial technology components, such as semiconductors, resulting in a significant decrease in domestic EV production in 2022 (Bogdanova, 2022).

The status of the region as a transit zone has led to cooperation between the Caspian countries in transport and logistics. The foremost region's objective is to establish contemporary logistics hubs to facilitate intermodal transportation and enhance the standard of service. Simultaneously, the notion of green logistics encompasses the reduction of air, water, and soil pollution, reduction of energy consumption, minimization of traffic congestion, and enhancement of safety.

The logistics in *Kazakhstan* are distinguished by a high level of energy consumption. In the year 2022, the combined cargo weight transported by water and land amounted to 157,940 tons, representing a rise of 5.4% in comparison to the preceding year. The Aktau port, which has a capacity of 12 million tons of cargo per year, comprising petroleum products and household goods, surpasses the largest Caspian port, Baku (Table 10). The sector, however, accounts for 23.7% of the total pollution emissions into the atmospheric air and for 4.7% of CO_2 emissions (WB, 2022a). The authorities are currently implementing measures to address air pollution issues,

Table 10 Cargo turnover in the Caspian Sea ports

State/ Port	Freight turnover in 2020, tons / to previous year growth, %	Port capacity, tons / Cargo turnover in 2022, tons / %
Azerbaijan/ Baku	4.8 mil / + 20%	15 mil / 6.3 mil / + 13.6%
Kazakhstan/ Aktau, Kuryk, Bautino	4.8 mil / - 9%	21 mil / 5.6 mil / + 19%
Russia/ Astrakhan, Olya, Makhachkala	8.1 mil / +9.5%	- / 6.0 mil /– 13.9%

Source: World Bank Reports 2018, 2022

[5] Concept for the development of the production and use of electric road transport in Russia until 2030 (2021).

such as enacting stricter fuel regulations, optimizing the transportation and distribution infrastructure, and encouraging the generation of non-carbon emissions. Nonetheless, according to the WB (2022a), the taken measures are insufficient.

To address the issue of air pollution and GGEs, a new priority for greening in *Azerbaijan* is the promotion of a modal transition from roads to railways. However, the proportion of rail transportation is steadily declining (from 20% in 2000 to 6% in 2019), while the proportion of passenger and cargo traffic by road is increasing (from 50% to 66%, respectively) (ITF, 2020). Hence, the execution of the modal transition will necessitate coordinated efforts from various government authorities toward the enhancement of railway infrastructure and logistics efficacy.

There are several transport corridors on the territory of *Turkmenistan*, such as TRACECA corridor, OSJD corridors VI and X, CAREC Corridors II and III. Turkmenistan appears as a crucial component of the "Great Silk Road" project within the global "One Belt—One Road" initiative. The primary directions for advancement encompass the modernization of the Turkmenbash international seaport, the development of internal roads and railways, and the enhancement of legal baking for international transportation.

The countries of the Caspian littoral, with the exception of Iran, are currently implementing the "Promoting Green Ports and Connectivity in the Caspian Region" (OSCE) project, which aims to develop green ports and transform them into green logistics hubs, as a part of the gradual decarbonization of the transportation system. The OSCE provides assistance to Caspian Sea ports in implementing the EcoPorts Port Environmental Review System (PERS) and also initiates working groups to investigate the renewable energy potential for electrifying ports and related logistics hubs (OSCE, 2021).

5 Conclusions and Recommendations

The Caspian states are currently in the initial stages of forming a sustainable and environmentally friendly economic strategy, gradually transforming the traditional export model of raw materials into a resource-saving one, taking into consideration national priorities. However, the economic dependence on fossil fuels and limited administrative capacity appear to pose obstacles in attaining the objectives of Agenda 21 (UN, 1992). However, the continued exploitation of fossil resources, which produce the highest level of GGEs (Gütschow et al., 2016), undermines all efforts to reduce emissions to 1.5 °C.

Despite the immense potential of alternative and sustainable energy (ASE) in the region, the economic conditions remain unfavorable to both ASE producers and consumers, owing to the high cost of construction and operation of installations in comparison to conventional thermal power facilities. The continued expansion of the energy and fuel sector would require the implementation of governing rules and guidelines, as well as the replacement of outdated, conventional approaches in the generation of electrical and thermal energy with the more modern ones. The energy transformation should be directed toward:

i) The application of innovative and advanced environmentally friendly technologies in oil and gas processing, fuel production in accordance with EURO-5 standards, modernization of gas pipelines, and gas distribution networks.
ii) The development and adoption of technical and regulatory documents will expand the use of innovative technologies, construction of hydroelectric power plants on small rivers, irrigation canals and reservoirs, and the use of biomass.

The subsequent economic greening of the Caspian littoral countries will necessitate the implementation of measures aimed at combining conventional command-administrative regulations with the novel ones, as well as market instruments for reorienting financing toward green investments, the development of environmentally friendly technologies, and the enhancement of resource efficiency.

Appendix

Appendix A Correlation matrix data

Country	SDG_I	GGI	CO2_TOT_EM	NDGAIN_VUL	REN_CONS	FUEL_EX	FUEL_REV	DI	GDP_GR	GDP_CAP	INDUST
Algeria	70.8	45.67	267.01	0.4	0.15	95.8	39.5	67.6	3.1	4274	42.3
Angola	50.8	44.26	119.41	0.5	61.02	93.5	53.5	73.5	3	2999	44.9
Azerbaijan	73.5	57.18	53.43	0.41	1.23	90.6	58.5	74.5	4.6	7737	55.9
Bahrain	63.74	39.49	54.15	0.45	N/A	49.7	N/A	49.7	4.9	30,152	47.9
Bolivia	68.85	58.71	131.43	0.47	16.86	34.1	18.4	26.3	3.1	3523	26.4
Brunei	65.7	52.6	12.23	0.38	0.01	91.4	72.3	81.9	−1.6	37,152	67.9
Cameroon	55.15	52.6	128.4	0.47	78.94	36.6	14.9	25.7	3.5	1588	25.5
Chad	45.34	N/A	114	0.62	73.74	82.7	25.5	54.1	2.2	717	49.5
Colombia	70.05	56.8	270.31	0.41	31.31	53.4	9.6	31.5	7.5	6630	26.7
Congo, Dem. Republic	48.58	N/A	688,06	0.59	96.16	52.3	53.2	52.8	8.9	586	48.6
Ecuador	70.4	57.23	94.16	0.45	20.1	37	N/A	37	2.9	6391	31.5
Egypt	69.6	37.74	299.96	0.44	6.51	20.4	15.9	18.1	6.6	4295	32.7
Gabon	63.09	61.48	21.16	0.42	90.12	68.1	35.8	52	3	8820	57.4
Iran	69.09	37.35	844.41	0.4	1.03	66.3	35.5	50.9	2.7	4388	40.2
Iraq	64.79	33.63	261.79	0.45	1.08	98	92.1	95	7	5937	62.7
Kazakhstan	71.65	54.29	291.82	0.35	1.78	65.8	32	48.9	3.2	11,244	34.5
Kuwait	64.4	39.35	135.89	0.41	0.1	92.8	69.4	81.1	8.2	43,233	45.4
Libya	N/A	33.82	75.37	0.42	3.12	92.2	91.5	91.9	−1.2	6716	80.3
Malaysia	69.85	59.84	376.76	0.37	5.84	15.2	21.1	18.2	8.7	11,971	39.2
Mongolia	64.69	48.6	76.38	0.4	3.98	36.3	15.7	26	4.8	4947	37
Mozambique	52.69	45.33	103.81	0.52	80.91	42	13	27.5	4.1	541	21.9
Nigeria	54.27	46.05	369.38	0.5	82.51	92.3	44.6	68.4	3.3	2184	30.8
Norway	82	65.74	31.33	0.27	61,0.29	57.2	9.5	33.3	3.3	104,149	48.7

Country	SDG_I	GGI	CO2_TOT_EM	NDGAIN_VUL	REN_CONS	FUEL_EX	FUEL_REV	DI	GDP_GR	GDP_CAP	INDUST
Oman	68.59	40.9	95.1	0.42	0.05	75.3	N/A	75.3	4.3	25,057	57
Papua New Guinea	53.6	43.82	51.65	0.52	53.48	44.2	7	25.6	4.6	3020	33
Qatar	64.24	46.52	119.61	0.39	0.06	86.1	33.8	59.9	4.8	88,046	65.6
Russian Federation	73.79	58.03	1799.98	0.34	3.72	53	23.5	38.2	−2.1	15,345	32.8
Saudi Arabia	67.69	40.77	712.59	0.4	0.06	73.6	68.7	71.1	8.7	30,436	53.3
South Sudan	38.68	N/A	70.61	N/A	33.22	96.1	85.1	90.6	−10.8	1072	33.1
Sudan	48.55	38.2	128.14	0.61	62.13	23.8	13.3	18.5	−1	1102	4.9
Trinidad and Tobago	62.98	49.7	26.61	0.39	0.47	51.4	33.8	42.6	2.5	18,222	42.2
Turkmenistan	68.47	N/A	194.09	0.41	0.06	89.8	N/A	89.8	−3.4	7297	43
United Arab Emirates	69.71	50.76	249.93	0.37	0.92	50.3	47.4	48.9	7.4	53,758	47.5
Uzbekistan	71.15	43.39	194.15	0.4	1.03	19.2	N/A	19.2	5.7	2255	31.1
Venezuela	62.88	54.86	237.63	0.38	23.33	86.7	14.5	50.6	−3.9	15,976	37.2
Yemen	46.85	32.42	24.39	0.55	3.53	47.6	36.5	42	0.8	677	15.2

Source: Composed by the author based on World Bank (WB) data, Sustainable Development Report, ND-GAIN, Climate Watch (CW) data, Global Green Growth Institute (GGGI) data, International Energy Agency (IEA) data, International Renewable Energy

References

Afshan, S., Yaqoob, T., Ho, W. K., & Leong, K. J. (2023). Achieving sustainable growth in emerging economies: Insights from advance method moment of quantile regression. *Gondwana Research., 127*, 182. https://doi.org/10.1016/j.gr.2023.08.003

Anser, M. K., Hanif, I., Alharthi, M., & Chaudhry, I. S. (2020). Impact of fossil fuels, renewable energy consumption and industrial growth on carbon emissions in Latin American and Caribbean economies. *Atmosfera, 3*(33). https://doi.org/10.20937/atm.52732

Arze del Granado, J. F., Coady, D., & Gilligham, R. (2012). The Unequal Benefits of Fuel Subsidies: A Review of Evidence for Developing Countries. *World Development, 40*(11), 2234–2248. https://doi.org/10.1016/J.WORLDDEV.2012.05.005

Barbier, E. B. (2020). Is Green Rural Transformation Possible in Developing Countries? *World Development, 131*, 104955. https://doi.org/10.1016/j.worlddev.2020.104955

Bogdanova, I. (2022). The role of technology sanctions in crippling Russia's War machine. *IISD*. Accessed Dec 13, 2023, from https://www.iisd.org/articles/policy-analysis/technology-sanctions-russia-war

Bouckley, E. (2022). Russia EV sales to grow 40%–45% in 2022, but mass adoption remains distant. *S&P global commodity insights*. Accessed Dec 13, 2023, from https://www.spglobal.com/commodityinsights/en/market-insights/latest-news/energy-transition/020722-feature-russia-ev-sales-to-grow-40-45-in-2022-but-mass-adoption-remains-distant#article0

CEP. (2003). *Framework convention for the protection of the marine environment of the Caspian Sea* … Accessed Oct 03, 2023, from https://faolex.fao.org/docs/pdf/mul63152E.pdf

CREA. (2022). *Financing Putin's war: Fossil fuel exports from Russia in the first six months of the invasion of Ukraine*. Accessed Dec 13, 2023, from https://energyandcleanair.org/wp/wp-content/uploads/2022/09/Final_-Financing-Putins-war_-Fossil-fuel-exports-from-Russia-in-the-first-six-months-of-the-invasion-of-Ukraine.pdf

Dual Citizen. (2022). *The global green economy index 2022 by Dualcitizen*. Accessed Oct 03, 2023, from https://dualcitizeninc.com/results-from-the-2022-global-green-economy-index-ggei/

ECE. (2022). *Implementing the Pan-European strategic framework for greening the economy and the Batumi initiative on green economy: Compilation of responses to the Survey on the implementation of BIG-E commitments*. Ninth Environment for Europe Ministerial Conference. Accessed Oct 03, 2023, from https://unece.org/sites/default/files/2022-10/ece.nicosia.conf_.2022.inf_.6.pdf

GGGI. (2019). *Green growth index: Concept, methods and applications*. Accessed Oct 03, 2023, from https://greengrowthindex.gggi.org/wp-content/uploads/2019/12/Green-Growth-Index-Technical-Report_20191213.pdf

GGGI. (2020). *Green growth index 2020 GGGI: Measuring performance in achieving SDG targets/Coordinating Lead Author: Lilibeth A. Acosta*. Technical Report No.1. Republic of Korea. p. 138. Accessed Oct 03, 2023, from https://greengrowthindex.gggi.org/wp-content/uploads/2021/01/2020-Green-Growth-Index.pdf

GGGI. (2022). Green growth index 2022 GGGI: Measuring performance in achieving SDG targets/Coordinating Lead Author: Lilibeth A. Acosta. Technical Report No.27. Republic of Korea. 196.

Gütschow, J., Jeffery, L. M., Gieseke, R., Gebel, R., Stevens, D., Krapp, M., & Rocha, M. (2016). The PRIMAP-hist national historical emissions timeseries. *Earth System Science Data, 8*(2), 571–603. https://doi.org/10.5194/essd-8-571-2016

IEA. (2022). *Kazakhstan 2022 energy sector review*. Accessed Dec 13, 2023, from https://iea.blob.core.windows.net/assets/fc84229e-6014-4400-a963-bccea29e0387/Kazakhstan2022.pdf

IRENA. (2019). Renewable capacity statistics 2019. *International Renewable Energy Agency (IRENA), 47*, 446. https://doi.org/10.1016/S2255-4971(15)30126-9

ITF. (2020). *Decarbonising Azerbaijan's transport system: Charting the way forward, international transport forum policy papers. No. 87*. OECD Publishing. p. 28. Accessed Oct 03, 2023, from https://www.itf-oecd.org/sites/default/files/docs/decarbonising-azerbaijan-transport-system.pdf

Jensen, L. (2023). *Global decarbonization in fossil fuel export-dependent economies–fiscal and economic transition costs*. UNDP. p. 57. Accessed Dec 13, 2023, from https://www.undp.org/sites/g/files/zskgke326/files/2023-05/Global%20Decarbonization%20in%20Fossil%20Fuel%20Export-Dependent%20Economies.pdf

Karanfil, F., & Omgba, L. D. (2023). The energy transition and export diversification in oil-dependent countries: The role of structural factors. *Ecological Economics, 204*, 107681. https://doi.org/10.1016/j.ecolecon.2022.107681

Mahoozi, S. (2021). *Iran's failure to tackle climate change—a question of priority*. Aljazeera. Accessed May 03, 2023, from https://www.aljazeera.com/news/2021/11/9/irans-failure-to-tackle-climate-change-a-question-of-priority

Naifar, N. (2023). Does Climate Change Affect Sovereign Credit Risk? International Evidence. *Borsa Istanbul Review, 23*, S84–S95. https://doi.org/10.1016/j.bir.2023.10.001

O'Byrne, D. (2022). *Ukraine war threatens Kazakh oil exports. Eurasianet*. Accessed Dec 13, 2023, from https://eurasianet.org/ukraine-war-threatens-kazakh-oil-exports

OECD. (2011). *Towards green growth: monitoring progress*. Accessed Oct 03, 2023, from https://www.oecd.org/greengrowth/48224574.pdf

OECD. (2013). *Moving towards a common approach on green growth Indicators*. Accessed Oct 03, 2023, from https://www.oecd.org/greengrowth/GGKP%20Moving%20towards%20a%20Common%20Approach%20on%20Green%20Growth%20Indicators%5B1%5D.pdf

OECD. (2022). *Green Economy Transition in Eastern Europe, the Caucasus and Central Asia: Progress and ways forward. Ninth Environment for Europe Ministerial Conference Nicosia*. Accessed Oct 03, 2023, from https://unece.org/sites/default/files/2022-09/ece.nicosia.conf_.2022.inf_.8.pdf

OSCE. (2021). *Promoting green ports and connectivity*. Accessed May 03, 2023, from https://unece.org/sites/default/files/2021-11/Session%202.11%20Daniel%20Kroos%20-%20OSCE%20activities%20in%20support%20of%20green%20trade%20in%20ports.pdf

PCGCC. (2009). *Climate change mitigation measures in the European Union*. International Brief 3. Pew Center on Global Climate Change. Accessed Oct 03, 2023, from https://www.c2es.org/wp-content/uploads/2009/12/european-union-climate-change-mitigation-measures.pdf

Sachs, J., Lafortune, G., Kroll, C., Fuller, G., & Woelm, F. (2022). *Sustainable Development Report 2022. From Crisis to Sustainable Development: the SDGs as Roadmap to 2030 and Beyond* (p. 493). Cambridge University Press.

Selomane, O., Reyers, B., Biggs, R., Tallis, H., & Polasky, S. (2015). Towards integrated social–ecological sustainability indicators: Exploring the contribution and gaps in existing global data. *Ecological Economics, 118*, 140–146. https://doi.org/10.1016/j.ecolecon.2015.07.024

Shaffer, B. (2010). Caspian energy phase II: Beyond 2005. *Energy Policy, 38*(11), 7209–7215. https://doi.org/10.1016/j.enpol.2010.07.051

Shear, F., Ashraf, B. N., & Butt, S. (2023). Sensing the heat: Climate change vulnerability and foreign direct investment inflows. *Research in International Business and Finance, 66*, 102005. https://doi.org/10.1016/j.ribaf.2023.102005

Shrestha, S., Kotani, K., & Kakinaka, M. (2021). The relationship between trade openness and government resource revenue in resource-dependent countries. *Resources Policy, 74*, 102332. https://doi.org/10.1016/j.resourpol.2021.102332

Solano-Rodríguez, B., Pye, S., Li, P.-H., Ekins, P., Manzano, O., & Vogt-Schilb, A. (2021). Implications of climate targets on oil production and fiscal revenues in Latin America and the Caribbean. *Energy and Climate Change, 2*, 100037. https://doi.org/10.1016/j.egycc.2021.100037

UN. (1992). *Agenda 21*. United Nations Conference on Environment & Development. Rio de Janerio, Brazil. Accessed Jun 1, 2022, from https://sustainabledevelopment.un.org/content/documents/Agenda21.pdf

UN. (2015). *Transforming our world: The 2030 agenda for sustainable development: A/RES/70/1*. Accessed May 19, 2022, from https://sustainabledevelopment.un.org/content/documents/21252030%20Agenda%20for%20Sustainable%20Development%20web.pdf
UN. (2021). *Renewables 2021 Global Status Report*. Accessed Dec 13, 2023, from https://www.ren21.net/wp-content/uploads/2019/05/GSR2021_Full_Report.pdf
UNCTAD. (2019). *Guidance on core indicators for entity reporting on contribution towards implementation of the Sustainable Development Goals*. Accessed Oct 03, 2023, from https://unctad.org/system/files/official-document/diae2019d1_en.pdf
UNDP. (2011). Lessons learnt from the UNDP-GEF project "Kazakhstan—Wind Power Market Development Initiative." Accessed April 13, 2024, from https://www.undp.org/sites/g/files/zskgke326/files/migration/kz/7075-25305.pdf
UNECE. (2022). *Green Economy Transition in Eastern Europe, the Caucasus and Central Asia: Progress and ways forward.* Ninth Environment for Europe Ministerial Conference Nicosia, 5–7 October 2022. Accessed Apr 27, 2023, from https://unece.org/sites/default/files/2022-09/ece.nicosia.conf_.2022.inf_.8.pdf
UNEP. (2011). *Towards a green economy: Pathways to sustainable development and poverty eradication: A synthesis for policy makers*. Department of Economic and Social Affairs. Accessed May 19, 2022, from https://sustainabledevelopment.un.org/content/documents/126GER_synthesis_en.pdf
UNFCCC. (2015). *United nations framework convention on climate: Paris agreement*. Accessed May 19, 2022, from https://unfccc.int/sites/default/files/english_paris_agreement.pdf
UNFCCC. (2022). *NDC synthesis report*. Accessed Oct 03, 2023, from https://unfccc.int/process-and-meetings/the-paris-agreement/nationally-determined-contributions-ndcs/ndc-synthesis-report/ndc-synthesis-report
US EIA. (2022). *Country analysis executive summary: Iran*. Accessed Dec 13, 2023, from https://www.eia.gov/international/content/analysis/countries_long/Iran/pdf/iran_exe.pdf
Wang, H., & Zhang, D. (2023). Examining the interplay between fossil fuel mining, sustainable growth, and economic prosperity. *Resources Policy, 87*, 104324. https://doi.org/10.1016/j.resourpol.2023.104324
WB. (2018). Green economy: realities & prospects in Kazakhstan. Accessed May 04, 2023, from https://sk.kz/upload/iblock/8d9/8d97878e7ec2466e04ab62e5d8f4c3a3.pdf
WB. (2022a). *Azerbaijan: towards green growth–Issues Note*. 70 p. https://doi.org/10.1596/38073.
WB. (2022b). Offshore Wind Development Program: Offshore Wind Roadmap for Azerbaijan. Accessed April 13, 2024, from https://documents1.worldbank.org/curated/en/099125006022242537/pdf/P1757160c9ba20078097880a6781b95d5eb.pdf
WCED. (1987). *Report of the world commission on environment and development: our common future*. Accessed Oct 03, 2023, from https://sustainabledevelopment.un.org/content/documents/5987our-common-future.pdf
WTO. (2022). *Trade and climate change*. Information brief # 6. Accessed Oct 03, 2023, from https://www.wto.org/english/news_e/news21_e/clim_03nov21-6_e.pdf
Wu, X., Fu, B., Wang, S., Liu, Y., Yao, Y., Li, Y., Xu, Z., & Liu, J. (2023). Three main dimensions reflected by national SDG performance. *The innovation, 4*(6), 100507. https://doi.org/10.1016/j.xinn.2023.100507
Zakharova, D. & Medas, P. A. (2009). *A primer on fiscal analysis in oil-producing countries*. IMF Working Paper, 09/56. 41. https://doi.org/10.5089/9781451872033.001.

Financial System Readiness to Achieve Carbon Neutrality

Umair Saeed Bhutta and Ahmar Qasim Qazi

Abstract This chapter examines the state of green finance in Oman and evaluates its capabilities to achieve environmental targets set by the country. This study analyzes the status of developing innovative green finance products, such as green Sukuk (green bonds) and other banking products in the Sultanate. Green finance is important to facilitate and encourage various stakeholders of society to promote financing for a sustainable future. The chapter also gauges the green financing initiatives initiated by different financial institutes of the country to promote the use of renewable resources in the Sultanate. The study uses the Content analysis methodology to ascertain the country's progress toward green finance. Several aspects of the data are identified and analyzed for content analysis. This analysis concludes that green finance is still in the emerging stage in the country. For commercial projects, green finance products have shown little progress. In the recent past, different financial institutes started green finance products for financing domestic projects. However, the scale of this activity is also lower than required to achieve the country's environmental goals. The study concludes that the government is trying to enhance green finance activities at various levels. Still, compared to other countries of the region, there is much to be done to achieve climate goals. The study recommends policymakers develop a mechanism to overcome challenges in promoting green finance. Also, it needs to support issuers in issuing advanced financial products like Green Bonds and Sukuk. It can be done by raising awareness about green finance, providing preference treatment for green financial products, and developing a clear regulatory framework.

U. S. Bhutta (✉)
Higher Colleges of Technology, Abu Dhabi, UAE

A. Q. Qazi
University of Technology and Applied Sciences, Ibri, Oman
e-mail: ahmar.qasimqazi@utas.edu.om

N. Naifar (ed.), *Climate Change and Finance*, Sustainable Finance,
https://doi.org/10.1007/978-3-031-56419-2_7

1 Introduction

The world is facing the issue of climate change, and it is one of the most pressing issues that needs the attention of world leaders. The recent COP26 UN climate conference is one of the forums where leaders discuss climate issues and ways to cope with the emerging challenge of climate change. Carbon dioxide (CO_2) is considered one of the primary causes of environmental pollution, and around 75% of greenhouse gas (GHG) emissions are due to it. Greenhouse gas emissions are one of the key factors of global warming, and it needs immediate world attention with concrete steps to improve the livability of humans on Earth (Abbasi et al., 2022). It is time to ensure that the increase in world temperature is not above 1.5 °C from the preindustrial level. It is necessary for human life, which faces a threat due to environmental degradation (Mahat et al., 2019).

The availability of finance is one of the key factors in facilitating a low-carbon and climate-friendly economy. The Paris Agreement also emphasizes that governments align financial flows toward those projects that help in reducing GHG emissions. It facilitates sustainable development (Fang, 2023). Literature shows that many countries need more finances to achieve climate objectives. It hurts the global goal of preserving the environment through sustainable business practices (Quynh et al., 2022). It is suggested that the world needs more funds for climate-related investments. The partnership between the public–private sector is required to minimize the budgetary gaps to facilitate climate finance and achieve sustainable development goals (Wang et al., 2022). The economies require a comprehensive green finance system to facilitate stakeholders' investment in climate-friendly projects.

Green finance is a financial system that supports the investment in sustainable projects to support the economy's transition toward using renewable resources. Economic systems worldwide face challenges in establishing a financial framework where funds are channeled to facilitate sustainable projects. The main challenge of such a system is ensuring fairness and effective utilization of such funds for intended projects (Guild, 2020). In the developed world, investment in green projects can be funded with certain guarantees, as the institutional framework in these countries is robust and ensures the legit utilization of the funds. However, developing economies are still in the phase of making a framework to facilitate green finance products.

The role of financial institutions is critical for economic development and to finance climate-friendly projects. It helps the inclusion of society to achieve sustainable development goals (Le et al., 2019). Financial inclusion of the community enables institutes to offer advanced innovative products. It ensures that all sectors of society are committed to muzzling up resources for a particular goal (Usman & Balsalobre-Lorente, 2022). Including the community through different financial instruments helps sustainably satisfy the needs of various segments of society (World Bank, 2018). Financial inclusion through a proper financial framework ensures that different segments of society can gather the funds to fulfill their financial needs (Ozili, 2021). The economic framework designed explicitly for climate-friendly projects provides opportunities and challenges to society. It offers an opportunity for more climate-friendly projects and to facilitate financing at attractive terms. It helps to build momentum to achieve SD goals (Lahiani et al., 2021). It is also a challenge as the economy, at the initial phase, has to divert resources for the negative NPV projects. These projects are considered risky at the initial stage and need a solid commitment from the different segments of society.

Developing a mechanism of green finance needs suitable resource allocation from the government. The government needs to develop a financial development framework that supports the financing of climate-friendly projects. Financial development and inclusion support overall economic development, and it also helps society achieve goals. Economic growth brings society prosperity. To sustain this economic growth, society uses non-renewable fuels that hurt societies' climate neutrality goals (Danish & Wang, 2019). It brings the debate to an exciting situation: whether the financial framework for sustainable projects is adequate for societies to achieve climate neutrality goals or aggravates the carbon emissions situation of the economy.

Gulf nations are also facing challenges due to climate change. Due to climate change, we can see that the region faces the problem of a rise in sea levels and extreme weather conditions. Consequently, the stated problems affect the natural habitats of the area. Gulf region requires around US$ 230 billion annually to facilitate the Paris Climate Change initiatives (IEA 2021). It is estimated that countries in the region need more than $100 billion. The situation requires leaders' attention throughout the region. These nations need to channel the funds for making their economies green and develop a mechanism to bring all stakeholders of society toward pro-environmental investments.

Oman has also committed itself to sustainable development through climate-friendly projects. As per Vision 2040 decreed by the Sultanate, sustainable development is a core objective; to achieve this, the role of the financial sector is essential. One of the main themes of the vision is to restructure the economy. The country's goal is to rely less on income from oil-based sources. The vision also emphasizes environmental issues and promotes using renewable resources to make the environment sustainable for future generations. The vision focuses on partnering with the private sector to promote social and economic development activities. Green finance is critical for the Sultanate to achieve the goals outlined by Vision 2040. Bank Muscat takes the initiative to provide loans to customers to use solar energy to fulfill their energy needs. Hyport Duqm is one of the first projects highlighting the Sultanate's move toward its green energy ambitions. These are the initial steps for different stakeholders in the Sultanate for sustainable projects. Still, much work needs to be done to achieve the goals formulated by Vision 2040 for achieving carbon neutrality.

Earlier studies on the Gulf region and Oman discuss the commercial bank's strategy for financing green projects (Miah et al., 2021). Khalil et al. (2023) discuss the role of green finance in promoting green innovation in the region. As per our understanding, there is a need for the study to comprehend the strategies implemented by the government and regulators in Oman for the development of green finance. The main objective of this study is to look at the state of the Sultanate's financial framework and evaluate its capabilities to achieve the net zero target by 2050. The focus of the study is not only to gauge the status of the financial structure but also to discuss the financial institutions' preparedness to meet the targets set by the Paris Agreement. Furthermore, through this study, we try to give some suggestions to improve the overall financial structure to achieve the targets of a green economy by comparing the status with the best practices of green finance worldwide. Lastly, we will also evaluate the possibility of launching innovative green finance products, like green bonds and green Sukuk, to facilitate and encourage private segments of society to promote financing for a sustainable future.

In the next part of this chapter, we will discuss the relevant literature related to the green financial system. Afterward, we will discuss the methodology of the current study in detail and then discuss the results of the study. Finally, we conclude the chapter with the conclusion and policy recommendations for the regulators to facilitate climate financing in the country.

2 Literature Review

In the recent past, the focus of the world shifted toward green or climate finance. The main objective is to allocate funds for climate-friendly projects and to achieve the objectives of the Paris Climate Conference. However, there needs to be more clarity about the meaning and interpretation of the term green finance or climate finance. People worldwide interpret it differently. Due to this ambiguity, the main objective of climate finance is yet to be fulfilled. Still, green finance is regarded as one of the critical tools to achieve sustainability goals (Bendarzsevszkij et al., 2021). To make green finance effective and objective-oriented, there is a need for one common definition of the term (Campiglio et al., 2018). Also, a commitment is required by all stakeholders to mitigate the effects of climate challenges faced by inhabitants of the planet Earth.

All nations should prepare a mechanism to achieve climate objectives collectively. It is necessary to ensure the livability of humans on Earth. The developed nations account for the major share of the world's greenhouse gas emissions (Shahbaz et al., 2016). Developed countries must participate actively in green finance projects and allocate resources exclusively for such projects. Participation only from the smaller countries is not sufficient to achieve climate objectives. The contribution to greenhouse emissions from the smaller countries is limited. Still, they are suffering badly from the climate crisis.

The previous research on the topic of financing and carbon emissions shows contradictory results. From the Pollution haven hypothesis perspective, the availability of finances allows firms to be involved extensively in climate-damaging activities (Al-Mulali et al., 2015). This behavior results in more carbon emissions and hurts sustainable growth objectives. On the other hand, other studies conclude that countries actively allocating funds for green finances help industries use innovative ways of production. These innovative ways of production are less carbon-intensive and help reduce carbon emissions (Bose & Kohli, 2018). Using advanced and environmentally friendly ways of doing business helps countries improve the quality of the environment and achieve sustainable growth goals.

As per the pollution halo hypothesis, the availability of the funds results in innovative production methods, which help reduce carbon emissions. In addition, some previous research on the topic indicates the non-linear relationship between the availability of funds and carbon emissions. According to that perspective, the availability of funds initially deteriorates the environment due to increased carbon emissions. However, as companies achieve a threshold, they reduce carbon emissions. This is due to changes in production methods to more environment-friendly business methods (Alshubiri & Elheddad, 2020). From the perspectives mentioned above, research on the topic is inconclusive, and the availability of funds has a different impact on the nations' carbon-producing activities.

Financial inclusion and development are crucial to facilitate the green growth of the economy to achieve sustainable development goals (SDGs). Gathering finances for climate-friendly projects is one of the critical tasks. Without the availability of finances for climate-friendly projects, achieving the SD goals looks like a distant dream. As climate-related issues arise and environmental issues are getting more attention, international financial institutes and governments advocate developing different financial instruments to finance climate-friendly projects (Krueger et al., 2020). The development of a comprehensive instrument is critical to financing climate-friendly projects. It helps firms to get finance for such projects at preferred rates, and climate-sensitive investors prefer to invest in these instruments (Bhutta et al., 2022). Green bonds are one such instrument that has gotten much attention in the recent past. Green bonds are specifically meant for climate-sensitive projects and support governments and corporations in financing them at affordable rates. As per academic research, green bonds, compared to conventional bonds, provide greenium, making financing cheaper. Financing with green instruments is aligned with the UN's sustainable development goals (SDGs).

The researchers examined the economic system's role in improving the firms' climate performance. Consequently, governments and financial institutes worldwide are developing innovative financial instruments. It facilitates climate-friendly projects and achieves the SDG goals outlined by the UN. The economic system facilitation for such finance is important to enhance the investment in green projects. Central banks worldwide are changing their policies to facilitate climate finance and mitigate the risks of greenhouse gas emissions (Dikau & Volz, 2021). The central banks' role is critical for developing green finance. It also boosts investors' confidence in seeking such investment opportunities and drives the funds toward green projects.

There is a clear distinction between green and conventional capital. Green capital is used for climate-friendly projects with minimal material intensity and also makes a mechanism for recycling. Furthermore, green projects use renewable resources compared to conventional projects (Dafermos et al., 2018). In the modern economic system, capital is raised by the financial system through gathering public savings. The people trust these systems and put their investments there for specific purposes.

General communities and firms invest their savings in different instruments offered by banks and other firms. The firms and banks used this capital for various projects as capital rationing is important to curtail environmental degradation. Institutional investors are critical for promoting green finance and also play an important role in promoting green finance (Ameli et al., 2020). Therefore, regulators and financers can restrict the financing for carbon-producing projects and encourage investment in green, environment-friendly projects. As the demand for such environment-friendly projects rises, it will help to lower the cost of borrowing as the supply of capital for such projects increases.

Central banks worldwide have started to assess the risk associated with climate-related issues and their impact on the economic environment of the economy. The carbon-intensive sectors of the economy are a potential systematic risk for the financial systems throughout the world (Chenet, 2021). The banking and insurance sectors may face numerous losses and insurance claims due to changes in weather patterns. Central banks and the government play a key role in enhancing green financing within society. They determine the base discount rate, provide funds for funding green within the community through commercial banks, and invest in different green instruments issued by the other firms. Creating and developing a green economy is the critical role of the country's central bank (Nasir et al., 2019). The researchers argue that the financial system's impact on economic transformation should be judged holistically (Nasir et al., 2015). The financial sector of any country should not be independently seen as an engine for economic growth. However, how the financial sector contributes to the development of society also needs to be gauged. In this case, in facilitating a green economy, the central bank's role is critical, and it can guide the different stakeholders to promote green activities. The problem is that users and fund providers are not incorporating climate change risks in their business activities. To overcome this problem.

The disclosure mechanism needs to be strengthened, and each stakeholder related to a project needs to have complete information about the project's risks.

The users of the funds' voluntary disclosures about climate issues can build trust among stakeholders, and financing can be enhanced by such disclosure (Urban & Wójcik, 2019). These voluntary disclosure requirements still need to be made obligatory for firms in many parts of the world. Therefore, there might be more effective ways to judge the effectiveness of such initiatives around the globe. Regulators need to develop a mechanism to ensure firms issue climate-related disclosures (Edenhofer et al., 2017). It will help financial institutes evaluate the riskiness of investment in certain projects and adjust their financing rates accordingly.

Green finance is gaining popularity in supporting a sustainable financial system and dealing with future financial challenges (Ziolo et al., 2019). The financial industry is required to fulfill the human economic needs in modern times. However, it also needs to take care of the environmental and social consequences of the financial decision. Regulators must develop a financial system that addresses the community's economic woes and integrates ESG factors to create a sustainable economic model. However, many challenges and limitations hinder the growth of the financial system and prevent it from achieving carbon neutrality targets. Financial Institutes must provide innovative financial instruments to support the SDG goals (Schoenmaker & Schramade, 2018). The mechanism is required to mitigate the climate challenges and to remove the stakeholders' concerns with green finance.

Financial institutes are developing innovative financial products and supporting firms' green initiatives (He et al., 2019). Developing effective green channels is one of the essential tools to provide finance for green projects. However, the effectiveness of such initiatives by the financial sector is still in the nascent stage.

To achieve the target of sustainable, economically friendly development, regulators need to develop a mechanism to divert funds toward the low-carbon productive sectors. Diversifying the green financial system requires converting the overall economy. Three sectors are significant for achieving the objectives of green development (Campiglio, 2016). These three sectors are (i) Energy production through renewable sources, (ii) Efficient energy use in transport and buildings, and (iii) Efficient utilization of natural capital. Keefe (2011) highlighted that non-financial factors like ecological, social, and governance issues need attention before making a financial decision. The main objective of a carbon-neutral society can only be achieved by investing in green projects and reducing financing for carbon-intensive projects. Investment in green sectors has grown faster recently, and more significant portions of those investments are channeled toward sustainable energy production units.

Financial stability and efficiency are critical for economic growth, and we can understand this relationship's ecological impact. As the economy grows due to the well-functioning financial system, it can result in more energy-consuming products. This phenomenon threatens sustainable development goals (SDGs) and may negatively affect the environment (Cetin & Ecevit, 2017). The literature provides mixed results on the impact of financial system viability and its economic consequences. Shahbaz et al. (2018) studied the context of France and reported that the financial system contributes positively to improving the quality of the environment. However, utilizing such resources for the intended purpose is still a question. Many projects are less green than they are advertised to investors.

Some studies show that financial development increases the use of carbon, which hurts the economy's environment. For instance, the study by Salahuddin et al. (2018) shows that economic development negatively impacts Kuwait's ecological system. The earlier study by Zhang (2011) also presents the same results in China: the environment is negatively affected due to the loopholes in the financial system. It shows that using finance for projects that help in economic development may not be good for the ecological system of that particular economy.

The other studies on this topic reported no impact of financial development on ecological footprints. They show that innovation in financial instruments is neutral to the environment (Bekhet et al., 2017). The studies provide mixed results that can be a guiding line for our future studies. We assume that a financial framework is vital to safeguard the environment in this context. The literature suggests that it may positively or negatively impact the production of carbon footprints. How the financial framework is developed and the type of instruments the system launches is critical in financing green projects.

In this study, we try to understand the development of the green financial framework in Oman. We study the current status of the Omani financial sector and its readiness to finance green projects. Through this study, we also try to understand the limitations and constraints that hamper the development of a green economy in Oman. Furthermore, we also check the viability of different green products within Oman's financial framework. Lastly, we also analyzed the feasibility of various financial instruments to be launched in Oman. It also gives ideas on mitigating the negative consequences of economic development on the environment of the Sultanate.

3 Methodology

This study applies the content analysis approach to answer the following research questions:

- How does the Sultanate of Oman at the government level promote green finance?
- What green finance products is the country's financial sector offering, and how are they aligned with the goals of the Paris Agreement?

The content analysis with qualitative aspects of the data is used to conduct this study. Qualitative content analysis is used for business and social science studies (Elo & Kyngäs, 2008). Content analysis is an important technique in business studies to study the data in a specific context (Boettger & Palmer, 2010). Krippendorff (1989) defined content analysis as a useful research tool for building a valid inference from the available data to the context. In this methodology, researchers use verbal, written, or visual data to make valid inferences about the data. By using this technique, the researchers make inferences from official or private communication (Kassarjian, 1977). Using this technique, researchers try to understand the policies. The content analysis should be objective, systematic, and conclusive. The inductive or deductive way of content analysis can be carried out (Krippendorff, 2013). As per Mehmetoglu and Dann (2003), content analysis is a dynamic technique and can be used in different forms of human communication in diverse formats.

The researchers' choice of data affects the interpretation. The main data sources are books, journals, government documents, and visual TV and social media programs (Salem et al., 2021). It is one of the innovative techniques, and it is popular in social sciences research (Orlikowski et al., 1991). Using this method, researchers understand the progress in the field by checking newspapers, websites, books, government publications, and other social media pages (Salem & Čavlek, 2016). For analysis, we are using the official documents from the Sultanate of Oman related to the promotion of Green Finance in this study. Further to the government documents, we also look at the important documents and news from the financial institutes of the Sultanate to understand the progress of green finance.

3.1 Sampling and Data Collection

For data collection, we used the defined criteria represented by the research question. Data for the study was collected from year 2016 to 2022. The reason for selecting 2016 as a starting point for the study is the world's reinforcement of instant climate action after the Paris Agreement. Governments worldwide allocate resources for sustainable projects with new enthusiasm and show more commitment to working together to minimize climate degradation. Climate change also affects Oman, and the government has started allocating more resources to promote sustainable projects.

For sampling purposes, we select the documents from official government papers from the Sultanate. We gathered climate change and green finance documents and selected pertinent ones for further review. The main document is the Vision 2040 document of the Sultanate. We searched the documents and then picked the topics related to our study for further analysis. To analyze further, we searched newspapers and journals around this period and checked the progress, i.e., green finance during this era in the Sultanate. All the related material is scanned, and any related document is selected for further analysis. Lastly, we checked the initiatives taken by financial institutions of the Sultanate by looking at their websites and annual reports. Any document showing progress in green finance is selected for further analysis. We used the steps prescribed by Baudendistel et al. (2015) to understand the initiative's impact on achieving sustainable development goals. The steps we used are as follows:

a. Transfer the relevant data from the main source to the Microsoft Excel file into logical sequencing.
b. Give the codes topic-wise to classify the text.
c. Sorting the text with the help of the codes to make logical reasoning out of it.
d. Logical inferences are made from the coded text.
e. Analysis of the text to understand the progress in green finance.

We gathered the data from official documents and other relevant sources for our analysis. Initially, we shortlisted a total of 63 documents for the analysis. Out of these 63 documents, relevant data is analyzed for further processing. Some of the information is duplicated. Furthermore, some of the initially selected documents are irrelevant as they discuss issues related to sustainability that are not related to our main area of study, i.e., sustainable finance. For this reason, 12 documents are discarded from the data. After all scanning, 31 documents are finally selected for analysis. All the documents are sourced from government publications, bank reports and websites, and newspaper reports related to our scope of the study.

3.2 Coding Framework

Coding is a process of labeling the original text for abstraction and interpretation. By giving specific labels and meaning to the text, we minimize the risk of missing the important content. For coding purposes, it is important to keep the context of the study and construct the keywords accordingly. In the case of our study, we keep the context of finances for sustainable projects in context and frame the keywords accordingly. It helps to include all the terms related to the field of green finance and minimize the chances of missing any important term.

To categorize the development of the Sultanate's green finance, we developed several keywords to complete the content analysis. We take the main data for content analysis from government documents, official government announcements, websites, annual reports of the financial institutes working in Oman, and news from the leading newspapers of the Sultanate. Reviewing the relevant literature ensures that all the keycode words are included in our analysis. It helps minimize the chances of being left out of any key concept related to our study.

The data must give all the required information to answer the research question. The code and keyword definitions are precise to get the right answers for our research questions (Lindgren et al., 2020). We select distinctive themes for getting the data to conduct our research. The list of keywords we selected for our analysis is broad terms. By reading the related material and looking at those terms, we ensure that all concepts encompass the topic are included in our review. The main keywords for this study are listed in Table 1.

We collected the documents required for this study and then searched for these keywords. All the websites of the banks are searched to conduct this content analysis. Also, newspaper articles are googled to get the required information. The data search process remains smooth, but it can be noted from the data collection process that green finance is still at a nascent stage in the Sultanate. The Sultanate has recently taken some initiatives to promote green finance. Furthermore, financial institutes also take some initiatives to promote green finance. All the data related to the keywords are recorded in an Excel file. The information, particularly related to our study, is further analyzed using the functions of data sorting and data analysis in Excel.

Table 1 Key terms for content analysis

– Green finance
– Green Sukuk
– Green bonds
– Climate finance
– Sustainable finance
– SDG development goals
– Vision 2040
– Finance for the future

4 Results and Discussion

To answer the first question of the study, our study goes through government documents from the Sultanate's official data sources to analyze the progress at the governmental level in the field of green finance. The country is in the initial stages of promoting green finance. The government also encourages activities to curtail climate degradation by incorporating a financial framework to promote a green economy in the Sultanate.

The most important government document that emphasizes the importance of a green economy and improving the country's environmental performance index is Vision 2040. It states that the sultanate wants to be in the top 20 countries of the world regarding environmental performance by 2040. In this document, the government is committed to providing funds for sustainable projects without compromising national growth. However, if we look deeply, the document does not mention sustainable or climate finance. That shows that the Sultanate still needs to make a clear policy about sustainable finance for the future. To achieve goals prescribed by the COP26 and further emphasized by the UN sustainable goals framework, it is the need of the hour for the Sultanate to come up with a clear green finance framework and give guidelines to different segments of the society to promote green activities and provide finance for such activities. To meet the challenges of climate change, the Sultanate announced that it aimed to reduce greenhouse gas emissions by 7% through 2030. For such an ambitious target to be achieved, a clear financial framework is needed to promote green finance to achieve the objectives stated by the government.

The Central Bank of Oman (CBO) shows some progress in green finance. CBO signed a memorandum of understanding with Oman Environmental Services Holding Company (be'ah) on March 16, 2022. Per this MoU, both parties agree to cooperate in a sustainable and circular economy. The main feature of this MoU is that both parties agree to cooperate in green finance and the development of fintech to promote ESG activities. The CBO's significant step is to promote green finance in the country. It also supports SMEs to get green finance for their projects and promotes the culture of green production in the Sultanate. CBO's recent financial stability report warns of significant risk to the Omani financial system due to tropical storms, floods, and rising sea levels. To mitigate those risks, CBO and the government must develop financing products that promote the culture of green projects to minimize the impact of carbon emissions on the country's financial system.

In the recent past, the Capital Market Authority of Oman organized a seminar on green finance in March 2022. The seminar's main objective is to promote green finance activities within the Sultanate and base the funding on ESG principles. The seminar is one of the major steps to attract the attention of stakeholders to work together toward the green economy. They are advocating and authorizing the financial products that help the growth of the green economy within the Sultanate as per the guidelines of Vision 2040. The CMA is trying to promote and support green finance products by making a system of accountability for such products. These financial products are specifically designed to promote the economy's growth in line with environmental sustainability and climate preservation. CMA is trying to introduce products such as green bonds, green loans, and green Sukuk to channel the funding for green projects to protect the environment.

Overall, there is still a lot to be done to promote green finance. From the official documents and newspapers, we realize that green finance initiatives in the Sultanate are at the initial stage, and there is a lot to be done to promote green activities within the country. In the region, some of the countries have taken steps to promote green culture within the respective countries. Oman can take help from those examples and promote green finance within the Sultanate. As per the warning from CBO, Oman's financial sector is facing vulnerability due to climate change. Therefore, it is high time for the regulators to provide a clear framework to the stakeholders in the country to promote green finance and take help from the other financial institutes operating in the country to streamline the flow of funds for such projects.

In the second part of the study, we search to answer questions regarding the role of financial institutes in the country in promoting green finance activities. To find answers to this question, we visited the websites, annual reports, and news reports for the country's financial institutes. From the information for the last 5 years, we can say that the country's financial sector is taking small steps to promote green finance. There is still a long way to go in promoting this financing. Some of the key initiatives taken by financial institutes of the country are discussed below for readers' understanding.

The country's financial institutions have been publishing sustainable reports for years. The main content of the sustainable report is the discussion on governance and environmental and social initiatives the financial institutes took to mitigate the negative impact of carbon emissions on the Sultanate's overall society. Bank Muscat, Oman Arab Bank, and Bank Nizwa have been issuing sustainable reports for quite some time now to promote the culture of sustainability within the banking industry and to promote green finance. However, the promotion and issuance of green finance instruments still need to be improved. As of now, a few products have been introduced in the market. We will discuss each financial institute's initiatives in detail in the following paragraphs:

Most financial institutions in the Sultanate provide green finance for domestic-level rather than commercial activities. For instance, recently, Bank Muscat launched a scheme to finance solar panels for domestic use. The main purpose of this financing is to enhance the sense of responsibility among customers about using sustainable renewable energy resources. It is an encouraging initiative by one of the leading banks of the Sultanate. However, launching these kinds of financial products for businesses is also important. It will help the country achieve the targets set by Vision 2040. More information about initiatives by financial institutes and the government is summarized in Table 2 of the study.

Another financial institute working actively toward sustainability and taking the initiative by offering green financing products is Bank Nizwa. One of the country's Islamic banks, Bank Nizwa, has started to offer sustainability-linked debit and credit cards. The main feature of these cards is that they are made of regeneratable

Table 2 Oman's progress toward green finance

Organization/ Govt department	Document and issuance date	Green finance activity
Government Vision 2040	Vision 2040 Sept 2019	Encourage companies to be involved in Green projects, but a clear green finance framework needs to be established.
Bank Muscat	Sustainability Report 2019	Domestic solar panel financing for the use of renewable energy
Bank Nizwa	News Report 2022	First green financing for industries in Oman
Oman Arab Bank	Sustainability Report 2020	The bank provided support for the environmental and social development project. Still no clear framework for green finance for industrial projects
Ahli Bank	News Report 2022	The bank is involved in some CSR initiatives but needs a product to support green financing
HSBC Oman	News Report 2022	HSBC in Oman launched a loan product for its customers where customers will repay their loans as they meet their sustainability targets. It is specially designed for the oilfield service business.
Sohar International Bank	Website 2022	Banks provide financing for some underprivileged segments of society, but no proper green finance mechanism has yet been established.
Bank Dhofar	News Report 2017	The bank promotes green activities by encouraging people to recycle and other green activities. A specific financing product needs to be launched by the bank to support green projects.
Oman Arab Bank	News Report 2021	The bank is committed to planting 10 million trees across the country with the help of environmental authorities. The initiative is to meet targets set by Oman Vision 2040. However, no exclusive financing for green projects were announced by the bank.

plastic. It is a small step, but it shows management's commitment to starting to work toward green finance products in the future. Bank Nizwa, as an ethical Islamic Bank, has also recently introduced a sustainability-linked financing initiative. It also shows the bank's commitment to the country's transition to Net Zero emissions. The bank is encouraging partner companies to adopt sustainable practices, and the bank is providing help by financing such projects. The bank's initiatives are commendable, but it is still a long way to launch sustainability-linked financial products. Bank management should look for alternative financial products for sustainable projects and give financial solutions to customers interested in launching sustainability-linked business solutions. It has been seen that entrepreneurs cannot change their business orientation toward sustainable procedures due to the lack of finance available for such projects. Therefore, the Bank must launch products to help firms gather finance for sustainable projects.

HSBC Oman is also one of the financial institutes that launched the scheme to finance solar panels for domestic users. With this finance, the customers can cover the cost of solar panels, solar water heaters, and other energy-saving domestic products. Customers can avail of financing without any down payment. The bank's interest rate on such loans is lower than the other conventional loans. The length of such a loan is up to 48 months, and there are no processing fees for such loans. Getting such a loan is simple. Citizens and residents of the country can avail of this loan if their salary is OMR 400. It is a nice initiative by one of the leading international banks to encourage customers to purchase green products and make them realize the importance of green products to achieve sustainability-linked goals. However, this kind of financing is still limited to domestic customers. There is a need to make some products for commercial customers to achieve the sustainability-linked goals prescribed by Vision 2040 of the Sultanate.

From the above analysis of the financial institutes of the Sultanate, we can determine that financial institutes in the country are providing sustainability-linked financing products to the domestic segment of society. It is a positive initiative and will help the Sultanate achieve its Vision 2040 of a sustainable environment. However, the country's banks and other financial institutes must issue contemporary financial products for the commercial segment of society to facilitate businesses' conversion of their business processes toward sustainable practices. With the help of such products, businesses can finance such projects that will help the country fulfill its commitment to green production. Government-level facilitation is also required in this regard, and the government must compile a comprehensive framework for green finance. The green finance framework should benefit financial institutes actively participating in green finance activities and launching related products. The government can take a leaf out of neighboring countries like Saudi Arabia and UAE, where sustainability-linked financial products are launched extensively for commercial projects and successfully implemented. The Dubai government created the Dubai Green Fund to finance environment-friendly projects in the UAE. Abu Dhabi launched a program to finance green projects through Green Bonds. Also, the UAE encourages businesses with good environmental practices with sustainability certificates that promote sustainability practices. The Saudi government is also in the process of preparing a green finance framework.

It is yet to be seen how the Sultanate will move forward in this direction and facilitate such products for commercial projects in the future. This transition will help Oman achieve its sustainability goals and progress toward Vision 2040 of the Sultanate.

5 Conclusion

In this chapter, we analyzed Oman's progress and future outlook for financing green projects to achieve sustainable development as the Climate Conference in Paris 2015 prescribes. Each country faces challenges to keep its climate livable for future generations as unsustainable human activities degenerate the climate badly. Oman is also facing climate challenges, and for sustainable growth, Oman needs to concentrate its resources on climate-friendly projects. The other countries in the region actively engage in activities to promote sustainable projects and form their framework to finance such projects. Climate financing is facing many challenges, as proper utilization of such resources is still a big question. Also, the Covid hampers the development of green finance worldwide. However, the world needs to keep its focus on providing sustainable projects as it is important for future growth and sustainability.

This chapter evaluates Oman's preparedness to facilitate green finance for sustainable projects. We concluded that so far, the green finance structure in the country is at the initial stage, and there are limited financing initiatives taken to finance sustainable projects. Through its Vision 2040 plan, the government encourages businesses to start green projects to achieve certain environmental targets mentioned in COP28. However, there needs to be a proper framework for financing such projects. Some financial institutes in the country are involved in limited activities to promote green financing. Still, the scale of such initiatives is narrow, and little is allocated for such projects.

We propose that financial institutes extend support for green financing. They should start financing industrial projects that help the Sultanate achieve sustainable environment targets. Also, we propose that the Sultanate encourage new financial products like Green Bonds, Green Sukuk, and green short-term financial products to extend the financing for such projects. Lastly, the government should work with the private sector to extend green financing at multiple levels in the country.

This study has some limitations that need to be addressed in future research. The study only used available public and private documents to check the green financing status in the Sultanate. We propose that future studies not only rely on documents for such studies but also conduct formal interviews with officials at the helm of affairs to understand their future outlook for green financing and get current information about the progress in this sector of the economy. Further, the chapter only analyses Oman's situation. In future studies, a geographical area study will be beneficial to understand the overall green finance initiatives in the region. The COP28 is held in the region, and it will help countries accelerate their green finance initiatives and make a mechanism to utilize the resources to achieve sustainable development goals effectively.

References

Abbasi, K. R., Hussain, K., Haddad, A. M., Salman, A., & Ozturk, I. (2022). The role of financial development and technological innovation towards sustainable development in Pakistan: Fresh insights from consumption and territory-based emissions. *Technological Forecasting and Social Change, 176*, 121444.

Al-Mulali, U., Ozturk, I., & Lean, H. H. (2015). Economic growth, urbanization, trade openness, financial development, and renewable energy influence European pollution. *Natural Hazards, 79*, 621–644.

Alshubiri, F., & Elheddad, M. (2020). Foreign finance, economic growth, and CO2 emissions nexus in OECD countries. *International Journal of Climate Change Strategies and Management, 12*(2), 161–181.

Ameli, N., Drummond, P., Bisaro, A., Grubb, M., & Chenet, H. (2020). Climate finance and disclosure for institutional investors: Why transparency is not enough. *Climatic Change, 160*, 565–589.

Baudendistel, I., Winkler, E., Kamradt, M., Brophy, S., Längst, G., Eckrich, F., Heinze, O., Bergh, B., Szecsenyi, J., & Ose, D. (2015). The patients' active role in managing a personal electronic health record: A qualitative analysis. *Supportive Care in Cancer, 23*, 2613–2621.

Bekhet, H. A., Matar, A., & Yasmin, T. (2017). CO2 emissions, energy consumption, economic growth, and financial development in GCC countries: Dynamic simultaneous equation models. *Renewable and Sustainable Energy Reviews, 70*, 117–132.

Bendarzsevszkij, A., Eszterhai, V., Gere, L., Klemensits, P., & Polyák, E. (2021). World economic forum 2017. *International Journal of Research in Engineering, Science and Management, 4*(2).

Bhutta, U. S., Tariq, A., Farrukh, M., Raza, A., & Iqbal, M. K. (2022). Green bonds for sustainable development: Review literature on development and impact of green bonds. *Technological Forecasting and Social Change, 175*, 121378.

Boettger, R. K., & Palmer, L. A. (2010). Quantitative content analysis: It is used in technical communication. *IEEE Transactions on Professional Communication, 53*(4), 346–357.

Bose, S., & Kohli, B. (2018). Study of FDI trends and patterns in BRICS economies from 1990–2015. *Emerging Economy Studies, 4*(1), 78–101.

Campiglio, E. (2016). Beyond carbon pricing: The role of banking and monetary policy in financing the transition to a low-carbon economy. *Ecological Economics, 121*, 220–230.

Campiglio, E., Dafermos, Y., Monnin, P., Ryan-Collins, J., Schotten, G., & Tanaka, M. (2018). Climate change challenges for central banks and financial regulators. *Nature Climate Change, 8*(6), 462–468.

Cetin, M., & Ecevit, E. (2017). The impact of financial development on carbon emissions under the structural breaks: Empirical evidence from Turkish economy. *International Journal of Economic Perspectives, 11*(1).

Chenet, H. (2021). *Climate change and financial risk* (pp. 393–419). Springer International Publishing.

Dafermos, Y., Nikolaidi, M., & Galanis, G. (2018). Climate change, financial stability and monetary policy. *Ecological Economics, 152*, 219–234.

Danish, & Wang, Z. (2019). Investigation of the ecological footprint's driving factors: What we learn from the experience of emerging economies. *Sustainable Cities and Society, 49*.

Dikau, S., & Volz, U. (2021). Central bank mandates, sustainability objectives and the promotion of green finance. *Ecological Economics, 184*, 107022.

Edenhofer, O., Knopf, B., Bak, C., & Bhattacharya, A. (2017). Aligning climate policy with finance ministers' G20 agenda. *Nature Climate Change, 7*(7), 463–465.

Elo, S., & Kyngäs, H. (2008). The qualitative content analysis process. *Journal of Advanced Nursing, 62*(1), 107–115.

Fang, Z. (2023). Assessing the impact of renewable energy investment, green technology innovation, and industrialization on sustainable development: A case study of China. *Renewable Energy, 205*, 772–782.

Guild, J. (2020). The political and institutional constraints on green finance in Indonesia. *Journal of Sustainable Finance & Investment, 10*(2), 157–170.
He, L., Liu, R., Zhong, Z., Wang, D., & Xia, Y. (2019). Can green financial development promote renewable energy investment efficiency? A consideration of bank credit. *Renewable Energy, 143*, 974–984.
International Energy Agency. (2021). *Net Zero by 2050 A roadmap for the global energy sector.*
Kassarjian, H. H. (1977). Content analysis in consumer research. *Journal of Consumer Research, 4*(1), 8–18.
Keefe, J. (2011). *From socially responsible investing to sustainable investing*. Green Money J.
Khalil, R. G., Damrah, S., Bajaher, M., & Shawtari, F. A. (2023). Unveiling the relationship of ESG, fintech, green finance, innovation and sustainability: Case of gulf countries. *Environmental Science and Pollution Research, 30*, 1–14.
Krippendorff, K. (1989). Content analysis. In E. Barnouw, G. Gabner, W. Schramm, T. L. Worth, & L. Gross (Eds.), *International encyclopedia of communication* (Vol. 1, pp. 403–407).
Krippendorff, K. (2013). *Content analysis: An introduction to its methodology* (3rd ed.). Sage.
Krueger, P., Sautner, Z., & Starks, L. T. (2020). The importance of climate risks for institutional investors. *The Review of Financial Studies, 33*(3), 1067–1111.
Lahiani, A., Mefteh-Wali, S., Shahbaz, M., & Vo, X. V. (2021). Does financial development influence renewable energy consumption to achieve carbon neutrality in the USA? *Energy Policy, 158*, 112524.
Le, T. H., Chuc, A. T., & Taghizadeh-Hesary, F. (2019). Financial inclusion and its impact on financial efficiency and sustainability: Empirical evidence from Asia. *Borsa Istanbul Review, 19*(4), 310–322.
Lindgren, B. M., Lundman, B., & Graneheim, U. H. (2020). Abstraction and interpretation during the qualitative content analysis process. *International Journal of Nursing Studies, 108*, 103632.
Mahat, T. J., Bláha, L., Uprety, B., & Bittner, M. (2019). Climate finance and green growth: Reconsidering climate-related institutions, investments, and priorities in Nepal. *Environmental Sciences Europe, 31*(1), 1–13.
Mehmetoglu, M., & Dann, G. (2003). Atlas/ti and content/semiotic analysis in tourism research. *Tourism Analysis, 8*(1), 1–13.
Miah, M. D., Rahman, S. M., & Mamoon, M. (2021). Green banking: The case of commercial banking sector in Oman. *Environment, Development and Sustainability, 23*, 2681–2697.
Nasir, M. A., Ahmad, M., Ahmad, F., & Wu, J. (2015). Financial and economic stability as 'two sides of a coin' non-crisis regime evidence from the UK based on VECM. *Journal of Financial Economic Policy, 7*(4), 327–353.
Nasir, M. A., Huynh, T. L. D., & Tram, H. T. X. (2019). Role of financial development, economic growth & foreign direct investment in driving climate change: A case of emerging ASEAN. *Journal of Environmental Management, 242*, 131–141.
Orlikowski, W. J., Markus, M. L., & Lee, A. S. (1991, December). A workshop on two techniques for qualitative data analysis: Analytic induction and hermeneutics. In *Proceedings of the Twelfth International Conference on Information Systems* (pp. 390–391).
Ozili, P.K., 2021. Financial inclusion research around the world: A review. In Forum for social economics (Vol. 50, 4, pp. 457–479). Routledge.
Quynh, M. P., Van, M. H., Le-Dinh, T., & Nguyen, T. T. H. (2022). The role of climate finance in achieving Cop26 goals: Evidence from N-11 countries. *Cuadernos De Economía, 45*(128), 1–12.
Salahuddin, M., Alam, K., Ozturk, I., & Sohag, K. (2018). The effects of electricity consumption, economic growth, financial development and foreign direct investment on CO2 emissions in Kuwait. *Renewable and Sustainable Energy Reviews, 81*, 2002–2010.
Salem, I. E. B., & Čavlek, N. (2016). Evaluation of hotel website contents: Existence-importance analysis. *Journal of Hospitality and Tourism Technology, 7*(4), 366–389.

Salem, I. E., Elbaz, A. M., Elkhwesky, Z., & Ghazi, K. M. (2021). The COVID-19 pandemic: The mitigating role of government and hotel support of hotel employees in Egypt. *Tourism Management, 85*, 104305.

Schoenmaker, D., & Schramade, W. (2018). *Principles of sustainable finance*. Oxford University Press.

Shahbaz, M., Shahzad, S. J. H., Ahmad, N., & Alam, S. (2016). Financial development and environmental quality: The way forward. *Energy Policy, 98*, 353–364.

Shahbaz, M., Nasir, M. A., & Roubaud, D. (2018). Environmental degradation in France: The effects of FDI, financial development, and energy innovations. *Energy Economics, 74*, 843–857.

Urban, M. A., & Wójcik, D. (2019). Dirty banking: Probing the gap in sustainable finance. *Sustainability, 11*(6), 1745.

Usman, M., & Balsalobre-Lorente, D. (2022). Environmental concern in the era of industrialization: Can financial development, renewable energy and natural resources alleviate some load? *Energy Policy, 162*, 112780.

Wang, Z., Shahid, M. S., Binh An, N., Shahzad, M., & Abdul-Samad, Z. (2022). Does green finance facilitate firms in achieving corporate social responsibility goals? *Economic research-Ekonomska istraživanja, 35*(1), 5400–5419.

World Bank. (2018). *Financial inclusion*. Retrieved on January 26, 2022.

Zhang, Y. J. (2011). The impact of financial development on carbon emissions: An empirical analysis in China. *Energy Policy, 39*(4), 2197–2203.

Ziolo, M., Filipiak, B. Z., Bąk, I., & Cheba, K. (2019). How to design more sustainable financial systems: The roles of environmental, social, and governance factors in the decision-making process. *Sustainability, 11*(20).

Part II
Climate Change, Policy, and Economic Governance

Emerging Trends in Climate Change and Global Economic Governance

Achyutananda Mishra, Ananya Pandey, and Joseph Varghese Kureethara

Abstract Climate change is one of the most significant challenges of our time leading humanity towards a catastrophic future. Its adverse impacts undermine the ability of all countries to achieve sustainable development. Increases in global temperature, sea level rise, ocean acidification, and other climate change impacts seriously affect coastal areas and low-lying coastal countries, including many least-developed countries and small island developing states. The survival of many societies and the biological support systems of the planet is at risk.

The global economic system and policy planning need to be fully prepared to face the challenges of climate change, although it can play a pivotal role in addressing it. Green finance has emerged as an effective alternative to address the challenges of global climate change. Green finance refers to using private and governmental funds for initiatives that protect the environment from harm and its associated effects, such as climate change and air pollution, produce a wide range of social advantages, and provide investors with adequate financial returns. In addition to tools for reducing greenhouse gas emissions and adapting to climate change, green finance has emerged as a strategy that includes financial services and products that address a more comprehensive range of environmental issues, such as industrial pollution control, waste management, sanitation, and hygiene, and ecological protection.

This chapter explores the idea of green finance in the present scenario and its potential to achieve the goals of climate change mitigation and adaptation. Further, it will examine how green finance can be integrated into global economic governance. The study finds a deep correlation between global economic governance and the achievement of sustainability. Further, it is found that economic governance and policy planning are crucial in achieving a carbon-neutral state across the globe.

A. Mishra (✉) · A. Pandey · J. V. Kureethara
Christ University, Bangalore, India
e-mail: achyuta.mishra@christuniversity.in

N. Naifar (ed.), *Climate Change and Finance*, Sustainable Finance,
https://doi.org/10.1007/978-3-031-56419-2_8

1 Introduction

Anthropogenic activities causing the rise in global greenhouse emissions have led to biodiversity loss and ecosystem changes worldwide. It has resulted in climate change, seriously affecting the human habitat and the planet's future. Due to the extreme effects of climate change, the survival of many societies and the biological support systems of the planet are at risk. Climate change is a global challenge that knows no boundaries. It presents "an urgent and potentially irreversible threat to human societies and the planet" (Paris, 2015). It requires concerted efforts by all countries and efficacious and suitable international response to mitigate its ill effects. The international community has been attempting successively through legislative and policy means to address the problem. Finance has a significant role to play in furthering climate action. The global economic system needs to be activated to decarbonise rapidly to reach the target set by the Paris Agreement, 2015 (Kocharekar, 2023). This chapter is inspired by the idea of exploring the role of global economic governance in attaining sustainable development goals to mitigate and adapt to climate change.

2 Global Economic Governance Vis-à-Vis Climate Change

"Global economic governance" refers to the framework, institutions, and mechanisms through which nations and international organisations collaborate to manage and regulate the global economy. It encompasses many issues, including trade, finance, monetary policy, development, and economic stability. "The foundations of the contemporary economic order were established in the late 1940s by the financial and trade institutions system set up at an international conference in Bretton Woods, New Hampshire, in 1944" (Gereffi, 2005). It is regulated by the World Trade Organisation (WTO) and the International Monetary Fund (IMF) (Wolf, 2004). This architecture does not have a clear definition. Although there is no consensus as to it, it consists of major international financial institutions like the World Bank Group, WTO, and IMF (Kocharekar, 2023). It has promoted global economic governance by promoting international trade, finance, and investment.

Recognising the importance of the environment and sustainable development, the agreement establishing WTO aims at achieving the goals of sustainable development, protecting and preserving the environment while promoting production of goods and services. This has resulted in the creation of the Trade and Environment Committee which looks into the environment when environmental policies have a substantial impact on trade. It does not look into anything else beyond that (WTO). So far, global economic architecture has paid little or no attention to environmental and climatic concerns which is having dire consequences for the planet. It is widely felt that the global economy must change its focus from trade and investment to a model based more on climate sustainability. It is ill-equipped to address the environmental and climate change issues. It is yet to centralise such issues in economic policy planning and is unable to deliver on it (Kocharekar, 2023).

The UNFCCC recognises the need for significant financial resources to combat climate change effectively. It established funding mechanisms, such as the Green Climate Fund (GCF) and the Global Environment Facility (GEF), to support developing countries' climate mitigation and adaptation projects. Green finance sources, such as green bonds and climate funds, contribute to these financing mechanisms, enabling the mobilisation of funds for climate-related projects and initiatives. The United Nations Framework Convention on Climate Change (UNFCCC) mandates the state parties to cooperate in promoting an open and supportive international economic system to achieve sustainable economic growth and development for all countries, specifically developing countries, enabling them to address the issues of climate change (UNFCCC, 2023a, 2023b, 2023c).

According to the Intergovernmental Panel on Climate Change (IPCC), investments in climate action totalling $1.6 trillion–3.8 trillion annually will be needed within the next several years to avoid global temperatures from increasing by more than 1.5 °C above pre-industrial levels (Kocharekar, 2023 p. 12). To achieve this financing, climate action can be considered an effective method. Cooperation and collaboration between international organisations and financing bodies focusing on climate action is the need of the hour. To fill this gap, green finance has emerged as an instrument to cater to the need for climate action.

UNFCCC calls for encouraging an open, accommodating, and supportive global economic system that would result in sustainable economic growth and development for all Parties, especially those from developing nations, allowing them to more effectively tackle the issues associated with climate change (UNFCCC, 2023a, 2023b, 2023c).

3 Global Climate Agreements: Strengthening Cooperation

In recent years, nations worldwide have recognised the urgent need to unite and take collective action to combat this existential threat. Global climate agreements have emerged as a critical tool for fostering economic cooperation while addressing climate change (Hoegh-Guldberg et al., 2019). To mitigate the adverse effects of climate change, the community of nations under the aegis of the United Nations (UN) have framed conventions and treaties such as the United Nations Framework Convention on Climate Change (UNFCCC), Kyoto Protocol, and Paris Agreements.

These agreements, such as the Paris Agreement, are about reducing greenhouse gas emissions and reshaping economic systems and forging a sustainable future for all. International agreements like the Paris Agreement have been crucial in addressing climate change. The Paris Agreement, adopted in 2015 under the United Nations Framework Convention on Climate Change (UNFCCC), marked a turning point in global climate governance (Zhang et al., 2017). It emphasised the need for nations to cooperate on climate action and set a global goal to limit global warming to well below 2 °C above pre-industrial levels, with an aspiration to limit it to 1.5 °C. This requires financial resources, large-scale investment, and spirited cooperation among all stakeholders.

According to Article 9 of the Paris Agreement, developed country Parties must continue to fulfil their current commitments under the Convention by contributing financial resources to help developing country Parties with mitigation and adaptation. It is encouraged for other Parties to offer or maintain this kind of voluntary assistance (UNFCCC, 2023a, 2023b, 2023c).

The goal of the Kyoto Protocol is to help nations adapt to the negative consequences of climate change. It makes it easier to create and implement technology that can strengthen resistance to the effects of climate change. The purpose of the Adaptation Fund is to provide funding for adaptation initiatives and programmes in developing nations that are Kyoto Protocol Parties. The Fund was primarily funded during the first commitment term by a portion of the money earned from CDM project activities. It was resolved in Doha in 2012 that the Adaptation Fund will receive a 2% share of earnings for the second commitment period from international emissions trading and cooperative implementation (UNFCCC, 2023a, 2023b, 2023c).

The UNFCCC has a climate finance data portal that serves as a resource for information on initiatives sponsored in developing nations to carry out climate action. It also provides useful explanations, pictures, and figures to help comprehend the climate finance process. The financial mechanism's running entities and Parties have made information available in each of the three modules that make up the finance site. The Convention's "common but differentiated responsibility and respective capabilities" principle states that wealthy country Parties must contribute financial resources to help developing country Parties carry out the UNFCCC's goals (UNFCCC, 2023a, 2023b, 2023c).

There is a growing emphasis on climate finance to support developing nations in their mitigation and adaptation efforts. This financial assistance is crucial for global economic stability and equity. The mobilisation of climate finance is a key aspect of global climate agreements. Developed countries commit to providing financial support to developing nations to help them mitigate and adapt to climate change. This financial cooperation has several economic implications (Aykut et al., 2022):

- Global Economic Stability: Climate finance assists vulnerable nations in building resilience, which, in turn, contributes to global economic stability by reducing the risk of climate-induced conflicts and displacement.
- Market Opportunities: Climate finance often supports projects in developing countries, creating new business market opportunities and promoting global economic growth.
- It aligns financial markets with climate objectives and supports the transition to a more sustainable and climate-resilient global economy. Emerging trends in global climate governance include:

    a. Enhanced Commitments: Countries are pressured to increase their emissions reduction targets. The ambition to limit global warming to 1.5 °C is driving more aggressive climate actions.

b. Climate Diplomacy: Climate change is becoming a central topic in diplomatic relations. Nations cooperate and negotiate climate-related agreements, fostering greater international collaboration (Downie & WilliamGaiser, 2003). Climate change is now a central topic in international diplomacy. Nations must collaborate and negotiate on various aspects of climate action. This diplomacy fosters economic cooperation by:

- Facilitating Trade Agreements: Climate agreements often lead to trade partnerships centred around sustainable technologies and products, boosting global economic integration.
- Encouraging Joint Ventures: Countries increasingly collaborate on research and development initiatives to advance clean technologies, leading to shared economic benefits.

4 Climate Change Mitigation and Green Finance

To achieve the climate change goals set out in the Paris Agreement, green finance can play a significant role by generating financial resources. Green finance is critical in mobilising and channelling funds toward projects and initiatives that reduce greenhouse gas emissions and enhance climate resilience. Green finance encompasses various financial instruments and strategies that direct capital towards projects with environmental benefits. It arranges finance for projects that are environmentally sustainable or adopt measures of climate action, which include "production of energy from renewable sources like solar, wind, biogas, etc.; clean transportation that involves lower greenhouse gas emission; energy-efficient projects like green building; waste management that includes recycling, efficient disposal and conversion to energy, etc. (RBI, 2021; Sharma et al., 2022). Green finance aims to minimise the negative impacts of economic activities on the environment, drive innovation, and create new opportunities for economic growth. It involves engagement of 'traditional capital markets in creating and distributing a range of financial products and services that deliver both investable returns and environmentally positive outcomes" (ADB, 2018).

4.1 The Significance of Green Finance

The significance of climate change mitigation and economic governance lies in their mutually reinforcing relationship. Effective financial management can enable and support climate mitigation efforts, which, in turn, can lead to a more resilient, equitable, and sustainable economy. Addressing climate change is an environmental imperative and a critical component of responsible economic governance in an increasingly interconnected world. Climate change threatens ecosystems, economies, and societies worldwide (Petersen & Chia, 2003). Rising sea levels, extreme weather events, and disruptions to food and water supplies are just a few of the consequences. Green finance is pivotal in mitigating these impacts by redirecting investments away from activities contributing to carbon emissions and environmental degradation (Nhamo & Swart, 2012). Instead, it focuses on initiatives that facilitate the transition to a more sustainable, carbon-neutral future. Every person benefits from green finance's economic and environmental benefits. It equalises the transition to a low-carbon society and increases access to environmentally friendly products and services for individuals and businesses, leading to more equitable social growth. Certain benefits arising out of climate finance are as follows (Stoddart & Tindall, 2015):

4.1.1 Auto Styling Mitigating Climate Risks

Green finance is crucial in mitigating climate risks by promoting investments in technologies and practices that reduce greenhouse gas emissions. This is vital to meet the targets set by international agreements such as the Paris Agreement, which aims to limit global warming to below 2 °C above pre-industrial levels.

4.1.2 Transitioning to a Low-Carbon Economy

Traditional industries heavily reliant on fossil fuels face increasing pressure to transition towards cleaner alternatives. Green finance provides the necessary funds for these industries to adopt sustainable practices and technologies, ensuring a smoother transition to a low-carbon economy.

4.1.3 Enhancing Resilience

Climate change brings various challenges, including extreme weather events and resource scarcity. Green finance supports projects that enhance the resilience of communities and businesses against such challenges, such as investing in resilient infrastructure and disaster preparedness.

4.1.4 Attracting Investment Opportunities

The growing demand for environmentally friendly products and services creates a market ripe for investment. Green finance allows investors to capitalise on these opportunities, driving economic growth in sustainable sectors.

4.2 *The Role of Financial Institutions and Governments*

Financial institutions, including banks, asset managers, and insurance companies, are critical in advancing green finance. They can develop innovative financial products tailored to sustainable investments and incorporate environmental, social, and governance (ESG) criteria into their decision-making processes (Kristijono et al., 2022). Governments also play a crucial role by establishing regulatory frameworks, incentives, and standards that encourage green finance initiatives. Policies such as carbon pricing and tax incentives for green investments can significantly influence the flow of capital towards sustainable projects. Businesses are taking climate change seriously on many levels. Many people have begun incorporating climate change into their strategies to mitigate global warming and climate change hazards. Business reaction to climate change has primarily been assessed using voluntary benchmarks like the Carbon Disclosure Project (CDP) and the Johannesburg Stock Exchange Socially Responsible Index (JSE SRI) (Derevyanko et al., 2020). It is argued that these benchmarks establish the absolute minimal standards for criteria related to climate change. The initial replies have demonstrated that the corporate sector is susceptible to thorough and academic reporting on business responses to climate challenges as part of CSR (Ayton, 2013). The entities in charge of the reporting frameworks also voiced concern over the expansion of reporting requirements globally.

5 Mechanism of Green Finance

One of the most significant emerging trends is the global shift towards a green economy. Nations increasingly recognise the need to reduce carbon emissions and invest in renewable energy sources. This transition towards sustainability is changing the landscape of global economic governance in several ways (Sodano & Hingley, 2013):

5.1 *Green Investments*

Governments are incentivising and subsidising green investments, including renewable energy projects, electric vehicles, and sustainable agriculture. This trend is creating new opportunities for businesses while reshaping traditional economic sectors. Green bonds are debt securities between governments, municipalities, and corporations issued to finance environmentally friendly projects (Tran et al., 2020). The proceeds from these bonds are exclusively allocated to projects with positive environmental impacts.

5.2 *Carbon Pricing*

Implementing a price on carbon emissions through mechanisms like carbon markets or taxes encourages industries to reduce their emissions, as it becomes more cost-effective to invest in green technologies. Carbon pricing mechanisms, such as carbon taxes and cap-and-trade systems, are becoming more prevalent. These policies encourage businesses to reduce emissions and invest in cleaner technologies (Green, 2021).

5.3 *Sustainable Finance*

The financial sector adapts to climate risks and opportunities by incorporating Environmental, Social, and Governance (ESG) criteria into investment decisions. Sustainable finance is increasingly influencing how companies operate and access capital like green bonds and sustainable loans fund projects with environmental benefits (Schütze & Stede, 2021). However, in this case, the funds come from financial institutions rather than the capital markets.

5.4 *Venture Capital and Private Equity*

These investment strategies channel funds into innovative startups and businesses that develop sustainable technologies and solutions.

6 Need for a Transition to a Green Economy

The transition to a low-carbon economy, as outlined in the Paris Agreement, necessitates fundamental global economic system changes. This transition involves (World Bank, 2015):

- Shifting Investments: As countries redirect investments from fossil fuels to renewable energy and sustainable infrastructure, economic cooperation occurs among various industries, driving innovation and job creation.
- Inclusive Growth: A green economy can promote inclusive growth by focusing on social equity, job creation, and poverty.
- The Paris Agreement encourages countries to enhance their emissions reduction commitments regularly. This ongoing process strengthens economic cooperation in several ways (Cranston & Hammond, 2010):
- Investment Certainty: Clear and ambitious climate targets provide businesses with investment certainty, encouraging long-term planning and innovation.

- Technological Advancement: As nations strive to meet their climate goals, they invest more in green technologies, such as renewable energy and energy-efficient infrastructure. This drives economic growth and job creation.
- Innovation and Adaptation of Technology: Technology plays a pivotal role in addressing climate change while shaping economic governance.

6.1 Renewable Energy Breakthroughs

Advancements in renewable energy technologies, such as solar and wind, are making clean energy more accessible and cost-effective. This is reducing dependence on fossil fuels. The global transition toward renewable energy sources represents a significant shift in energy production, consumption, and the broader economic landscape. This transition addresses environmental concerns and has profound implications for global financial management. Renewable energy technologies, including solar, wind, and hydropower, rapidly reshape financial strategies, risk management, and investment decisions worldwide (Newell, 2019). Renewable energy is not just a solution to climate change but a transformative force in global financial management.

The sector offers attractive investment opportunities, drives innovations in risk management, and introduces new financial instruments. As the world continues its shift towards cleaner energy sources, financial managers and institutions will play an increasingly vital role in driving the sustainable, low-carbon economy of the future. Successful financial management in this evolving landscape will require agility, innovation, and a deep understanding of the complex interplay between renewable energy and global finance (Barbir, 2009).

6.2 Climate Adaptation Technologies

Innovations in climate adaptation technologies, like drought-resistant crops and flood mitigation systems, are helping communities and industries adapt to changing climate patterns. Climate adaptation starts with effective monitoring and early warning systems. Investment in technology to predict extreme weather events, rising sea levels, and changing precipitation patterns is essential for preparedness and risk reduction. Water scarcity and flooding are exacerbated by climate change. Innovative water conservation, purification, and efficient distribution technologies are necessary for adapting to changing hydrological conditions. Environmental, Social, and Governance (ESG) criteria have gained prominence in investment decisions (Adom & Amoani, 2021). Financial institutions increasingly consider climate adaptation and resilience key to their ESG assessments.

6.3 *Digitalisation and Climate Data*

Big data, artificial intelligence, and satellite technology improve climate modelling and disaster management, aiding in risk assessment and mitigation. Digitalization, driven by technological advancements, data collection, and analysis, is revolutionising how we understand, address, and cooperate on climate change. When leveraged through digital tools and platforms, climate data enhances cooperation among nations, businesses, and stakeholders, ultimately contributing to more effective climate action. Digital technologies like remote sensing, satellites, and IoT sensors enable real-time data collection on various climate parameters, including temperature, sea-level rise, and greenhouse gas emissions (Santarius, 2022). This data is essential for monitoring climate trends and impacts. Digital platforms aid in efficiently allocating climate finance by identifying high-impact projects and tracking funding flows. This ensures that resources are directed towards the most effective climate mitigation and adaptation initiatives, fostering international cooperation in supporting vulnerable regions.

- Inclusive Decision-Making: The Role of Stakeholders Emerging trends in global economic governance involve more inclusive decision-making processes:
- Multi-Stakeholder Engagement: Governments, businesses, civil society, and academia are increasingly shaping climate policies and strategies. This inclusivity fosters diverse perspectives and solutions.

Bottom-Up Approaches: Local and regional authorities are taking a proactive role in climate action. This decentralisation of power is vital for effective climate governance.

7 Climate Change and Economic Governance: International Perspective

7.1 *Europe*

Climate change and economic governance in Europe are two interconnected and critical issues that have garnered significant attention in recent years. Europe has been at the forefront of efforts to address climate change as a region. It has also implemented various economic governance mechanisms to navigate the challenges posed by this global crisis (Mahat et al., 2019). Therefore, how these two aspects intersect and their implications for the continent can be described as follows:

- Emissions Reduction Targets: Europe has set ambitious targets to reduce greenhouse gas emissions. The European Union (EU) has committed to becoming carbon-neutral by 2050, with intermediate targets for 2030 (Vestreng et al., 2020). Achieving these targets requires significant economic restructuring and investments in clean energy, transportation, and industrial processes. Economic governance is crucial to allocate resources efficiently and ensure that member states adhere to these targets.

- Green Investment: To combat climate change, Europe has adopted policies and financial instruments that promote green investments. This includes the European Green Deal, which aims to mobilise over €1 trillion in assets to support sustainability and decarbonization (Pimonenko et al., 2019). Economic governance in the form of budget allocations, grants, and subsidies plays a pivotal role in directing funds towards green projects.
- Carbon Pricing: European countries have implemented carbon pricing mechanisms like the Emissions Trading System (ETS). These mechanisms put a price on carbon emissions and create economic incentives for industries to reduce their emissions. Effective economic governance is essential to manage and expand these carbon pricing mechanisms.
- Regulation and Standards: The EU has introduced stringent rules and standards for various sectors, including energy, transportation, and agriculture, to reduce emissions and promote sustainability. These regulations are enforced through economic governance mechanisms that penalise non-compliance and reward sustainable practices.
- Transitioning Energy Systems: Europe is transitioning from fossil fuel-based energy systems to renewables. This transition involves substantial investments in renewable energy infrastructure, grid improvements, and storage solutions. Economic governance ensures that the costs and benefits of this transition are distributed fairly among member states.
- Just Transition: Managing climate policy's economic and social impacts is crucial. Economic governance in Europe includes efforts to ensure a "just transition" for regions heavily reliant on fossil fuels or carbon-intensive industries (Brescia et al., 2023). This involves retraining workers, supporting affected communities, and promoting new economic opportunities in green sectors.
- International Climate Agreements: Europe is leading in international climate negotiations. Economic governance is vital for fulfilling commitments like the Paris Agreement. This includes funding climate adaptation and mitigation efforts in developing countries and coordinating efforts with other nations.
- Financial Institutions: European financial institutions like the European Central Bank and the European Investment Bank have incorporated climate considerations into their policies and practices (González & Núñez, 2021). Economic governance within these institutions influences how they support climate-related projects and investments.

Therefore, addressing climate change and ensuring economic stability in Europe are intertwined challenges that require effective governance and policy coordination. The region's commitment to reducing emissions and transitioning to a sustainable, green economy is a testament to its recognition of the urgent need to combat climate change while ensuring economic prosperity. However, achieving these goals requires ongoing efforts in economic governance, investment strategies, and international cooperation.

7.2 Brazil

Brazil, one of the world's largest and most biodiverse countries, is critical in addressing climate change due to its vast forests, including the Amazon rainforest, and its contributions to greenhouse gas emissions. The critical aspects related to climate change and economic governance in Brazil are:

7.2.1 Climate Change Challenges

- *Deforestation*: Brazil has faced ongoing issues with deforestation, particularly in the Amazon rainforest. Deforestation contributes to increased greenhouse gas emissions and disrupts local and global ecosystems.
- *Carbon Emissions*: Brazil is one of the world's largest emitters of greenhouse gases, primarily due to deforestation, agriculture, and energy production (Reuters, 2022).
- *Extreme Weather Events*: The country experiences more frequent and severe extreme weather events, such as droughts, floods, and storms, linked to climate change.

7.2.2 Economic Governance

- *Economic* Structure: Brazil has a diverse economy including agriculture, mining, manufacturing, and services. However, it faces challenges related to income inequality and poverty.
- *Political* Stability: Brazil has had political instability and corruption scandals that have impacted economic governance and policy implementation.
- *Environmental* Regulations: The Brazilian government has implemented various environmental regulations and policies to address deforestation and curb greenhouse gas emissions.

7.2.3 Policy Responses

- *Paris Agreement*: Brazil is a signatory to the Paris Agreement, committing to reduce its greenhouse gas emissions and enhance its efforts to combat climate change.

- *Forestry Protection*: Efforts to combat deforestation have included government initiatives like the Amazon Fund and international collaborations to monitor and protect the rainforest.
- *Renewable Energy*: Brazil has made strides in renewable energy production, particularly in hydropower and biofuels, reducing its reliance on fossil fuels.

7.2.4 Challenges and Controversies

- *Political Will*: The effectiveness of climate change and environmental policies in Brazil can be hindered by political and economic interests prioritising development over conservation.
- *International Pressure*: Brazil faces international scrutiny and pressure regarding its environmental policies, particularly the Amazon rainforest.
- *Economic Trade-offs*: Balancing economic growth with environmental preservation is a complex challenge for Brazil's economic governance.

Brazil's climate change and economic governance approach is globally significant due to its vast natural resources and environmental impact. Striking a balance between economic development and ecological preservation remains a pressing challenge for the country and has implications for the broader global effort to combat climate change.

7.3 South Africa

Climate change and legal and economic governance in South Africa are intertwined issues that significantly affect sustainable development and environmental protection efforts.

7.3.1 Climate Change in South Africa

- *Impacts:* Like many other countries, South Africa is experiencing the adverse effects of climate change. These impacts include frequent and severe droughts, heat waves, flooding, and changing rainfall patterns (Calzadilla et al., 2014). These events have far-reaching consequences for agriculture, water resources, and public health.
- *Contributions:* While South Africa is not one of the largest global greenhouse gas emitters, it is one of the biggest on the African continent. The country's major contributors to emissions are energy production, mining, and transportation sectors.
- *Vulnerabilities:* South Africa's socio-economic landscape makes it particularly vulnerable to climate change. The country faces poverty, inequality, and unem-

ployment challenges exacerbated by climate-induced livelihood disruptions (Xie et al., 2022).

7.3.2 Legal Economic Governance in South Africa

- *Constitutional Framework:* South Africa's legal economic governance is primarily guided by its constitution. The country's progressive constitution enshrines various socioeconomic rights, including the right to a clean environment, access to water, and sustainable development (Cilliers & Aucoin, 2016).
- *Environmental Legislation:* South Africa has established a comprehensive legal framework to address environmental and climate change issues. Vital ecological laws include the National Environmental Management Act (NEMA), the National Climate Change Response Policy, and the Carbon Tax Act.
- *Economic Policies:* South Africa aims to balance economic growth with environmental sustainability. The country has been exploring ways to transition to a green economy, including investing in renewable energy and sustainable agriculture.

7.3.3 Intersection of Climate Change and Economic Governance

- *Policy Coherence:* South Africa's challenge is ensuring policy coherence between its economic development goals and climate change mitigation and adaptation efforts (Du Plessis, 2008). Striking this balance requires effective legal and financial governance that supports sustainable practices.
- *International Commitments:* South Africa is a signatory to international agreements like the Paris Agreement, which commits countries to limit global warming. Legal frameworks and economic governance must align with these commitments.
- *Socio-economic Equity:* Legal and economic governance should also address issues of socio-economic equity, ensuring that vulnerable communities benefit from climate change adaptation and mitigation measures.

In conclusion, addressing climate change in South Africa requires a multifaceted approach that integrates legal and economic governance into its climate action strategies. This involves aligning economic policies with environmental goals, complying with international commitments, and prioritising socio-economic equity in climate adaptation and mitigation efforts.

7.4 USA (United States of America)

The intersection of climate change and legal and economic governance in the USA is complex and multifaceted. It involves how the legal and regulatory framework in the USA addresses and responds to the challenges posed by climate change while also considering the economic implications of such policies. Here are some key points to consider:

- *Regulatory Framework*: The USA has a variety of federal and state laws and regulations that govern environmental protection and climate change mitigation. Key federal agencies, such as the Environmental Protection Agency (EPA), play a significant role in implementing and enforcing these regulations (Aglietta et al., 2015).
- *Climate Change Mitigation*: Efforts to combat climate change in the USA involve regulations to reduce greenhouse gas emissions. This can include industry standards, energy efficiency requirements, and incentives for developing renewable energy sources.
- *Economic Implications*: Climate change policies can have significant financial implications. For example, regulations requiring industries to reduce emissions increase compliance costs. However, they can also stimulate the growth of green initiatives and create new economic opportunities.
- *Legal Challenges*: Climate change policies often face legal challenges from various stakeholders, including businesses and advocacy groups. These challenges may revolve around the interpretation of existing laws or the constitutionality of specific regulations.
- *International Agreements*: The USA's approach to climate change is influenced by international agreements, such as the Paris Agreement. These agreements shape the country's climate policies and commitment to reducing global emissions.
- *State vs. Federal Authority*: There can be conflicts between state and federal governments regarding climate change policy. Some states have taken more aggressive measures to address climate change, while the federal government has different priorities under different administrations.
- *Market Mechanisms*: Economic governance related to climate change can involve market-based mechanisms, such as carbon pricing or cap-and-trade systems. These mechanisms aim to create economic incentives for emission reductions.
- *Innovation and Investment*: The legal framework can encourage innovation and private sector investment in clean energy technologies and climate adaptation measures. Tax incentives and subsidies are examples of such mechanisms.
- *Litigation and Liability*: There have been cases where individuals, communities, or even municipalities have pursued legal action against companies for climate-related damages. These cases raise questions about liability and legal responsibility.

- *Changing Political Landscape*: Climate change policies in the USA can vary significantly depending on the political climate and the current administration's and Congress's priorities. Leadership changes can lead to shifts in regulatory focus.

In recent years, climate change has become an increasingly prominent issue in legal and economic discussions in the USA. Policymakers, businesses, and activists continue to debate the best approaches to addressing climate change while considering financial implications and the legal framework within which these efforts take place.

7.5 India

In terms of both economic growth and climate change, India is in an intriguing situation. India's trip to a net-zero carbon economy is a marathon despite having abundant resources (forests, solar energy, etc.). Still, it lacks the funding and technical know-how to make it happen. The economy is typically stressed by immediate climate change legislation like COP-26 (Ahluwalia & Patel, 2021). The climatic aspects of climate change have unavoidably begun to change, severely harming the Indian economy. However, failing to strive for an emission-free economy will impede future advancement. Energy needs frequently conflict with climate change policies. Climate change also has an impact on agriculture, leaving the majority of Indians vulnerable.

Several examples of how catastrophic weather occurrences have impacted India's daily life have surfaced recently. According to exposure and vulnerability to climate risk events, India was rated seventh among the most affected nations in the Global Climate Risk Index 2021. One of the greatest issues of the twenty-first century is climate change, which poses serious threats to not just the environment but also human health, food security, and economic growth. Both the supply side of the economy (the potential for production) and the demand side (consumption and investment) are susceptible to the negative effects of climate change (Nawaz et al., 2021). Additionally, it has cross-regional and cross-sectoral spill-over effects, cross-border repercussions, and contagion concerns.

The impact of climate change on the Indian Economy and its governance can be enunciated as follows:

- Reduced Agricultural Output: Due to temperature changes, patterns of precipitation, infestations of pests, deterioration of soil, shortages in water supply, and extreme weather events like floods and droughts, climate change can seriously interrupt crop cycles and result in low agricultural productivity. The main source of income in India and a major economic contributor is agriculture and related industries. Low yields hurt rural economies and can increase inflation in cities as well.

- Risen sea surface temperatures resulting from climate change wreak havoc on the fisheries industry by altering fish species' distribution and migratory patterns. The presence of fish in a given area changes due to some species migrating or moving to colder waters. Fish catch composition and abundance altar, affecting fishermen's livelihoods.
- Expensive Medical Care: Malaria, dengue fever, cholera, heat stroke, respiratory infections, and stress are just a few illnesses that become more common and severe due to climate change. It also impacts the health and nutrition of vulnerable populations like the elderly, destitute, and vulnerable mothers and children. In addition to lowering labour productivity and raising public spending, health expenditures can also affect disposable income. The WHO estimates that between 2030 and 2050, climate change will result in an additional 2,50,000 deaths per year from starvation, malaria, diarrhoea, and heat stress (Roberts & Weikmans, 2017).
- Infrastructure Damage: Due to sea level rise, coastal erosion, landslides, storms, floods, and heat waves, climate change can harm physical infrastructure, including roads, bridges, trains, ports, airports, power plants, water supply systems, and buildings. In addition to raising maintenance and replacement costs, damaged infrastructure can interfere with commerce, connectivity, and economic activities. For instance, India has spent USD 3 billion on flood-related economic losses over the past 10 years, accounting for 10% of the total (Kumar & Parikh, 2001).
- Reduced Industrial Output: Due to variables including new climate-friendly laws, lower stock utilisation, and relocation of production processes and activities due to climate-related losses, climate change can raise operational costs and diminish profitability in the industrial sector. By 2030, India may be responsible for 34 million of the 80 million jobs lost worldwide due to heat stress-related productivity reduction.
- Energy Crisis: India's primary energy demand is expected to triple by 2030, according to the International Energy Agency (IEA). Because of the special relationship between energy and climate, rising temperatures necessitate increased energy use to help mitigate the consequences of the heat.
- Impact on Financial Services: Financial services may be under pressure because banks and other financial institutions are taking on more credit risk due to climate change. Climatic disasters like floods, hurricanes, or droughts may impact the ability of borrowers to repay loans. These occurrences can potentially cause property damage, supply chain disruptions, and profitability issues for enterprises, which could result in loan defaults and credit losses. Due to decreased demand, cancellations and safety worries might also increase insurance claims and cause travel and hospitality services disruptions.

7.5.1 Legal Framework in India

India has enacted several laws and policies to address climate change and promote sustainable economic development. Some critical legal frameworks include:

- *National Action Plan on Climate Change (NAPCC)*: India's comprehensive strategy to combat climate change, consisting of eight national missions addressing sectors like solar energy, water, and agriculture.
- *Environmental Laws*: India has a range of environmental laws and regulations that cover aspects related to climate change, air and water pollution, forest conservation, and wildlife protection.
- *Renewable Energy Policies*: India has ambitious renewable energy targets and policies to promote the development of clean energy sources like solar and wind power.
- *Carbon Pricing*: India has been exploring carbon pricing mechanisms, such as a carbon tax or emissions trading schemes, to incentivize emissions reductions.

7.5.2 Challenges and Opportunities

- *Economic Growth vs. Sustainability*: India faces the challenge of balancing its economic growth aspirations with environmental sustainability. Rapid industrialization and urbanisation can lead to increased emissions.
- *Climate Adaptation*: Legal economic governance needs to address climate adaptation strategies, particularly for vulnerable communities and sectors.
- *International Commitments*: India is a signatory to international agreements like the Paris Agreement. Legal frameworks must align with these commitments and reporting requirements.
- *Technology Transfer*: Legal mechanisms can facilitate the transfer of climate-friendly technologies to India.

7.5.3 Enforcement and Implementation

The effectiveness of climate and environmental laws in India often depends on enforcement and implementation at various levels of government. Strengthening institutional capacities and ensuring compliance are essential for achieving desired outcomes.

In conclusion, addressing climate change within the framework of legal economic governance is a complex and multifaceted challenge for India. Balancing economic growth with environmental sustainability, adapting to changing climate patterns, and fulfilling international commitments are key considerations in this context. Effective legal frameworks and their rigorous enforcement are critical to mitigating climate change impacts and promoting sustainable economic development in India.

8 Challenges and Future Prospects

1.1 *Standardisation*: Establishing consistent criteria for what constitutes a "green" project is essential to avoid "greenwashing," where projects with limited environmental benefits are labelled as such. So far there is no universally accepted criteria specifying as to what constitutes green finance, which is giving way to greenwashing. To achieve the goal of green finance only the EU has come up with the taxonomy prescribing sustainability and finance.
1.2 *Information Gaps*: Investors need accurate and transparent information about the environmental impacts of projects to make informed decisions.
1.3 *Risk Assessment*: Green projects sometimes involve higher upfront costs or longer payback periods, requiring innovative risk assessment models considering long-term environmental and financial impacts.
1.4 *Policy Support*: Governments and regulatory bodies are critical in incentivizing green finance through policies, regulations, and incentives.

9 Conclusion and Suggestions

To successfully address the challenges posed by climate change, continued economic cooperation among the nations is essential. In contemporary times, green finance has emerged as an effective tool having the potential to drive a significant shift towards sustainability by directing financial resources towards environmentally beneficial projects. As the urgency of addressing climate change grows, integrating green finance into mainstream financial practices becomes more important than ever (Ionescu, 2021). By addressing challenges and fostering collaboration between governments, financial institutions, and businesses, the power of finance can be harnessed to create a resilient, low-carbon, and environmentally sound future.

The future of our planet and global economic governance depends on our ability to respond to emerging trends effectively. A shift in the paradigm of global economic architecture towards climate action will reduce greenhouse gas emissions released from implementing a certain investment and increase its resilience to climate change impacts. Further, it will help develop more resilient and carbon-efficient policies like those of the International Solar Alliance and the National Clean Energy Fund. Emerging trends in climate change and global economic governance are intertwined and evolving rapidly. The transition to a global green economy, technological innovations and sharing, and inclusive decision-making will reshape how we tackle climate change. Economic collaboration between governments, international organisations, businesses, and civil society is crucial to achieving meaningful progress towards climate change action.

References

Adom, P. K., & Amoani, S. (2021). The role of climate adaptation readiness in economic growth and climate change relationship: An analysis of the output/income and productivity/institution channels. *Journal of Environmental Management, 293*, 112923.

Aglietta, M., Hourcade, J. C., Jaeger, C., & Fabert, B. P. (2015). Financing transition in an adverse context: Climate finance beyond carbon finance. *International Environmental Agreements: Politics, Law and Economics, 15*, 403–420.

Ahluwalia, M. S., & Patel, U. (2021). *Getting to net zero: An approach for India at CoP-26*. Centre for Social and Economic Progress.

Asian Development Bank. (2018). *Green finance, explained*. https://development.asia/explainer/green-finance-explained

Aykut, S. C., Morena, E., & Foyer, J. (2022). Incantatory governance: Global climate politics performative turn and its wider significance for global politics. In *The Climatization of global politics* (pp. 19–40). Springer International Publishing.

Ayton, N. (2013). *Profitable, ethical, sustainable: Investigating corporate responses to climate change*. University of Victoria (Canada).

Barbir, F. (2009). Transition to renewable energy systems with hydrogen as an energy carrier. *Energy, 34*(3), 308–312.

Brescia, V., Esposito, P., Amelio, S., & Biancone, P. P. (2023). Rethinking green investment and corporate sustainability: The south European countries experiences during the COVID-19 crisis. *EuroMed Journal of Business*.

Calzadilla, A., Zhu, T., Rehdanz, K., Tol, R. S., & Ringler, C. (2014). Climate change and agriculture: Impacts and adaptation options in South Africa. *Water Resources and Economics, 5*, 24–48.

Cilliers, J., & Aucoin, C. (2016). Economics, governance and instability in South Africa. *Institute for Security Studies Papers, 2016*(293), 1–24.

Cranston, G. R., & Hammond, G. P. (2010). North and south: Regional footprints on the transition pathway towards a low carbon, global economy. *Applied Energy, 87*(9), 2945–2951.

Derevyanko, Y., Lukash, O., Shkarupa, O., Melnyk, V., & Simonova, M. (2020). Greening economy vs. greening business: Performance indicators, driving factors and trends. *International Journal of Global Environmental Issues, 19*(1–3), 217–230.

Downie, C., & WilliamGaiser, T. (2003). Global change and regional impacts: Water availability and vulnerability of ecosystems and society in the semiarid northeast of Brazil. Springer Science & Business Media.S, M. After the Paris agreement: What role for the BRICS in global climate governance? *Global Policy, 9*(3), 398–407.

Du Plessis, W. (2008). Legal mechanisms for cooperative governance in South Africa: Successes and failures. *SA Publiekreg SA Public Law, 23*(1), 87–110.

Gereffi, G. (2005). *The global economy: Organization, governance, and development*. https://www.researchgate.net/publication/281870242

González, C. I., & Núñez, S. (2021). Markets, financial institutions and central banks in the face of climate change: Challenges and opportunities. *Banco de España. Occasional Paper, 2126*.

Green, J. F. (2021). Does carbon pricing reduce emissions? A review of ex-post analyses. *Environmental Research Letters, 16*(4), 043004.

Hoegh-Guldberg, O., Jacob, D., Taylor, M., Guillén Bolaños, T., Bindi, M., Brown, S., et al. (2019). The human imperative of stabilising global climate change at 1.5 C. *Science, 365*(6459), eaaw6974.

Ionescu, L. (2021). Transitioning to a low-carbon economy: Green financial behaviour, climate change mitigation, and environmental energy sustainability. *Geopolitics, History, and International Relations, 13*(1), 86–96.

Kocharekar, R. (2023). *The response of the international economic architecture to climate change: Reforming global economic governance to meet the Paris goals*. https://www.chathamhouse.

org/sites/default/files/2023-03/2023-03-29-response-economic-architecture-climate-change-kocharekar
Kristijono, N. H., Pramono, R., Supratikno, H., & Sudibjo, N. (2022). Ethical decisions of leaders in sustainable investing to promote sustainable development. *International Journal of Sustainable Development & Planning, 17*(1), 13.
Kumar, K. K., & Parikh, J. (2001). Indian agriculture and climate sensitivity. *Global Environmental Change, 11*(2), 147–154.
Mahat, T. J., Bláha, L., Uprety, B., & Bittner, M. (2019). Climate finance and green growth: Reconsidering climate-related institutions, investments, and priorities in Nepal. *Environmental Sciences Europe, 31*(1), 1–13.
Nawaz, M. A., Seshadri, U., Kumar, P., Aqdas, R., Patwary, A. K., & Riaz, M. (2021). Nexus between green finance and climate change mitigation in N-11 and BRICS countries: Empirical estimation through difference in differences (DID) approach. *Environmental Science and Pollution Research, 28*, 6504–6519.
Newell, P. (2019). Trasformismo or transformation? The global political economy of energy transitions. *Review of International Political Economy, 26*(1), 25–48.
Nhamo, G., & Swart, R. (2012). *A proposed scholarly framework for measuring business responsibility to climate change in South Africa* (Vol. 6, p. 50).
Petersen, A., & Chia, K. (2003, November 25) *Regulatory, financial and legal drivers for climate change corporate responsibility*. 65.
Pimonenko, T. V., Lieonov, S. V., & Ibragimov, Z. (2019). *Green investing for SDGS: EU experience for developing countries*.
Reuters. (2022). *Brazil's 2021 climate emissions highest since 2005*. https://www.reuters.com/world/americas/brazils-2021-climate-emissions-highest-since-2006-report-says-2022-11-01/#:~:text=Brazil%2C%20 despite%20having%20a%20 relatively,agriculture%20and%20other%20land%20use
Reserve Bank of India. (2021). *Green finance in India: Progress and challenges*. 04AR_2101202185D9B6905ADD465CB7DD280B88266F77.PDF (rbi.org.in)
Roberts, J. T., & Weikmans, R. (2017). Postface: Fragmentation, failing trust and enduring tensions over what counts as climate finance. *International Environmental Agreements: Politics, Law and Economics, 17*, 129–137.
Santarius, T. (2022). Climate justice and digitalization: A plea to consider broader socio-economic implications of digitalization and climate change. *GAIA-Ecological Perspectives for Science and Society, 31*(3), 146–150.
Schütze, F., & Stede, J. (2021). The EU sustainable finance taxonomy and its contribution to climate neutrality. *Journal of Sustainable Finance & Investment*, 1–33.
Sharma, M., Singh, R., & Kathuria, A. (2022). Climate change and the Indian economy–a review. *Current World Environment: An International Research Journal of Environmental Sciences, 17*(1), 20. https://doi.org/10.12944/CWE.17.1.3
Sodano, V., & Hingley, M. (2013). The food system, climate change and CSR: From business to government case. *British Food Journal, 115*(1), 75–91.
Stoddart, M. C., & Tindall, D. B. (2015). Canadian news media and the cultural dynamics of multilevel climate governance. *Environmental Politics, 24*(3), 401–422.
Tran, T., Do, H., Vu, T., & Do, N. (2020). The factors affecting green investment for sustainable development. *Decision Science Letters, 9*(3), 365–386.
UNFCCC. (2023a). *Climate action*. United Nation Climate Change. https://unfccc.int/
UNFCCC. (2023b). *Climate finance in the negotiations*. https://unfccc.int/topics/climate-finance/the-big-picture/climate-finance-in-the-negotiations#:~:text=In%20addition%2C%20Article%209%20states,particularly%20vulnerable%20to%20the%20adverse
UNFCCC. (2023c). *What is the Kyoto protocol?*. https://unfccc.int/kyoto_protocol#:~:text=In%20short%2C%20the%20Kyoto%20Protocol,accordance%20with%20agreed%20individual%20targets

Vestreng, V., Myhre, G., Fagerli, H., Reis, S., & Tarrasón, L. (2020). Twenty-five years of continuous Sulphur dioxide emission reduction in Europe. *Atmospheric Chemistry and Physics, 7*(13), 3663–3681.

Wolf, M. (2004). Globalisation and global economic governance. *Oxford Review of Economic Policy, 20*(1), 72.

World Bank. (2015, June). *Global economic prospects*. The Global Economy in Transition.

Xie, Z., Liu, X., Najam, H., Fu, Q., Abbas, J., Comite, U., & Miculescu, A. (2022). Achieving financial sustainability through revenue diversification: A green pathway for financial institutions in Asia. *Sustainability, 14*(6), 3512.

Zhang, Y. X., Chao, Q. C., Zheng, Q. H., & Huang, L. (2017). The withdrawal of the US from the Paris agreement and its impact on global climate change governance. *Advances in Climate Change Research, 8*(4), 213–219.

Climate-Related Financial Regulations and their Impact on Geotechnical Engineering

Ali Akbar Firoozi and Ali Asghar Firoozi

Abstract This chapter explores the dynamic interplay between Climate-Related Financial Regulations and Geotechnical Engineering, focusing on the resultant impacts, challenges, and opportunities. In an era marked by escalating climate concerns, these regulations have increasingly begun to shape practices across various sectors, particularly in geotechnical engineering. This chapter aims to provide a thorough overview of this nexus, equipping engineering firms with the knowledge to navigate the evolving regulatory landscape adeptly. Initially, it offers a concise overview of climate-related financial regulations, highlighting their global relevance and implications. Subsequently, it examines these regulations' direct and indirect impacts on geotechnical engineering, specifically focusing on their influence on risk management, project design, and financial decision-making. The discussion then pivots to strategic approaches for geotechnical firms, emphasizing how compliance with these regulations can drive innovation, enhance competitiveness, and bolster sustainability. The chapter concludes by underscoring the necessity of continuous stakeholder engagement, adopting adaptive business strategies, and a sustained commitment to sustainability. It advocates for proactive involvement from geotechnical engineering firms and other stakeholders in embracing these regulations, thereby fostering a culture of sustainability, innovation, and resilience.

1 Introduction

The dawn of the twenty-first century has ushered in an era where the once distinct realms of climate science, financial stability, and engineering practice are increasingly intertwined. Climate change has emerged as a pervasive force, challenging the

A. A. Firoozi (✉)
Department of Civil Engineering, Faculty of Engineering & Technology, University of Botswana, Gaborone, Botswana

A. A. Firoozi
Department of Civil Engineering, University Kebangsaan Malaysia (UKM), Bangi, Malaysia

N. Naifar (ed.), *Climate Change and Finance*, Sustainable Finance,
https://doi.org/10.1007/978-3-031-56419-2_9

siloed approach of various sectors, and demanding a cohesive, interdisciplinary response. The financial world, traditionally seen as separate from environmental stewardship, has begun to grapple with the stark reality of climate risks and their significant economic, societal, and environmental costs (Batten et al., 2016).

Parallel to this financial awakening, geotechnical engineering—a discipline deeply rooted in the interplay between human-made structures and the natural world—faces a paradigm shift. Geotechnical engineers, custodians of the earth's materials, are at the forefront of adapting to climatic uncertainties. Fluctuations in weather patterns, sea-level rise, and extreme meteorological events challenge the foundational tenets of geotechnical endeavors, demanding innovative solutions that are both financially sound and climatically robust (Vahedifard et al., 2018).

Responding to this need, climate-related financial regulations have emerged as a critical bridge linking economic and environmental resilience. These policies encourage organizations to disclose climate risks, driving a movement toward transparency, informed risk management, and sustainable investment behaviors that support the ethos of long-term viability (Alestra et al., 2022).

The ripples of these regulatory shifts are felt acutely within the geotechnical sector. The discipline must now reconcile the conventional approach with a new climate-informed paradigm, incorporating resilience and sustainability into the core of its practice. This necessitates a reevaluation of traditional methodologies and an embrace of innovation to meet the dual demands of regulatory compliance and environmental responsibility.

As shown in Fig. 1, this chapter endeavors to chart the complex interplay between climate-related financial regulations and geotechnical engineering, elucidating their mutual influence and the broader implications thereof. By dissecting the evolution, application, and repercussions of these regulations, the text will illuminate how geotechnical engineering is adapting to the changing regulatory climate, spotlighting the challenges, strategies, and case studies that exemplify this transformation.

2 Understanding Climate Change

Understanding climate change is fundamental to comprehending the context within which climate-related financial regulations impact geotechnical engineering. "Climate change" signifies the long-term alterations in temperatures and weather patterns, often emphasizing the notable shifts since the early twentieth century. The scientific consensus holds that human activities, particularly the emission of greenhouse gases (GHGs) from burning fossil fuels, are the main drivers of the rapid climatic changes we are witnessing today. This anthropogenic influence has led to a notable increase in global average temperatures, with a cascade of adverse effects.

Rising sea levels, changing precipitation patterns, and more frequent extreme weather events like hurricanes and heat waves exemplify the climate's volatility. These changes pose direct threats to human health, critical infrastructure, and the

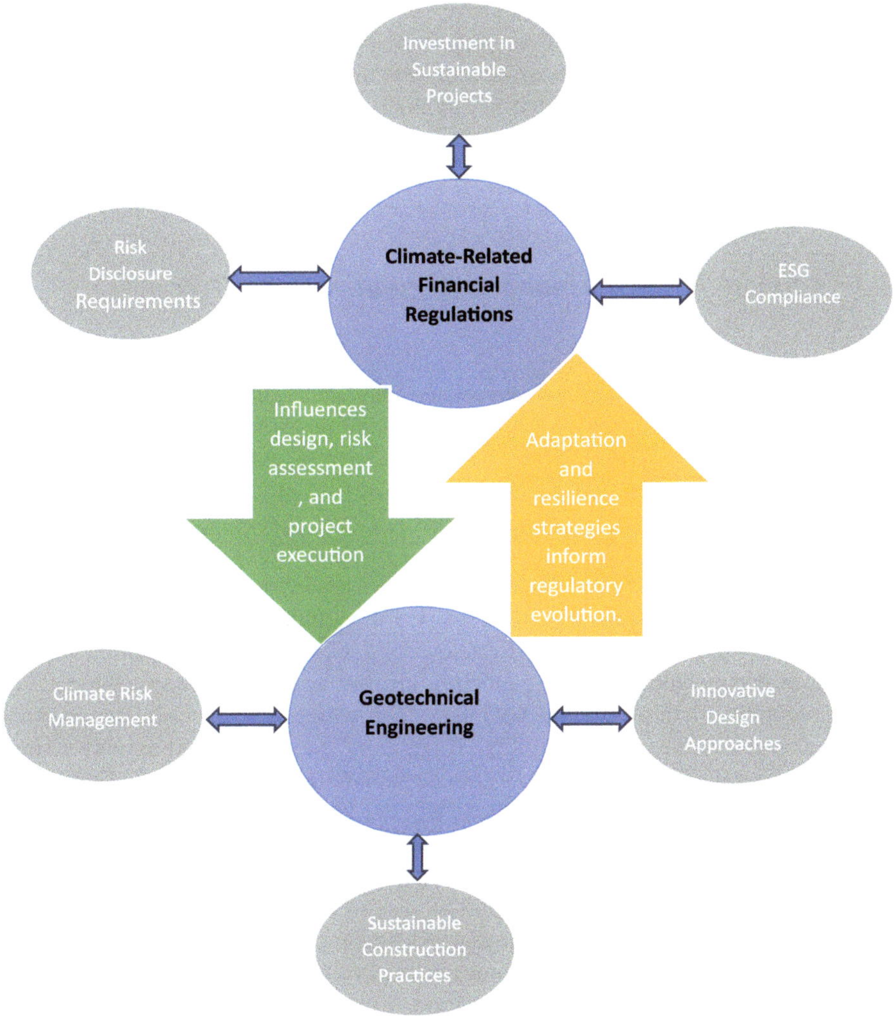

Fig. 1 A conceptual framework for climate-adaptive geotechnical engineering

natural systems upon which society relies (Abbass et al., 2022; Kemp et al., 2022; Meerow & Keith, 2022; Watts et al., 2015).

2.1 *Climate Change and the Financial System*

The financial system, historically centered on economic metrics, now increasingly acknowledges the environmental dimensions due to climate change's far-reaching impacts. Physical risks emerge from climatic and weather disturbances, while transition risks arise as economies strive for lower-carbon models. The financial implications are profound: insurers grapple with rising claims, banks face credit risks on properties and enterprises affected by climate events, and asset managers confront the possibility of stranded assets. Such systemic risks necessitate a transformative approach to integrating climate considerations into financial analyses and decision-making (Veraksa & Basseches, 2022; Boneva et al., 2022).

2.2 *Climate-Related Financial Regulations: An Overview*

The recognition of climate change as a destabilizing force in financial stability has catalyzed the development of climate-related financial regulations. These measures aim to embed environmental risks within financial decision-making, safeguarding financial systems against climatic uncertainties. Such regulations' primary goals are enhancing transparency, fortifying risk management, and stimulating sustainable investments.

A landmark in these efforts is the establishment of the Task Force on Climate-related Financial Disclosures (TCFD) by the Financial Stability Board (FSB) in 2015. The TCFD has crafted a framework for voluntary, consistent disclosures related to climate financial risks, empowering businesses to comprehend and respond to these challenges while meeting investor needs. Although adherence to TCFD recommendations remains voluntary, their widespread adoption underscores a shift toward greater accountability. Regulatory encouragement is also rising, exemplified by the UK government's commitment to mandatory TCFD-aligned disclosures by 2025 (Board, 2017; Ripple et al., 2022).

At the regional level, the European Union's actions, including the EU Taxonomy Regulation and the Sustainable Finance Disclosure Regulation, build a classification system for sustainable activities and set disclosure mandates for market participants, reflecting the EU's broader sustainable finance strategy (Kumar et al., 2022).

Central banks, too, are recognizing the importance of climate considerations, with the Network for Greening the Financial System (NGFS) serving as a pivotal collective, advocating for robust climate and environmental risk management within the financial sector (Sun et al., 2022).

In the framework of Institutional Theory, climate-related financial regulations are seen as constructs that not only mandate compliance but also foster a shift in organizational norms and behaviors. These regulations act as mechanisms that embed environmental consciousness into corporate strategies, particularly in sectors like geotechnical engineering where environmental impact is significant. Institutional Theory underscores the role of such regulations in shaping organizational structures and practices, compelling firms to integrate sustainability into their core operations (Meyer & Rowan, 1977; DiMaggio & Powell, 1983). By aligning with these regulatory frameworks, geotechnical engineering firms not only adhere to legal requirements but also reflect evolving societal values toward environmental stewardship.

2.3 Climate-Related Financial Regulations and Geotechnical Engineering

Geotechnical engineering, as a specialized discipline of civil engineering, centers on the interaction between human-made structures and earth materials. It is instrumental in designing and constructing essential components of the built environment, such as dams, tunnels, highways, and airports. The advent of climate change, with its pronounced shifts in weather patterns and increased incidence of natural disasters, escalates challenges within geotechnical engineering. The introduction of climate-related financial regulations amplifies these challenges, necessitating a redefined approach to project planning and execution.

One significant implication is the augmented necessity for comprehensive climate risk assessments. With the financial sector's growing emphasis on climate-related disclosures, geotechnical projects are now scrutinized for both physical and transitional climate risks. For example, projects in flood-prone areas must account for these risks in their planning and secure financing accordingly. Conversely, projects misaligned with a transitioning low-carbon economy risk becoming stranded assets, unsuitable for future climate realities (Kandalai et al., 2023; Allman, 2022).

Concurrently, there is an opportunity for geotechnical engineers to lead in sustainable project innovation. The push for investments in climate change mitigation and adaptation could favor projects that contribute positively to these objectives. Geotechnical expertise is pivotal in renewable energy infrastructure and in enhancing the resilience of communities through projects like advanced flood defenses. Additionally, the field is central to developing CCUS technologies, with geotechnical knowledge crucial for underground CO_2 storage assessments (Pelling et al., 2022; Qerimi & Sergi, 2022).

Carbon footprint quantification is integral to this new focus, as illustrated in Eq. 1. This equation offers a methodological approach to calculating the emissions footprint of geotechnical projects, encompassing all significant activities and inputs.

$$C = \Sigma\left(Q_i \times EF_i\right) \tag{1}$$

where:

C: represents the total carbon footprint of the project,
Q_i: denotes the quantity of each specific activity or input in the project,
EF_i: signifies the emission factor corresponding to each activity or input, reflecting the CO_2 equivalent emissions per activity unit.

This emphasis on carbon accounting underscores the transformative impact climate-related financial regulations have on geotechnical engineering. Such regulations, while presenting challenges, also unveil opportunities for innovation and alignment with global climate action efforts. Table 1, which follows, encapsulates the specific impacts of these regulations across various geotechnical project types, detailing the unique climate-related financial risks each entails.

Table 1 Impacts of climate-related financial regulations on geotechnical projects

Project type	Potential climate-related financial risks	Regulation impacts
Infrastructure projects (bridges, roads, tunnels)	Infrastructure deterioration due to extreme weather events; increased maintenance and repair costs.	Climate-related financial regulations may necessitate additional financial allocations for climate risk mitigation strategies and integration of sustainable practices.
Coastal engineering projects	Increased flooding risks due to sea-level rise, erosion, and infrastructure instability.	Regulations may influence project design to include adaptation strategies such as natural infrastructure and resilient design elements.
Building foundations	Changes in soil properties due to temperature increases or changes in precipitation; structural integrity risks.	Regulations may require rigorous climate risk assessment and the implementation of climate-resilient design features.
Dams and levees	Increased precipitation leads to heightened risk of overtopping and failure.	Regulations may necessitate the inclusion of more robust safety measures and the consideration of worst-case climate scenarios in the design phase.
Mining projects	Water scarcity or flooding affects operational safety and efficiency; instability risks due to thawing permafrost in certain regions.	Regulations may influence site selection, operational practices, and post-closure plans to ensure long-term sustainability and safety.

2.4 Adapting Geotechnical Engineering to Climate-Related Financial Regulations

The advancement of climate-related financial regulations necessitates a transformative approach to geotechnical engineering practices. As the financial sector increasingly mandates the incorporation of climate risks into decision-making, geotechnical engineers are called upon to align their methodologies with these evolving expectations. This alignment involves developing and implementing strategies to effectively manage both the potential physical and transition risks associated with geotechnical projects.

Physical risks pertain to the tangible impacts of climate change, such as extreme weather events, that can affect the economic performance of investments. For geotechnical professionals, this translates into integrating future climate projections into their project designs and risk assessments, ensuring resilience to forthcoming climate conditions (Gilchrist et al., 2021; Zhou et al., 2023). Table 2 illustrates the multifaceted implications of climate change scenarios for geotechnical engineering, alongside their financial consequences.

Furthermore, calculating the net present value (NPV) of geotechnical projects, which considers the time value of money, is now more critical than ever. This

Table 2 Climate change scenarios and their geotechnical engineering implications

Climate change scenario	Expected geotechnical challenges	Possible financial implications
Increased frequency of extreme precipitation events	Soil instability, landslides, increased erosion	Increased project costs due to necessary mitigation measures, potential for costly project delays, increased maintenance, and repair costs
Sea-level rise	Coastal erosion, increased load on coastal infrastructure, groundwater salinization	Need for investment in adaptive infrastructure, potential loss of property, and associated financial resources
Increased temperature and heatwaves	Changes in soil properties, permafrost thaw, increased evaporation rates affecting soil moisture	Increased costs for climate-resilient design, potential for costly remediation of damage, increased water management costs
Increased frequency and intensity of storms	Wind loading on structures, storm surge impacts on coastal infrastructure, debris impact on infrastructure	Increased design and construction costs for resilience, potential for costly repair and recovery after events

calculation enables engineers to compare a project's future cash inflows against the initial investment, effectively gauging its financial viability:

$$NPV = \sum \left(\frac{C_t}{(1+r)^t} \right) - C_0 \tag{2}$$

where:

NPV is the net present value of the project,
C_t is the net cash inflow during the period t,
r is the discount rate, and,
C_0 is the total initial investment cost.

Professional bodies in geotechnical engineering are encouraged to embed climate-related financial risks into their standards and guidelines. They should offer training and resources to help engineers navigate these challenges and capitalize on emerging opportunities. Such initiatives ensure that the profession remains robust and responsive to the demands of a changing climate and evolving financial landscape (Quorning, 2023; Chua et al., 2022; Anderson et al., 2022). Table 3 summarizes various financial assessment tools that geotechnical engineers can utilize to evaluate the financial aspects of climate adaptation in their projects.

The Risk-Adjusted Return on Capital (RAROC) provides a framework for analyzing financial performance by adjusting for risk, an increasingly important metric as the industry faces new challenges:

Table 3 Financial assessment tools for climate adaptation in geotechnical engineering

Financial assessment tool	Brief description	Application in geotechnical engineering
Cost-benefit analysis (CBA)	A systematic approach for comparing the advantages and disadvantages of a project or decision.	CBA can be used to assess the financial feasibility of incorporating climate adaptation measures in geotechnical projects.
Life cycle cost analysis (LCCA)	Evaluates the total cost of a project over its lifespan, including initial costs and future costs such as maintenance and operation.	LCCA can help determine the long-term financial viability of climate-resilient infrastructure designs.
Environmental, social, and governance (ESG) metrics	Standards for measuring the sustainability and societal impact of a project or investment.	ESG metrics can be used to assess and communicate the sustainability of geotechnical projects, potentially influencing funding decisions.
Climate risk assessment (CRA)	Evaluation of potential financial losses due to climate change impacts.	CRA can help geotechnical engineers anticipate potential financial losses due to climate-related events and incorporate mitigation strategies.

$$RAROC = \frac{R - L - O - CC}{EC} \tag{3}$$

where:

R = Expected Revenue: the total income anticipated from the project.
L = Expected Loss: accounts for potential risks such as project failures or damages.
O = Operating Costs: cover all expenses related to the operations of the project.
CC = Capital Charge: the cost of capital, representing the return expected by investors.
EC = Economic Capital: the risk capital required to cover the risks associated with the project.

2.5 Innovation and Sustainability in Geotechnical Engineering

Amidst the challenges posed by climate-related financial regulations, geotechnical engineering is at the cusp of a revolution, embracing sustainability and innovation. These regulations, while necessitating adherence to new standards, concurrently open the market for sustainable investments. Investors' growing environmental consciousness propels demand for projects that not only mitigate climate change effects but also adapt to its realities. Geotechnical engineers can capitalize on this trend, leading the charge in sustainable project development.

Innovations within geotechnical engineering are multifaceted. Alternative materials with a reduced environmental impact are being explored, such as recycled, locally sourced, or newly developed materials that minimize carbon emissions. Project design innovation is another avenue, with engineers striving to create an infrastructure that stands resilient against climate change and contributes to mitigation efforts, like renewable energy infrastructure or carbon capture and storage facilities (Thomä & Chenet, 2017; Tang et al., 2020; Ohene et al., 2023).

Furthermore, the discipline's potential extends to addressing broader societal objectives, including the United Nations Sustainable Development Goals (SDGs). Geotechnical projects can advance SDG 13 (Climate Action) through the reduction of greenhouse gas emissions, contribute to SDG 11 (Sustainable Cities and Communities) by enhancing infrastructure resilience, and support SDG 9 (Industry, Innovation, and Infrastructure) through promoting sustainable industrial processes (Biswas et al., 2021).

Thus, climate-related financial regulations are not just regulatory hurdles but also beacons guiding geotechnical engineering toward a sustainable future. Embracing this transformation, geotechnical engineers can become pivotal players in the global response to climate change and in the pursuit of a sustainable future.

Figure 2, as it stands, showcases the alignment of sustainable and climate-aware geotechnical practices with the Sustainable Development Goals, highlighting the positive societal and environmental ripple effects of these practices.

Create a flowchart or diagram that visually connects geotechnical engineering practices with relevant SDGs. For example:

Place a globe at the center to represent the Earth, with icons of geotechnical projects (like a dam, tunnel, and wind turbine).

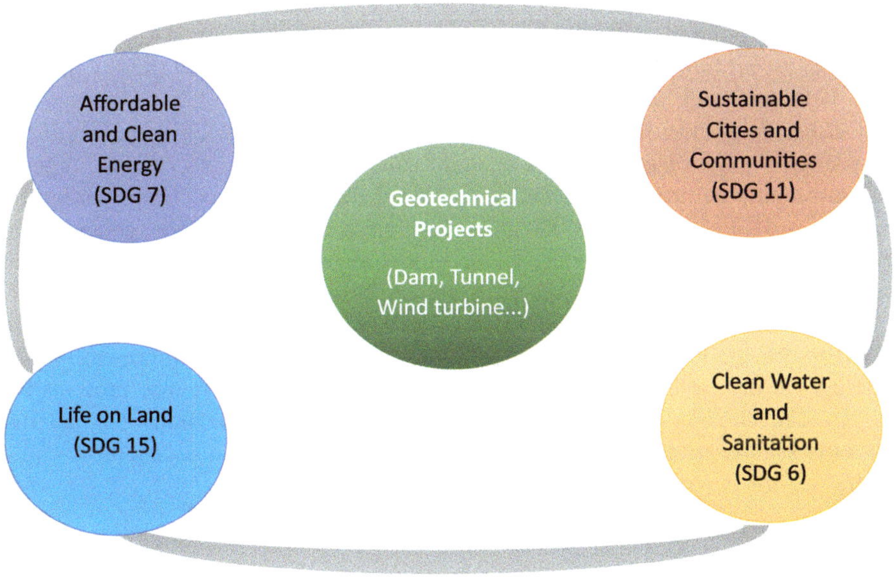

Fig. 2 The Role of geotechnical engineering in advancing sustainable development goals

Surrounding the globe, include symbols for various SDGs, such as a windmill for renewable energy, a city skyline for sustainable cities, or a tree for life on land.
Draw lines or arrows from the geotechnical project icons to the SDGs symbols, illustrating the contribution of these projects to each goal.
Underneath the diagram, provide a legend explaining the symbols and their corresponding SDGs.

2.6 Conclusion and Future Outlook

As the paths of geotechnical engineering and climate-related financial regulations intersect, the impetus for evolution within the field becomes increasingly clear. These regulations are not merely guidelines but catalysts for integrating financial considerations into geotechnical practices, ensuring the long-term sustainability and resilience of projects. This integration is poised to redefine the future trajectory of geotechnical engineering.

Looking ahead, it is crucial for geotechnical engineers to deepen their understanding of climate-related financial risks and their implications on project dynamics. This extends beyond technical acumen, encompassing a comprehensive grasp of financial and economic factors, and necessitates the development and application of tools for risk quantification and management (Romanello et al., 2022; Overland et al., 2022).

The need for collaboration and knowledge exchange is paramount. Establishing strong communication channels between geotechnical engineers, financial institutions, and policymakers will pave the way for standardized methodologies and enhance the reliability of climate-related financial disclosures. This synergy is essential for bridging knowledge gaps, fostering best practices, and driving innovation within the field.

Furthermore, there lies a substantial opportunity for pioneering sustainability in geotechnical engineering. Climate-related financial regulations serve as an impetus for developing cutting-edge technologies and sustainable approaches, steering the field toward greener practices. By embracing these regulatory trends, geotechnical engineers can actively contribute to the global agenda on climate mitigation and resilience building (Wing et al., 2022; Quemin, 2022).

This chapter has endeavored to provide a holistic view of the convergence between climate-related financial regulations and geotechnical engineering. It has highlighted the critical challenges and opportunities this convergence presents for the field. In adapting to the emerging financial landscape, geotechnical engineering not only secures its relevance but also affirms its crucial role in advancing societal welfare in the era of climate change.

3 Understanding Climate-Related Financial Regulations

The financial regulatory landscape is undergoing a significant transformation as a result of the increased awareness of climate change risks and their potential to impact the economy. Emerging climate-related financial regulations represent a concerted effort to mitigate these risks by promoting sustainable practices and enhancing climate resilience across business operations. These regulations have their roots in international climate agreements and green policies initiated by governments, reflecting a pivotal shift in how financial markets address the risks posed by climate change, including integrating climate risk assessments into financial decision-making processes.

Such regulations have direct relevance to the geotechnical engineering sector, influencing aspects such as project financing, risk management, and operational protocols. In subsequent sections, we will explore the specifics of these regulations, their historical development, key international examples, and their implications for geotechnical engineering (Alestra et al., 2022; Risi et al., 2023). Figure 3 underscores the expanding scope of climate-related financial regulations. The timeline starts sparsely in the early 2010s and ascends sharply, illustrating an escalating acknowledgment of climate change's financial repercussions by governments and regulatory bodies.

Table 4 provides a concise overview of pivotal climate-related financial regulations pertinent to geotechnical engineering. These regulations, mainly targeting financial institutions and public entities, also exert substantial effects on sectors closely tied to financial operations, such as geotechnical engineering.

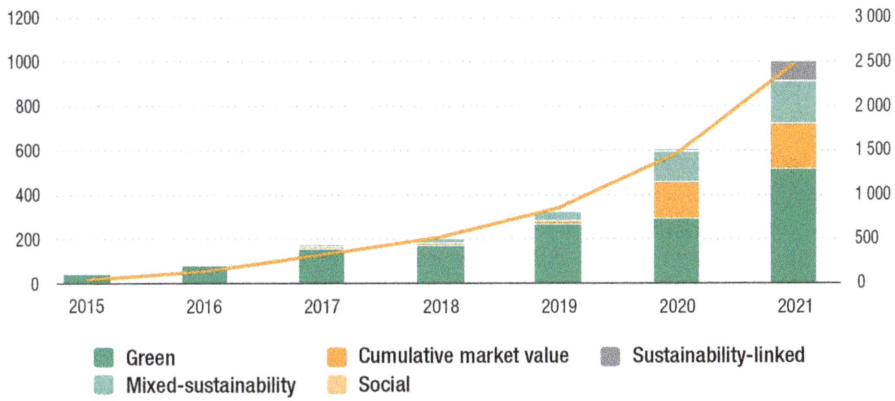

Fig. 3 Evolution of climate-related financial regulations (Source: UNCTAD based on data from Environmental Finance and Climate Bonds Initiative)

Table 4 Key Climate-related financial regulations and their influence on geotechnical engineering

Regulation name	Year enacted	Jurisdiction	Key provisions	Implications for geotechnical engineering
Sustainable Finance Disclosure Regulation (SFDR)	2021	European Union	Requires financial market participants to disclose how they integrate ESG risks into their risk processes.	Geotechnical projects funded by these entities may be required to demonstrate compliance with ESG risk integration processes.
Task Force on Climate-related Financial Disclosures (TCFD)	2017	International	Encourages businesses to voluntarily disclose climate-related financial risks.	Geotechnical firms may be required to disclose their climate-related financial risks, especially in projects with significant public interest or environmental impact.
Network for Greening the Financial System (NGFS)	2017	International	A network of central banks and supervisors willing to share best practices and contribute to the development of environment and climate risk management in the financial sector.	Influences the practices of financial institutions, which could impact the funding and risk assessment of geotechnical projects.
Climate Risk Disclosure Bill	2023	United States	Requires public companies to disclose information about their exposure to climate-related risks.	May affect geotechnical firms that are public companies or those working with public companies.
Green bond principles	2018	International	Encourage transparency, disclosure, and integrity in the development of the Green bond market.	Geotechnical projects funded through green bonds may be required to meet certain sustainability criteria.

3.1 Detailed Explanation of What Climate-Related Financial Regulations Are

Climate-related financial regulations are an integral component of environmental, social, and governance (ESG) guidelines, comprising a series of norms, rules, and standards developed by authorities to ensure that financial institutions and corporate entities incorporate climate risk assessments into their operational, investment, and risk management protocols, as well as their reporting processes. The overarching intent of these regulations is to redirect financial flows toward more sustainable and environmentally conscious pathways, aligning with objectives such as those set forth by the Paris Agreement.

At their core, these regulations confront two principal risk categories: physical and transition. Physical risks involve the potential for property damage or other financial losses due to climate change-induced events, such as severe weather occurrences or persistent shifts in climate patterns. Transition risks are associated with the economic and financial changes necessitated by the global shift to a low-carbon economy, encompassing policy alterations, technological advancements, and evolving market dynamics.

A critical element of climate-related financial regulations is the emphasis on financial disclosures related to climate. These disclosures are designed to enlighten investors, lenders, and insurers, equipping them with the necessary information to evaluate how climate risks might influence a company's strategic direction and financial performance, thereby fostering a culture of transparency and enabling stakeholders to gauge a company's climate resilience (Veraksa & Basseches, 2022; Board, 2017; Smoleńska, 2023).

Table 5 enumerates climate-related financial regulations from various global jurisdictions and the potential impacts these regulations may have on the field of geotechnical engineering. Each regulation is contextualized with a brief description and assessed for its influence on the practice and business of geotechnical engineering, demonstrating the widespread and varied implications of these regulations.

Table 5 Global climate-related financial regulations and their impact on geotechnical engineering

Country/ Region	Regulation	Brief description	Impact on geotechnical engineering
European Union	Sustainable Finance Disclosure Regulation (SFDR)	Requires financial market participants to disclose how they integrate ESG risks into their investment decisions.	Forces investment in infrastructure projects to consider environmental impacts and climate change risks, influencing geotechnical project design, planning, and execution.
United States	Climate Risk Disclosure Act	Aims to direct the SEC to issue rules requiring every public company to disclose more information about its exposure to climate-related risks.	Promotes transparency in disclosing climate risks associated with infrastructure projects, affecting funding decisions for geotechnical projects.
China	Green Bond Endorsed Project Catalogue	Lists the projects that can be funded by issuing green bonds, including those that aid in adaptation to climate change.	Encourages geotechnical projects that help in climate change adaptation to receive funding from green bonds.
Australia	Climate Active Carbon Neutral Standard	Offers a way to certify that a business, building, event, or product is carbon neutral.	Impacts geotechnical engineering by promoting designs and methods that minimize carbon footprints.

3.2 A Historical Perspective on the Evolution of Climate-Related Financial Regulations

The perception of climate risk as a financial risk has crystallized in the consciousness of regulatory bodies only in recent decades. The genesis of climate-related financial regulations is rooted in the burgeoning awareness of global environmental issues that emerged toward the end of the twentieth century. During the 1990s, the financial sector began to tentatively address environmental concerns, primarily through nascent "green investing" initiatives and corporate social responsibility (CSR) frameworks. These early efforts laid the groundwork for integrating environmental considerations into financial decision-making.

The financial turmoil experienced in 2007–2008 served as a catalyst, highlighting the necessity of robust risk management strategies within the financial sector. This period brought to the forefront the realization that non-traditional risks, such as those posed by climate change, bore the potential for systemic repercussions (Shipton & Dauvergne, 2022; Tao et al., 2022; Sgambati, 2022).

The adoption of the Paris Agreement in 2015 represented a critical inflection point, underscoring the integral role of financial regulations in mitigating climate risk. The Agreement's ambitious targets to curtail global warming to well below 2 degrees Celsius necessitated a substantial overhaul of economic structures and investment patterns. The subsequent establishment of the Task Force on Climate-related Financial Disclosures (TCFD) in 2017 by the Financial Stability Board symbolized a pivotal step in the maturation of climate-related financial regulations, advocating for standardized and transparent climate-related financial reporting.

In the wake of the TCFD's recommendations, the Network for Greening the Financial System (NGFS) has been instrumental in advancing the discourse on climate risk. Comprising a network of central banks and financial supervisors, the NGFS has been pivotal in championing the assimilation of climate-related risks into financial stability monitoring and the supervisory framework (Board, 2017; Smoleńska, 2023; Naylor & Ford, 2023).

3.3 Examples of Major Global Climate-Related Financial Regulations and their Implications

Internationally, the impetus for climate action has given rise to various climate-related financial regulations. The European Union (EU) has been at the forefront of initiatives such as the Sustainable Finance Disclosure Regulation (SFDR) and Taxonomy Regulation. The SFDR mandates that financial market participants disclose sustainability risks and impacts, fostering informed investment decisions. The Taxonomy Regulation, meanwhile, categorizes sustainable economic activities to guide investment toward more sustainable technologies and sectors (Cremasco & Boni, 2022).

China has also been proactive, with the Green Bond Endorsed Project Catalogue steering investments toward "green" projects, and the Green Financial System Guidelines pushing banks to enhance their environmental risk management (Lin & Hong, 2022). In the USA, the focus has been on enhancing climate-related risk disclosures for publicly traded companies, as pursued by the Securities and Exchange Commission (Bremner, 2023).

These diverse regulations have a significant bearing on geotechnical engineering. They affect how finances are allocated, driving the sector toward sustainable and climate-resilient projects, and necessitating heightened transparency and environmental risk disclosure.

Table 6 delineates the contrast between traditional risk management and a climate-informed approach within geotechnical engineering. The latter method expands upon the conventional by factoring in the potential impacts of future climate scenarios on geotechnical endeavors.

3.3.1 Task Force on Climate-Related Financial Disclosures

The Task Force on Climate-Related Financial Disclosures (TCFD) recommendations have set a new precedent for financial transparency, with a particular focus on climate change impacts. Formed under the aegis of the Financial Stability Board (FSB), the TCFD provides a structured framework for entities to disclose clear, comprehensive information about their climate-related risks and opportunities. This framework is built around four thematic pillars: governance, strategy, risk management, and metrics and targets, which allow businesses to evaluate and communicate their preparedness for climate change. For geotechnical engineering, this means that firms now need to scrutinize how climate change may affect their long-term viability, including the assessment of geotechnical risks enhanced by climate variability.

Table 6 Contrast of risk management approaches in geotechnical engineering

	Traditional risk management	Climate-informed risk management
Focus	Primarily on geotechnical risks, such as soil stability and foundation integrity.	Includes traditional geotechnical risks and integrates climate risks, such as increased precipitation, sea-level rise, and temperature changes.
Time horizon	Short to medium-term, typically aligned with a project lifecycle.	Medium to long-term, considering potential future climate scenarios and their impacts.
Risk assessment	Based on historical data and current conditions.	Utilizes climate projections and scenario analysis in addition to historical data.
Risk mitigation	Traditional engineering solutions are typically focused on addressing immediate geotechnical challenges.	Incorporates climate adaptation measures and seeks to increase the resilience of infrastructure to future climate changes.
Regulatory compliance	Compliance with traditional engineering standards and regulations.	Compliance with both traditional engineering regulations and emerging climate-related financial regulations.

This is not just about immediate operational risks but also about how climate change can alter project landscapes, supply chains, and the longevity of infrastructure. By adhering to TCFD guidelines, geotechnical firms can provide a higher degree of confidence to investors and stakeholders who are increasingly sensitive to climate-related issues (Board, 2017; Barkemeyer et al., 2017).

3.3.2 Network for Greening the Financial System

The Network for Greening the Financial System (NGFS) is a collective commitment by Central Banks and Supervisors to acknowledge and address the financial risks posed by climate change. Since its inception, the NGFS has been a driving force in encouraging the financial industry to account for environmental risks. The geotechnical sector is impacted by NGFS initiatives as the banks and financial institutions that fund infrastructure projects are now assessing environmental risks as part of their financing decisions. Consequently, geotechnical engineering projects must demonstrate their environmental credibility and how they minimize climate risks to secure funding. Moreover, the NGFS's focus on mobilizing capital for green investments can lead to increased funding opportunities for geotechnical projects that support sustainable economic transitions, such as those involving green infrastructure or ecological restoration (Sun et al., 2022; Dikau & Volz, 2018).

3.3.3 European Union's Sustainable Finance Disclosure Regulation

The Sustainable Finance Disclosure Regulation (SFDR) requires financial market participants within the EU to make detailed disclosures on how sustainability risks are integrated into their investment decisions and the expected impact on the returns of their products. This regulation is reshaping the investment landscape, with direct consequences for geotechnical engineering projects. Projects will need to demonstrate their sustainability credentials and the positive impact they have on ESG factors to attract investment. For the geotechnical field, this could translate into a greater emphasis on sustainable construction materials, eco-friendly design, and mitigation of environmental impacts through innovative engineering solutions (van Oostrum, 2021).

3.3.4 China's Green Finance Guidelines

China's Green Finance Guidelines are among the most comprehensive national frameworks for green finance. By defining what qualifies as a 'green' project, these guidelines have significant implications for geotechnical engineering projects in China. For instance, projects that reduce carbon emissions, improve energy efficiency, or enhance environmental protection are more likely to receive funding. Geotechnical engineers in China must now consider these guidelines in the planning and implementation phases of their projects to ensure alignment with national green finance criteria, potentially influencing everything from site selection to the choice of construction techniques (Tian et al., 2022).

3.3.5 UK's Mandatory TCFD-Aligned Disclosures

The UK's commitment to implementing mandatory TCFD-aligned disclosures is indicative of a broader international trend toward increased climate governance. This will require public companies and certain private companies to integrate climate-related risks into their financial reporting. For geotechnical engineering firms, particularly those in the UK or those that partner with UK-based companies, this means that climate risk analysis and disclosure must become a routine part of financial and project reporting. It also suggests that there may be a growing market for climate risk assessment services within the geotechnical sector, as firms seek to comply with these new regulations (James & Quaglia, 2023).

3.4 Implications of Climate-Related Financial Regulations

Climate-related financial regulations hold significant consequences for a wide array of sectors, not least of which is geotechnical engineering. These regulations underpin a concerted effort to instill transparency, invigorate sustainable investments, and embed climate risk management into the fabric of financial and corporate activities.

Transparency is pivotal within these regulations. Mandating entities to openly report their climate risks allows for a more informed decision-making process by investors and stakeholders. For geotechnical engineering, this transparency translates to rigorous environmental impact assessments and clear communication regarding project risks and mitigation strategies. Consequently, geotechnical firms may be required to enhance their reporting mechanisms to align with these expectations (Sullivan & Gouldson, 2017).

Investment stimulation is another key facet of these regulations. They incentivize projects that align with green objectives while discouraging those that are deemed environmentally detrimental. The EU Taxonomy Regulation, for example, fosters investments in projects that contribute to climate change mitigation or adaptation, prompting geotechnical firms to orient their project designs toward sustainability (Cremasco & Boni, 2022).

Climate risk management is also reinforced, to bolster the economy's resilience against the impacts of climate change. In geotechnical engineering, this necessitates the integration of comprehensive climate risk assessments into project design, encouraging the construction of infrastructure capable of withstanding future climatic stresses (Smoleńska, 2023).

The influence of these regulations on *global financial flows* is substantial. They drive capital toward sustainable investments, as seen in the burgeoning green bond market, which flourishes under the banner of regulation-mandated disclosure and transparency (Flammer, 2021; Diffenbaugh, 2022).

Financial stability gains reinforcement from these regulations by acknowledging and addressing climate change as a tangible financial risk. Financial entities adhering to these guidelines are more adept at managing physical and transition risks, positioning them to better endure climate-related shocks (Schoenmaker & Van Tilburg, 2016).

In terms of *climate policy goals*, financial regulations are becoming increasingly aligned with climate objectives. By necessitating disclosures on climate impacts, these regulations can motivate companies toward greener practices, thereby aiding in the fulfillment of national and international climate targets (Tan, 2022).

The *Social Cost of Carbon (SCC)* is an economic metric that encapsulates the estimated damages from an incremental increase in CO_2 emissions. It is a critical tool in the valuation of the climate impact of policy decisions, calculated as follows:

$$SCC = \frac{D}{(1+r)^t} \tag{4}$$

where:

D represents the economic damages from additional CO_2 emissions,
r is the discount rate, and,
t is the time horizon over which the damages are assessed.

This quantification provides a monetized value of the detriments associated with an additional ton of carbon dioxide, facilitating the economic evaluation of climate policies.

4 The Intersection of Geotechnical Engineering with Climate-Related Financial Regulations

Geotechnical engineering plays a pivotal role in the construction and infrastructure sector, and its integration with climate-related financial regulations is driving transformative change in the industry. The rise of these regulations has catalyzed a shift in the field, urging it toward sustainable and resilient practices.

Projects within geotechnical engineering are diverse, ranging from foundational work and earthworks to managing geological hazards. The environmental footprint of these projects is significant, and with the advent of climate change, factors like sea-level rise, increased rainfall, and temperature fluctuations become critical considerations. As a result, climate-related financial regulations are necessitating thorough climate risk assessments and the integration of climate-resilient designs to ensure projects align with environmental sustainability standards.

Increased demands for transparency and the requirement to disclose climate-related risks significantly affect project financing. Financial backers, including banks and investors, are inclined to favor projects that exemplify sustainable practices and resilience to climate risks. This preference influences geotechnical firms to transition toward adopting greener technologies, using sustainable materials, and implementing efficient operational practices (Ottesen et al., 2022; Kinsella et al., 2023).

Furthermore, the regulations are a boon for innovation in the field of geotechnical engineering. They encourage the development of novel techniques and technologies that aim to minimize environmental impacts and bolster resilience to the effects of climate change. Examples include advanced modeling for climate risk assessment, innovative designs that provide foundational resilience, and sustainable approaches to waste management (Parmenter & Losleben, 2023).

In essence, climate-related financial regulations are not just influencing but actively reshaping geotechnical engineering, guiding it toward a path of sustainable development and enhanced climate resilience. The intersection of these regulations with geotechnical practices is forging a new landscape where the ability to adapt and innovate is paramount.

4.1 The Role of Geotechnical Engineering in Understanding Climate Risks

Geotechnical engineering is at the forefront of discerning and addressing climate risks. Central to this discipline is the analysis of earth materials and their interaction with human-made structures. With the progression of climate change resulting in altered weather patterns and a rise in extreme weather events, geotechnical engineering's role becomes even more critical. The sector is responsible for ensuring the structural integrity of buildings and infrastructure, which are now under increased threat from the climate crisis.

For example, heightened rainfall and rising sea levels can intensify soil erosion and heighten the risk of landslides, undermining the stability of critical infrastructure. Conversely, escalating temperatures and extended periods of drought can cause soil desiccation, leading to ground subsidence and the potential compromise of building foundations. Geotechnical engineers, with their profound understanding of such climate-induced risks, are instrumental in devising strategies to protect infrastructural integrity against the vagaries of climate change. These measures are increasingly valued by financial stakeholders, with risk mitigation strategies being recognized as crucial financial assets. This recognition effectively bridges the gap between geotechnical engineering and climate-related financial regulations, highlighting the financial value of robust engineering solutions to climate risks (Jain et al., 2022; Svennevig et al., 2023; Das, 2021).

Applying Neo-Institutional Theory to geotechnical engineering reveals how these firms adapt to regulatory environments characterized by climate-related financial regulations. This theory explains organizational changes as responses to external pressures for legitimacy and survival, emphasizing the concept of isomorphism—mimetic, coercive, and normative (DiMaggio & Powell, 1983). Mimetic isomorphism occurs as firms emulate successful practices in response to uncertainties posed by climate change. Coercive isomorphism results from formal and informal pressures exerted by regulatory bodies, influencing firms to align their practices with environmental regulations. Normative isomorphism, driven by professionalization, leads to standardization in approaches to climate risk assessment and sustainability practices. This theoretical lens helps to dissect the processes through which geotechnical engineering firms incorporate climate risk considerations in line with financial regulations, shaping their strategies to align with broader industry norms and expectations (Scott, 2008).

By incorporating climate risk mitigation into their projects, geotechnical engineers not only contribute to the resilience of the built environment but also to the attractiveness and financial viability of these projects from an investor and insurer perspective. This alignment with climate-related financial regulations demonstrates the essential role of geotechnical engineering in the broader context of sustainable development and climate risk management.

4.2 Implications and Opportunities for Geotechnical Engineering

The advent of climate-related financial regulations heralds significant changes in geotechnical engineering. These regulations require firms to incorporate sustainable practices and rigorous climate risk assessments, affecting everything from material selection to construction techniques. While this might introduce initial costs, it sets the stage for long-term benefits and innovation (Cao et al., 2022).

Conversely, these challenges pave the way for opportunities. A stringent regulatory environment acts as a catalyst for innovation, leading to the development of new materials and technologies that contribute to carbon reduction and resource efficiency. Aligning with these regulations opens up new markets, attracts investment, and enhances a firm's reputation as a sustainable and resilient business leader, which can be a significant draw for both community engagement and talent acquisition (Girard-Guerraud et al., 2023; Bassen & Kovács, 2020).

Table 7 outlines various climate adaptation strategies that geotechnical engineering projects can implement, each with distinct financial implications ranging from initial costs to potential long-term savings and investment opportunities.

Table 7 Climate adaptation measures and financial implications in geotechnical engineering

Adaptation measures	Brief description	Financial implications
Climate-resilient design	Incorporating design features that increase infrastructure's resilience to future climate changes, e.g., designing foundations to withstand changes in soil properties due to climate change.	Can increase upfront project costs but may result in long-term savings through reduced repair and maintenance costs.
Climate risk assessment	A systematic process to identify, analyze, and evaluate the climate risks associated with a geotechnical project.	Requires investment in data analysis and risk modeling but can inform the more efficient allocation of resources and mitigate financial risks from unanticipated climate events.
Green infrastructure	Use of natural or semi-natural systems for the management of water, temperature, or other environmental variables.	Can sometimes be more cost-effective than traditional gray infrastructure. Also, may open up opportunities for project financing through green bonds or other ESG investment instruments.
Implementation of early warning systems	Use of technology and data analysis to provide advance warning of potential climate-induced issues, e.g., monitoring soil moisture levels to predict landslides.	Requires investment in technology and ongoing operational costs but can prevent costly infrastructure failures and associated financial liabilities.
Insurance and financial hedging	Transfer of climate risks to third parties through insurance or financial instruments.	Incurs premiums or costs but can provide financial protection against unanticipated climate events.

4.2.1 Enhanced Role and Demand

The integration of climate risks into financial decision-making elevates the role and demand for geotechnical engineering expertise. Firms are now tasked with conducting detailed climate risk assessments, which demand the specialized skills of geotechnical engineers to analyze physical risks and develop risk management strategies. This broadened role solidifies the importance of geotechnical engineering in constructing climate-resilient infrastructure (Ng et al., 2018).

4.2.2 Incorporation of Financial Knowledge

As the link between geotechnical engineering and financial regulations strengthens, engineers must become adept in the financial aspects of climate risks and regulatory compliance. This knowledge is becoming a vital part of the skill set required in the modern geotechnical engineering landscape (Semieniuk et al., 2021).

4.2.3 Innovation and New Services

The regulatory shift is spurring innovation within the geotechnical sector. There is a burgeoning demand for new services that aid businesses in regulatory compliance. Developing specialized risk assessment tools that cater to the specifics of climate-related financial disclosure, such as those recommended by the TCFD, represents a new avenue for service expansion within geotechnical firms (Abhayawansa & Adams, 2022).

In summary, climate-related financial regulations are reshaping the geotechnical engineering sector, steering it toward sustainability and resilience. Firms that can adeptly navigate this new landscape will not only contribute positively to the climate agenda but also reap the benefits of long-term sustainability and competitive advantage.

5 Practical Examples of the Intersection of Geotechnical Engineering and Climate-Related Financial Regulations

In the context of climate change and evolving financial regulations, geotechnical engineering emerges as a critical field, demonstrating adaptability and innovation. This section explores real-world scenarios that exemplify the intersection of geotechnical engineering with climate-related financial regulations. From infrastructure projects grappling with sea-level rise and extreme weather to renewable energy developments and sustainable real estate ventures, these examples highlight how geotechnical engineering is pivoting toward sustainability and resilience. They showcase the field's crucial role in adhering to and advancing climate-related financial regulations, underlining the importance of incorporating environmental considerations into every stage of engineering projects. Through these cases, we gain valuable insights into the synergistic relationship between geotechnical engineering practices and the financial sector's focus on climate risk management, illustrating the transformative impact of these regulations on the field.

5.1 Infrastructure Projects

Climate-related financial risks, accentuated by phenomena such as sea-level rise and increased rainfall, present significant challenges to infrastructure projects. A pertinent example is the Florida Department of Transportation's (FDOT) proactive approach to these challenges. FDOT's comprehensive analysis of sea-level rise implications on transportation infrastructure underscores the critical role of geotechnical engineering in both risk identification and cost evaluation. This initiative is closely aligned with the TCFD recommendations, which advocate for robust risk disclosures. The geotechnical studies not only inform infrastructure resilience strategies but also provide a framework for financial disclosures that meet regulatory requirements. The data gathered from such studies can inform investment decisions, ensuring that the financial implications of climate risks are appropriately accounted for and disclosed, in line with the TCFD guidelines (Rowan et al., 2013).

5.2 Renewable Energy Projects

In renewable energy ventures, geotechnical engineering is indispensable. For offshore wind farms, geotechnical site investigations are crucial for assessing soil conditions, evaluating the impact of climate change on those conditions, and ensuring the stability of the structures. In terms of financial regulations, such as those promulgated by the TCFD, these assessments fulfill the dual purpose of risk management and regulatory compliance. They not only provide investors with the confidence needed to finance these projects but also align the projects with regulatory expectations for climate risk disclosure and management (Liu et al., 2020).

5.3 Real Estate Development

The role of geotechnical engineering in real estate is foundational, especially in assessing the effects of climate change on land suitability and structural stability. Take, for instance, Miami's real estate sector, where geotechnical engineers are integral to addressing and mitigating the risks posed by sea-level rise. By conducting thorough soil analyses and designing structures that can withstand climatic stresses, geotechnical engineers contribute to the climate resilience of real estate developments. These efforts are vital in satisfying the demands of climate-related financial regulations, which require the disclosure of such risks to investors and lenders, thereby influencing the flow of capital into real estate projects that prioritize climate resilience (Wdowinski et al., 2016).

5.4 Hazard Mitigation and Insurance

The increasing instances of natural disasters emphasize the importance of geotechnical engineering in hazard mitigation. Geotechnical engineers design infrastructures that can endure disasters, thereby mitigating potential damages. Insurance companies, which are directly impacted by these risks, depend on geotechnical assessments for accurate risk pricing. The precision of these assessments is crucial in an era of climate-related financial regulations that stress the need for transparent risk management. By providing detailed risk evaluations, geotechnical engineering facilitates the insurance sector's compliance with regulations that demand clarity on how climate risks are quantified and managed (Martinez Salgueiro & Tarrazon-Rodon, 2021).

In each of these examples, the relevance of climate-related financial regulations is underscored by the role geotechnical engineering plays in ensuring that projects not only comply with these regulations but also become more attractive to risk-aware investors. The regulations drive the demand for more sophisticated geotechnical assessments, and in turn, these assessments inform the financial sector, ensuring that climate risks are integrated into the economic evaluation of projects.

6 Future Trends and Considerations in Geotechnical Engineering and Climate-Related Financial Regulations

The nexus of geotechnical engineering and climate-related financial regulations is poised to play an increasingly significant role in an era marked by escalating climate change and a shift toward sustainable finance. This section explores the key trends and considerations that will shape the future of this intersection.

6.1 *The Rise of Sustainable Finance*

Sustainable finance, with its focus on integrating environmental, social, and governance (ESG) factors into investment decisions, is witnessing a notable surge. This trend is not just a paradigm shift in investment strategies but also a clear indication of the heightened emphasis on environmental impact and sustainability in the financial world. For geotechnical engineering, this translates into an expanded scope of work where engineers must provide in-depth analyses of climate-related risks to support informed investment decisions. This development also places geotechnical engineers at the forefront of helping firms meet regulatory demands for transparency and sustainability in their projects (Amel-Zadeh & Serafeim, 2018).

6.2 *Advancements in Geotechnical Technologies*

The field of geotechnical engineering is rapidly evolving with advancements in technologies like remote sensing, GIS, and machine learning. These tools are revolutionizing how climate risks are assessed, leading to more accurate and efficient risk management strategies. The integration of these technologies with traditional geotechnical expertise is set to become a key differentiator in the industry. For geotechnical engineers, staying updated with these technological trends and effectively incorporating them into their practices is essential for maintaining relevance and compliance with financial regulations (Benbouras, 2022).

6.3 Regulatory Changes and their Impact

The landscape of climate risk disclosure and sustainable finance regulations is in constant flux. Geotechnical engineers must stay informed about these changes and understand their potential impacts on the industry. For example, initiatives like the International Financial Reporting Standards (IFRS) Foundation's proposal to establish a Sustainability Standards Board underscore the growing emphasis on climate risk disclosure. Such regulatory developments could introduce new compliance requirements and opportunities for geotechnical professionals, necessitating a proactive and informed approach to regulatory developments (Adams & Mueller, 2022).

In conclusion, the interplay between geotechnical engineering and climate-related financial regulations is set to deepen, driven by the rise of sustainable finance, technological advancements, and ongoing regulatory changes. These factors collectively underscore the need for geotechnical engineering to adapt and evolve, positioning it as a key contributor to the global pursuit of sustainability and resilience in the face of climate change.

7 Strategies for Navigating the Intersection of Geotechnical Engineering and Climate-Related Financial Regulations

As climate-related financial regulations increasingly influence global financial and investment practices, geotechnical engineering firms are compelled to adapt and refine their strategies. This necessity paves the way for a range of strategic approaches, encompassing capability building, regulatory engagement, research advancement, and collaborative endeavors. A key financial tool relevant to this context is the Capital Asset Pricing Model (CAPM), which assesses the expected return on investment concerning its risk:

$$E(R) = R_f + \beta \times (R_m + R_f) \tag{5}$$

where:

E(R) is the Expected Return on the investment.
R_f is the Risk-free Rate, typically the return on a government bond.
β is a measure of the investment's volatility compared to the market.
R_m is the Expected Market Return.

7.1 Building Capabilities to Address Climate Risks

The primary strategy for geotechnical engineering firms in this evolving context is to develop robust capabilities for assessing and managing climate-related risks. This entails a thorough understanding of how climate change impacts geotechnical projects, such as alterations in soil properties, heightened landslide risks, and potential effects on infrastructure stability.

Geotechnical engineers should be equipped with the latest knowledge and skills relevant to these challenges. This involves training in climate science, environmental risk assessment, and climate modeling. Additionally, the integration of advanced tools and technologies like GIS, remote sensing, and machine learning into risk assessment processes can significantly enhance accuracy and efficiency.

Firms should also consider establishing specialized climate risk teams or appointing climate risk champions to spearhead the integration of climate risk considerations into project planning, execution, and management. This might include conducting comprehensive climate risk assessments for projects, engaging proactively with clients and stakeholders on climate risk issues, and ensuring adherence to pertinent regulations. Such proactive measures not only prepare firms to meet regulatory demands but also position them as leaders in risk management and sustainability within the geotechnical engineering sector (Benbouras, 2022; Kelman et al., 2016; Aitsi-Selmi et al., 2015).

By fostering these capabilities, geotechnical engineering firms can navigate the challenges posed by climate-related financial regulations, turning potential obstacles into opportunities for growth, innovation, and leadership in sustainable engineering practices.

7.2 Engaging with Climate-Related Financial Regulations

For geotechnical engineering firms, adeptly navigating climate-related financial regulations is imperative. These regulations, designed to enhance transparency around climate risks and stimulate sustainable finance, play a critical role in project financing, insurance, and investment decisions. Firms must actively engage with these regulations, which involves understanding their intricacies, implications, and the opportunities they present.

Staying updated with regulatory changes, both nationally and internationally, is essential. This requires an understanding of how these changes impact projects and clients. Participation in industry discussions and providing expert input during consultations around new regulations can also be beneficial (Busch & Lewandowski, 2018).

At the project level, aligning risk assessments and disclosures with regulatory requirements is crucial. This might include implementing specific risk metrics, reporting formats, or adherence to prescribed disclosure standards. Conducting regular audits and reviews ensures compliance and helps in identifying areas needing improvement. Additionally, geotechnical engineers should explore ways their services can assist clients in complying with these regulations, such as through detailed climate risk assessments for project financing or designing mitigation strategies for insurance purposes (Minutiello & Tettamanzi, 2022).

7.3 Advancing Research on Climate Risk and Geotechnical Engineering

Advancing research in the area of climate risk and geotechnical engineering is another strategic approach. Research can enhance our understanding of climate risks, refine risk assessment methods, and develop more effective mitigation strategies. It can also guide regulatory developments and establish best practices in climate risk management. Geotechnical firms, with their technical expertise and practical experience, are uniquely positioned to contribute to this research domain.

Engagement with academic institutions, research bodies, and industry associations is vital in this endeavor. This could involve participating in collaborative research projects, sponsoring academic studies, or providing opportunities for student internships and research. Disseminating research findings is also crucial, which can be achieved through publishing in academic journals, presenting at conferences, or sharing insights via company reports and websites. These efforts not only enhance the firm's reputation but also contribute to industry knowledge and global understanding of climate risks in geotechnical engineering (Zhang et al., 2021; Hijma & Cohen, 2019).

A crucial aspect of aligning with global climate-related financial regulations is accounting for the carbon footprint of geotechnical projects. A common approach to estimating Greenhouse Gas (GHG) emissions is through Eq. 6:

$$GHG = AD \times EF \tag{6}$$

where:

GHG represents the total Greenhouse Gas emissions, expressed as carbon dioxide equivalent (CO_2e).

AD stands for Activity Data, which is the magnitude of human activity resulting in emissions over a given period.

EF is the Emission Factor, a coefficient quantifying the emissions of a gas per unit of activity.

This formula aids in quantifying the environmental impact of geotechnical projects, aligning them with the growing focus on sustainability and climate responsibility in financial regulations.

7.4 Forming Strategic Collaborations

In the realm of geotechnical engineering, strategic collaborations are becoming increasingly pivotal for navigating the intricacies of climate-related financial regulations. These collaborations, spanning a wide array of partnerships, provide numerous benefits and opportunities for firms to adapt and thrive in this evolving landscape.

Partnerships with Other Engineering Firms: Collaborating with peer firms can lead to a fruitful exchange of knowledge, resources, and best practices regarding climate risk management and adherence to financial regulations. These partnerships might involve joint ventures on projects, shared learnings from individual experiences, and a collective approach to engaging with and influencing regulatory developments (Bekessy et al., 2018).

Collaborations with Financial Institutions: Partnerships with financial entities can offer invaluable insights into the financial aspects of climate risks and regulations. Such interactions can enlighten engineering firms on how the financial sector perceives and manages climate risks, influencing the development of risk assessment and mitigation strategies. They also present opportunities for securing project financing and attracting investments (Homsy & Warner, 2015).

Engagement with Regulatory Bodies: Regular interaction with regulatory authorities can keep firms up-to-date with regulatory changes, influence regulatory frameworks, and ensure compliance. This engagement can also bolster a firm's credibility and reputation, fostering trust among clients and stakeholders.

Collaborations with Research Institutions: As mentioned earlier, aligning with academic and research institutions can advance research in climate risk and geotechnical engineering. This collaboration is essential for driving innovation and developing new methodologies.

Client Engagement: Open dialogues with clients about climate risks and regulatory requirements can align project goals, establish effective mitigation strategies, and ensure client satisfaction (Adamkiewicz et al., 2022).

In addition to these collaborative strategies, Cost-Benefit Analysis (CBA) is a crucial financial tool for evaluating the feasibility and value of investments in climate change adaptation or mitigation measures in geotechnical projects. The basic formula is:

$$NB = B - C \tag{7}$$

where:

NB represents the Net Benefit of the project.

B is the Total Benefits of the project, encompassing all expected revenue or gains, including both direct and indirect benefits.

C represents the Total Costs of the project, including all expenses involved in its implementation, such as direct, indirect, and opportunity costs.

The use of CBA in the context of geotechnical projects helps in making informed decisions by weighing the economic advantages against the costs involved, particularly in scenarios where climate change adaptation or mitigation measures are being considered. This approach ensures that investments are not only environmentally sound but also economically viable.

8 Case Studies of Geotechnical Engineering in the Context of Climate-Related Financial Regulations

To enrich the theoretical discussions and strategies previously outlined, let us examine real-life case studies where geotechnical engineering firms have adeptly navigated climate-related financial regulations. These case studies demonstrate the varied approaches firms have taken to manage climate risk and comply with regulatory demands in their projects, offering practical insights into the integration of geotechnical challenges, climate adaptation measures, and financial considerations.

Table 8 showcases how the interplay of geotechnical engineering, climate change, and finance manifests globally. These case studies span a range of geotechnical challenges and climate adaptation strategies, emphasizing the financial aspects of each scenario.

Table 8 Case studies highlighting the intersection of geotechnical engineering, climate change, and finance

Case study	Geotechnical challenges	Climate adaptation measures	Financial considerations
Dutch "Room for the River" project	River flooding due to climate change	Widening of river channels and floodplains, relocation of dikes	Funded through public resources; increased upfront costs, but avoided future flood damage costs
Thawing permafrost in Alaska's infrastructure	Thawing permafrost causes ground instability	Use of thermosyphons to keep ground frozen, climate-resilient infrastructure design	Significant upfront costs, but long-term savings from reduced repair and maintenance
Coastal erosion Management in New Zealand	Coastal erosion due to sea-level rise	Use of seawalls, beach nourishment, managed retreat	A mix of public and private funding; costs varied based on adaptation measure
Landslide mitigation in Italy	Increased landslides due to heavy rainfall events	Implementation of early warning systems, soil nailing, slope terracing	Funded through EU, national, and regional funds; upfront investment in mitigation measures reduced costs from potential future landslides

8.1 Smith Engineering

Smith Engineering, a leader in global geotechnical engineering, has been proactive in adapting to climate-related financial regulations. Recognizing the increasing emphasis on climate risks, the firm has implemented advanced tools for climate risk assessment. These tools not only assess the potential impact of climate change on their projects but are also aligned with the guidelines of the Task Force on Climate-related Financial Disclosures (TCFD).

In response to TCFD recommendations, Smith Engineering has restructured its risk metrics and reporting formats to ensure comprehensive and transparent disclosure of climate risks. They have incorporated climate scenario analysis into their risk assessment framework, which allows them to evaluate the resilience of their projects under various future climate conditions. This approach is crucial in understanding the long-term sustainability and viability of their projects under shifting climatic scenarios.

Furthermore, Smith Engineering has established a robust internal audit system to maintain compliance with the evolving regulatory landscape. They have designated a dedicated team responsible for regularly reviewing their risk assessments and disclosures. This team ensures that all projects comply with current regulatory standards, which is crucial for maintaining the firm's reputation for regulatory compliance and transparency. Their commitment to aligning with climate-related financial regulations has positioned Smith Engineering as a leader in sustainable geotechnical practices, demonstrating how proactive adaptation to regulatory changes can enhance a firm's competitiveness and resilience in the face of climate challenges (Martinez et al., 2022; Chenet et al., 2019).

8.2 GreenTech Geoengineering

GreenTech Geoengineering, a firm renowned for its sustainable geotechnical solutions, offers a compelling case study in integrating environmentally friendly methods with traditional geotechnical practices. Their approach aligns with the global push for sustainability, notably aligning with the United Nations' Sustainable Development Goals (SDGs). GreenTech has adopted innovative technologies like biopolymer treatments for soil stabilization, carbon-negative concrete for foundations, and advanced groundwater monitoring systems, all aimed at mitigating the impacts of climate change (Alamoush et al., 2021). These techniques not only address geotechnical challenges but also significantly reduce the financial risks associated with climate change.

GreenTech's commitment to sustainability extends to its risk assessment models, demonstrating the effectiveness of its green technologies in risk mitigation. The firm has also established a thorough disclosure system, adhering to the Task Force on Climate-related Financial Disclosures (TCFD) guidelines. This transparency has garnered attention from investors and clients who prioritize environmental responsibility. Additionally, their strategic collaborations with research bodies and other firms have positioned GreenTech at the forefront of sustainable geotechnical innovation (Alamoush et al., 2021; Paulose, 2022; Rimmer, 2022).

8.3 TerraFirm Geotechnical Solutions

TerraFirm Geotechnical Solutions exemplifies the integration of geotechnical engineering with climate-related financial regulations in the housing and commercial building sector. Embracing the TCFD principles, TerraFirm has woven these guidelines into its risk management and operational strategies. A standout feature of their approach is the collaboration with insurance companies, fostering a shared understanding of climate-related financial risks. This exchange has refined TerraFirm's risk assessment models, aligning them more closely with the insurance sector's perspectives on climate risk (Al Hawaj & Buallay, 2022).

In terms of risk management, TerraFirm has adeptly integrated climate risks into its business strategy. They employ advanced geospatial modeling to evaluate the susceptibility of their projects to various climate change scenarios and have devised adaptive design solutions to counter these risks. Furthermore, the company has invested in extensive training for its staff to ensure a thorough understanding and effective implementation of climate-related financial disclosures. This strategy not only ensures consistency and accuracy in their reporting but also bolsters their credibility among clients and investors (Frykler, 2023).

These case studies, GreenTech Geoengineering and TerraFirm Geotechnical Solutions showcase how geotechnical engineering firms can effectively respond to climate-related financial regulations. They demonstrate that a proactive approach to sustainable practices and regulatory compliance can lead to innovation, enhanced credibility, and new opportunities in the rapidly evolving field of geotechnical engineering.

9 Future Directions for Geotechnical Engineering and Climate-Related Financial Regulations

As we progress through the twenty-first century, the interplay between geotechnical engineering and climate-related financial regulations is set to intensify, bringing forth a spectrum of challenges and opportunities. Anticipating and preparing for future developments will be crucial for the long-term viability and sustainability of geotechnical firms.

9.1 Future Regulatory Developments

The landscape of climate-related financial regulations is dynamic and is expected to evolve rapidly. With an expanding understanding of climate change and its impacts, these regulations are likely to become more stringent and encompassing. Emerging areas such as carbon pricing and climate change litigation could profoundly impact geotechnical firms. There is also a trend toward increased global coordination on climate-related financial regulations. For instance, frameworks like the Task Force on Climate-related Financial Disclosures (TCFD) are gaining momentum and may evolve into mandatory standards globally. While this presents challenges in navigating different regulatory environments, it also opens doors for streamlined compliance and efficiency (Chenet et al., 2019; Taylor, 2023).

9.2 Future Challenges and Opportunities for Geotechnical Engineering

The evolving regulatory landscape presents various challenges for geotechnical firms, but these can be transformed into opportunities with the right strategies:

- Aligning geotechnical risk management with changing regulations will require ongoing vigilance and adaptability. Firms will need to invest in expertise and tools to ensure compliance and effectively manage risks (Cook, 2022).
- Managing the financial impacts of these regulations will be another challenge. Projects with high climate risks may incur additional costs. Therefore, adopting innovative, green geotechnical methods and adaptive designs will be vital for financial sustainability (Vardon, 2015).
- Conversely, the growing emphasis on sustainability offers substantial opportunities. Firms that proactively adhere to climate-related financial regulations and excel in transparent risk disclosures can attract environmentally conscious investors and clients. The increasing demand for climate risk assessments in the financial sector could also create new service avenues. Geotechnical firms with expertise in climate risk assessments could expand their offerings to financial institutions, insurance companies, and investors, tapping into new markets and revenue streams (Cook, 2022; Vardon, 2015).

In conclusion, the future trajectory for geotechnical engineering in the context of climate-related financial regulations points toward a landscape where adaptability, innovation, and proactive engagement with sustainability issues will be key differentiators. Geotechnical firms that embrace these elements will not only navigate regulatory challenges effectively but also carve out a niche in a more environmentally conscious market.

10 Conclusion: Geotechnical Engineering in the Era of Climate-Related Financial Regulations

As we advance into the twenty-first century, the convergence of geotechnical engineering with climate-related financial regulations is becoming increasingly pivotal. This nexus presents a spectrum of challenges but also unlocks significant opportunities for innovation and growth in the geotechnical sector. Embracing these changes is not only a compliance necessity but also a strategic imperative for sustainable business practice.

Key Takeaways

- Climate-related financial regulations offer a unique opportunity for geotechnical firms to drive sustainable practices and integrate climate risks into their financial assessments, promoting industry-wide positive change.
- The role of geotechnical engineers is more critical than ever in the face of profound climate implications. Their expertise is essential in leading climate change mitigation and adaptation efforts, with financial regulations underscoring the value of these endeavors.
- Leading geotechnical firms are already harnessing these regulations to boost their competitiveness and sustainability, showcasing the potential for industry-wide transformation.
- Success in this landscape requires a holistic approach, blending technical geotechnical knowledge with financial and regulatory understanding.
- Innovation in geotechnical engineering, spurred by these challenges, can lead to transformative industry solutions and practices.

Summary and Discussion

The intersection of geotechnical engineering with climate-related financial regulations marks a significant shift toward sustainable and responsible business practices. This trend reflects a broader societal recognition of the role businesses play in climate change mitigation and adaptation. For geotechnical engineering firms, this means integrating sustainability into their strategies, operations, and stakeholder communications.

Future Work

Looking ahead, geotechnical engineering firms should:

- Embrace climate-related financial regulations as a catalyst for innovation and strategic differentiation.
- Develop the necessary skills and expertise to understand and navigate these complex regulations.
- Foster a culture of sustainability and innovation to align with regulatory demands and societal expectations.
- Communicate their sustainability efforts transparently to build trust and attract investment.

Final Thoughts
This chapter is a step in the ongoing journey toward understanding and navigating the complex landscape of climate-related financial regulations in geotechnical engineering. It highlights the need for continuous research, dialogue, and adaptation among all stakeholders. As the climate crisis unfolds, the role of geotechnical engineering firms in building a sustainable and resilient future becomes ever more critical. It is a journey that requires collective effort, innovative thinking, and a commitment to sustainable development.

Competing Interests The authors declare that they have no competing interests.

References

Abbass, K., Qasim, M. Z., Song, H., Murshed, M., Mahmood, H., & Younis, I. (2022). A review of the global climate change impacts, adaptation, and sustainable mitigation measures. *Environmental Science and Pollution Research, 29*(28), 42539–42559. https://doi.org/10.1007/s11356-022-19718-6

Abhayawansa, S., & Adams, C. (2022). Towards a conceptual framework for non-financial reporting inclusive of pandemic and climate risk reporting. *Meditari Accountancy Research, 30*(3), 710–738. https://doi.org/10.1108/MEDAR-11-2020-1097

Adamkiewicz, J., Kochanska, E., Adamkiewicz, I., & Łukasik, R. M. (2022). Greenwashing and sustainable fashion industry. *Current Opinion in Green and Sustainable Chemistry, 100710*, 100710. https://doi.org/10.1016/j.cogsc.2022.100710

Adams, C. A., & Mueller, F. (2022). Academics and policymakers at odds: The case of the IFRS foundation trustees' consultation paper on sustainability reporting. *Sustainability Accounting, Management and Policy Journal, 13*(6), 1310–1333. https://doi.org/10.1108/SAMPJ-10-2021-0436

Aitsi-Selmi, A., Egawa, S., Sasaki, H., Wannous, C., & Murray, V. (2015). The Sendai framework for disaster risk reduction: Renewing the global commitment to people's resilience, health, and Well-being. *International Journal of Disaster Risk Science, 6*, 164–176. https://doi.org/10.1007/s13753-015-0050-9

Al Hawaj, A. Y., & Buallay, A. M. (2022). A worldwide sectorial analysis of sustainability reporting and its impact on firm performance. *Journal of Sustainable Finance & Investment, 12*(1), 62–86. https://doi.org/10.1080/20430795.2021.1903792

Alamoush, A. S., Ballini, F., & Ölçer, A. I. (2021). Revisiting port sustainability as a foundation for the implementation of the United Nations sustainable development goals (UN SDGs). *Journal of Shipping and Trade, 6*(1), 1–40. https://doi.org/10.1186/s41072-021-00101-6

Alestra, C., Cette, G., Chouard, V., & Lecat, R. (2022). Growth impact of climate change and response policies: The advanced climate change long-term (ACCL) model. *Journal of Policy Modeling, 44*(1), 96–112. https://doi.org/10.1016/j.jpolmod.2021.10.001

Allman, E. (2022). Pricing climate change risk in corporate bonds. *Journal of Asset Management, 23*(7), 596–618. https://doi.org/10.1057/s41260-022-00294-w

Amel-Zadeh, A., & Serafeim, G. (2018). Why and how investors use ESG information: Evidence from a global survey. *Financial Analysts Journal, 74*(3), 87–103. https://doi.org/10.2469/faj.v74.n3.2

Anderson, C. C., Renaud, F. G., Hanscomb, S., & Gonzalez-Ollauri, A. (2022). Green, hybrid, or grey disaster risk reduction measures: What shapes public preferences for nature-based solutions? *Journal of Environmental Management, 310*, 114727. https://doi.org/10.1016/j.jenvman.2022.114727

Barkemeyer, R., Figge, F., Hoepner, A., Holt, D., Kraak, J. M., & Yu, P. S. (2017). Media coverage of climate change: An international comparison. *Environment and Planning C: Politics and Space, 35*(6), 1029–1054. https://doi.org/10.1177/0263774X16680818

Bassen, A., & Kovács, A. M. (2020). *Environmental, social and governance key performance indicators from a capital market perspective* (pp. 809–820). Springer Fachmedien Wiesbaden. https://doi.org/10.1007/978-3-658-16205-4_66

Batten, S., Sowerbutts, R., & Tanaka, M. (2016). *Let's talk about the weather: The impact of climate change on central banks*. https://doi.org/10.2139/ssrn.2783753.

Bekessy, S. A., Runge, M. C., Kusmanoff, A. M., Keith, D. A., & Wintle, B. A. (2018). Ask not what nature can do for you: A critique of ecosystem services as a communication strategy. *Biological Conservation, 224*, 71–74. https://doi.org/10.1016/j.biocon.2018.05.017

Benbouras, M. A. (2022). Hybrid meta-heuristic machine learning methods applied to landslide susceptibility mapping in the Sahel-Algiers. *International Journal of Sediment Research, 37*(5), 601–618. https://doi.org/10.1016/j.ijsrc.2022.04.003

Biswas, S. S., Ahad, M. A., Nafis, M. T., Alam, M. A., & Biswas, R. (2021). Introducing "α-sustainable development" for transforming our world: A proposal for the 2030 agenda. *Journal of Cleaner Production, 321*, 129030. https://doi.org/10.1016/j.jclepro.2021.129030

Board, F. S. (2017). *Recommendations of the task force on climate-related financial disclosures*.

Boneva, L., Ferrucci, G., & Mongelli, F. P. (2022). Climate change and central banks: What role for monetary policy? *Climate Policy, 22*(6), 770–787. https://doi.org/10.1080/14693062.2022.2070119

Bremner, C. (2023). The value of regulating climate disclosures. *Environmental Law, 53*(1), 107–145. https://www.jstor.org/stable/48724688

Busch, T., & Lewandowski, S. (2018). Corporate carbon and financial performance: A meta-analysis. *Journal of Industrial Ecology, 22*(4), 745–759. https://doi.org/10.1111/jiec.12591

Cao, Y., Xu, C., Kamaruzzaman, S. N., & Aziz, N. M. (2022). A systematic review of green building development in China: Advantages, challenges and future directions. *Sustainability, 14*(19), 12293. https://doi.org/10.3390/su141912293

Chenet, H., Ryan-Collins, J., & Van Lerven, F. (2019). *Climate-related financial policy in a world of radical uncertainty: Towards a precautionary approach* (p. 13). UCL Institute for Innovation and Public Purpose WP. https://ssrn.com/abstract=3520224

Chua, W. F., James, R., King, A., Lee, E., & Soderstrom, N. (2022). Task force on climate-related financial disclosures (TCFD) implementation: An overview and insights from the Australian accounting standards Board dialogue series. *Australian Accounting Review, 32*(3), 396–405. https://doi.org/10.1111/auar.12388

Cook, J. R. (2022). Reducing future climate impact on rural transport infrastructure in developing countries; the role of engineering geology. *Quarterly Journal of Engineering Geology and Hydrogeology, 55*(4), qjegh2021-052. https://doi.org/10.1144/qjegh2021-052

Cremasco, C., & Boni, L. (2022). Is the European Union (EU) sustainable finance disclosure regulation (SFDR) effective in shaping sustainability objectives? An analysis of investment funds' behaviour. *Journal of Sustainable Finance & Investment, 1-19*, 1. https://doi.org/10.1080/20430795.2022.2124838

Das, B. M. (2021). *Principles of geotechnical engineering*. Cengage learning.

Diffenbaugh, N. S. (2022). COVID-19 and the environment: Short-run and potential long-run impacts. *Annual Review of Environment and Resources, 47*, 65–90. https://doi.org/10.1146/annurev-environ-120920-125207

Dikau, S., & Volz, U. (2018). *Central banking, climate change and green finance*. https://eprints.soas.ac.uk/id/eprint/26445

DiMaggio, P. J., & Powell, W. W. (1983). The iron cage revisited: Institutional isomorphism and collective rationality in organizational fields. *American Sociological Review, 147-160*, 147. https://doi.org/10.2307/2095101

Flammer, C. (2021). Corporate green bonds. *Journal of Financial Economics, 142*(2), 499–516. https://doi.org/10.1016/j.jfineco.2021.01.010

Frykler, A. (2023). *Business innovation in complex ecosystems a case from the construction industry*. http://hdl.handle.net/20.500.12380/306236

Gilchrist, D., Yu, J., & Zhong, R. (2021). The limits of green finance: A survey of literature in the context of green bonds and green loans. *Sustainability, 13*(2), 478. https://doi.org/10.3390/su13020478

Girard-Guerraud, C., Goodman, J., & Louche, C. (2023). Transformations in shareholder activism: Past, present, and future. In *Recent trends in financial engineering: Towards more sustainable social impact* (pp. 133–164). https://doi.org/10.1142/9789811260483_0007

Hijma, M. P., & Cohen, K. M. (2019). Holocene Sea-level database for the Rhine-Meuse Delta, The Netherlands: Implications for the pre-8.2 ka sea-level jump. *Quaternary Science Reviews, 214*, 68–86. https://doi.org/10.1016/j.quascirev.2019.05.001

Homsy, G. C., & Warner, M. E. (2015). Cities and sustainability: Polycentric action and multilevel governance. *Urban Affairs Review, 51*(1), 46–73. https://doi.org/10.1177/1078087414530545

Jain, V., Guha, S., & Sagar, B. D. (2022). Quantitative geomorphology. In *Encyclopedia of mathematical geosciences* (pp. 1–18). Springer International Publishing. https://doi.org/10.1007/978-3-030-26050-7_27-1

James, S., & Quaglia, L. (2023). Differentiated de-Europeanisation: UK policy-making in finance after Brexit. *Journal of European Public Policy, 1-23*, 2445. https://doi.org/10.1080/13501763.2023.2183978

Kandalai, S., John, N. J., & Patel, A. (2023). Effects of climate change on geotechnical infrastructures—State of the art. *Environmental Science and Pollution Research, 30*(7), 16878–16904. https://doi.org/10.1007/s11356-022-24788-7

Kelman, I., Gaillard, J. C., Lewis, J., & Mercer, J. (2016). Learning from the history of disaster vulnerability and resilience research and practice for climate change. *Natural Hazards, 82*, 129–143. https://doi.org/10.1007/s11069-016-2294-0

Kemp, L., Xu, C., Depledge, J., Ebi, K. L., Gibbins, G., Kohler, T. A., et al. (2022). Climate endgame: Exploring catastrophic climate change scenarios. *Proceedings of the National Academy of Sciences, 119*(34), e2108146119. https://doi.org/10.1073/pnas.2108146119

Kinsella, E. L., Muldoon, O. T., Lemon, S., Stonebridge, N., Hughes, S., & Sumner, R. C. (2023). In it together?: Exploring solidarity with frontline workers in the United Kingdom and Ireland during COVID-19. *British Journal of Social Psychology, 62*(1), 241–263. https://doi.org/10.1111/bjso.12579

Kumar, S., Sharma, D., Rao, S., Lim, W. M., & Mangla, S. K. (2022). Past, present, and future of sustainable finance: Insights from big data analytics through machine learning of scholarly research. *Annals of Operations Research, 1-44*. https://doi.org/10.1007/s10479-021-04410-8

Lin, L., & Hong, Y. (2022). Developing a green bonds market: Lessons from China. *European Business Organization Law Review, 23*(1), 143–185. https://doi.org/10.1007/s40804-021-00231-1

Liu, B., Zhang, Y., Ma, Z., Andersen, K. H., Jostad, H. P., Liu, D., & Pei, A. (2020). Design considerations of suction caisson foundations for offshore wind turbines in southern China. *Applied Ocean research, 104*, 102358. https://doi.org/10.1016/j.apor.2020.102358

Martinez, A., DeJong, J., Akin, I., Aleali, A., Arson, C., Atkinson, J., et al. (2022). Bio-inspired geotechnical engineering: Principles, current work, opportunities and challenges. *Géotechnique, 72*(8), 687–705. https://doi.org/10.1680/jgeot.20.P.170

Martinez Salgueiro, A., & Tarrazon-Rodon, M. A. (2021). Weather derivatives to mitigate meteorological risks in tourism management: An empirical application to celebrations of Comunidad Valenciana (Spain). *Tourism Economics, 27*(4), 591–613. https://doi.org/10.1177/1354816619890751

Meerow, S., & Keith, L. (2022). Planning for extreme heat: A national survey of US planners. *Journal of the American Planning Association, 88*(3), 319–334. https://doi.org/10.1080/01944363.2021.1977682

Meyer, J. W., & Rowan, B. (1977). Institutionalized organizations: Formal structure as myth and ceremony. *American Journal of Sociology, 83*(2), 340–363. https://doi.org/10.1086/226550

Minutiello, V., & Tettamanzi, P. (2022). The quality of nonfinancial voluntary disclosure: A systematic literature network analysis on sustainability reporting and integrated reporting. *Corporate Social Responsibility and Environmental Management, 29*(1), 1–18. https://doi.org/10.1002/csr.2195

Naylor, A. W., & Ford, J. (2023). Vulnerability and loss and damage following the COP27 of the UN framework convention on climate change. *Regional Environmental Change, 23*(1), 38. https://doi.org/10.1007/s10113-023-02033-2

Ng, S. T., Xu, F. J., Yang, Y., Lu, M., & Li, J. (2018). Necessities and challenges to strengthen the regional infrastructure resilience within city clusters. *Procedia Engineering, 212*, 198–205. https://doi.org/10.1016/j.proeng.2018.01.026

Ohene, E., Chan, A. P., Darko, A., & Nani, G. (2023). Navigating toward net zero by 2050: Drivers, barriers, and strategies for net zero carbon buildings in an emerging market. *Building and Environment, 110472*. https://doi.org/10.1016/j.buildenv.2023.110472

Ottesen, D., Batchelor, C. L., Bjarnadóttir, L. R., Wiberg, D. H., & Dowdeswell, J. A. (2022). Glacial landforms reveal dynamic ice-sheet behaviour along the mid-Norwegian margin during the last glacial-deglacial cycle. *Quaternary Science Reviews, 285*, 107462. https://doi.org/10.1016/j.quascirev.2022.107462

Overland, I., Fossum Sagbakken, H., Isataeva, A., Kolodzinskaia, G., Simpson, N. P., Trisos, C., & Vakulchuk, R. (2022). Funding flows for climate change research on Africa: Where do they come from and where do they go? *Climate and Development, 14*(8), 705–724. https://doi.org/10.1080/17565529.2021.1976609

Parmenter, R. R., & Losleben, M. V. (2023). Influence of mixed conifer Forest thinning and prescribed fire on soil temperature and moisture dynamics in proximity to Forest logs: A case study in New Mexico, USA. *Forests, 14*(6), 1117. https://doi.org/10.3390/f14061117

Paulose, R. M. (2022). Death by a thousand cuts? Green tech, traditional knowledge, and genocide. *Genocide Studies and Prevention: An International Journal, 16*(1), 6. https://doi.org/10.5038/1911-9933.16.1.1881

Pelling, M., Chow, W. T., Chu, E., Dawson, R., Dodman, D., Fraser, A., et al. (2022). A climate resilience research renewal agenda: Learning lessons from the COVID-19 pandemic for urban climate resilience. *Climate and Development, 14*(7), 617–624. https://doi.org/10.1080/17565529.2021.1956411

Qerimi, Q., & Sergi, B. S. (2022). The case for global regulation of carbon capture and storage and artificial intelligence for climate change. *International Journal of Greenhouse Gas Control, 120*, 103757. https://doi.org/10.1016/j.ijggc.2022.103757

Quemin, S. (2022). Raising climate ambition in emissions trading systems: The case of the EU ETS and the 2021 review. *Resource and Energy Economics, 68*, 101300. https://doi.org/10.1016/j.reseneeco.2022.101300

Quorning, S. (2023). The 'climate shift'in central banks: How field arbitrageurs paved the way for climate stress testing. *Review of International Political Economy, 1-23*, 74. https://doi.org/10.1080/09692290.2023.2171470

Rimmer, M. (2022). Synthetic genomics: Intellectual property, innovation policy, and advanced biofuels. In *In 3rd generation biofuels* (pp. 847–868). Woodhead Publishing. https://doi.org/10.1016/B978-0-323-90971-6.00010-3

Ripple, W. J., Wolf, C., Gregg, J. W., Levin, K., Rockström, J., Newsome, T. M., … & Lenton, T. M. (2022). *World scientists' warning of a climate emergency* 2022. https://doi.org/10.1093/biosci/biac083.

Risi, D., Vigneau, L., Bohn, S., & Wickert, C. (2023). Institutional theory-based research on corporate social responsibility: Bringing values back in. *International Journal of Management Reviews, 25*(1), 3–23. https://doi.org/10.1111/ijmr.12299

Romanello, M., Di Napoli, C., Drummond, P., Green, C., Kennard, H., Lampard, P., et al. (2022). The 2022 report of the lancet countdown on health and climate change: Health at the mercy of fossil fuels. *The Lancet, 400*(10363), 1619–1654. https://doi.org/10.1016/S0140-6736(22)01540-9

Rowan, E., Evans, C., Riley-Gilbert, M., Hyman, R., Kafalenos, R., Beucler, B., et al. (2013). Assessing the sensitivity of transportation assets to extreme weather events and climate change. *Transportation Research Record, 2326*(1), 16–23. https://doi.org/10.3141/2326-03

Schoenmaker, D., & Van Tilburg, R. (2016). What role for financial supervisors in addressing environmental risks? *Comparative Economic Studies, 58*, 317–334. https://doi.org/10.1057/ces.2016.11

Scott, W. R. (2008). *Institutions and organizations: Ideas and interests*. Sage Publications.

Semieniuk, G., Campiglio, E., Mercure, J. F., Volz, U., & Edwards, N. R. (2021). Low-carbon transition risks for finance. *Wiley Interdisciplinary Reviews: Climate Change, 12*(1), e678. https://doi.org/10.1002/wcc.678

Sgambati, S. (2022). Who owes? Class struggle, inequality and the political economy of leverage in the 21st century. *Finance and Society, 8*(1), 1. https://doi.org/10.2218/finsoc.7115

Shipton, L., & Dauvergne, P. (2022). The influence of home country institutions on the adoption of corporate social responsibility policies by transnational mining corporations. *The Extractive Industries and Society, 10*, 101077. https://doi.org/10.1016/j.exis.2022.101077

Smoleńska, A. (2023). *Central bank supervisory role: Micro-prudential supervision and regulation of ESG risks*. https://ssrn.com/abstract=4446375

Sullivan, R., & Gouldson, A. (2017). The governance of corporate responses to climate change: An international comparison. *Business Strategy and the Environment, 26*(4), 413–425. https://doi.org/10.1002/bse.1925

Sun, L., Fang, S., Iqbal, S., & Bilal, A. R. (2022). Financial stability role on climate risks, and climate change mitigation: Implications for green economic recovery. *Environmental Science and Pollution Research, 29*(22), 33063–33074. https://doi.org/10.1007/s11356-021-17439-w

Svennevig, K., Koch, J., Keiding, M., & Luetzenburg, G. (2023). Assessing the impact of climate change to landslides using public data, a case study from Vejle, Denmark. *Natural Hazards and Earth System Sciences Discussions, 2023*, 1–26. https://doi.org/10.5194/nhess-2023-68

Tan, C. (2022). Private investments, public goods: Regulating markets for sustainable development. *European Business Organization Law Review, 23*(1), 241–271. https://doi.org/10.1007/s40804-021-00236-w

Tang, C. S., Paleologos, E. K., Vitone, C., Du, Y. J., Li, J. S., Jiang, N. J., et al. (2020). Environmental geotechnics: Challenges and opportunities in the post-COVID-19 world. *Environmental Geotechnics, 8*(3), 172–192. https://doi.org/10.1680/jenge.20.00054

Tao, H., Zhuang, S., Xue, R., Cao, W., Tian, J., & Shan, Y. (2022). Environmental finance: An interdisciplinary review. *Technological Forecasting and Social Change, 179*, 121639. https://doi.org/10.1016/j.techfore.2022.121639

Taylor, N. (2023). 'Making financial sense of the future': Actuaries and the management of climate-related financial risk. *New political economy, 28*(1), 57–75. https://doi.org/10.1080/13563467.2022.2067838

Thomä, J., & Chenet, H. (2017). Transition risks and market failure: A theoretical discourse on why financial models and economic agents may misprice risk related to the transition to a low-carbon economy. *Journal of Sustainable Finance & Investment, 7*(1), 82–98. https://doi.org/10.1080/20430795.2016.1204847

Tian, X., Zhang, Y., & Qu, G. (2022). The impact of digital economy on the efficiency of green financial investment in China's provinces. *International Journal of Environmental Research and Public Health, 19*(14), 8884. https://doi.org/10.3390/ijerph19148884

Vahedifard, F., Williams, J. M., & AghaKouchak, A. (2018). Geotechnical engineering in the face of climate change: Role of multi-physics processes in partially saturated soils. *IFCEE, 2018*, 353–364.

van Oostrum, C. (2021). Sustainability through transparency and definitions: A few thoughts on regulation (EU) 2019/2088 and regulation (EU) 2020/852. *European Company Law, 18*(1), 15. https://doi.org/10.54648/eucl2021003

Vardon, P. J. (2015). Climatic influence on geotechnical infrastructure: A review. *Environmental Geotechnics, 2*(3), 166–174. https://doi.org/10.1680/envgeo.13.00055

Veraksa, N., & Basseches, M. (2022). Dialectical thinking. In *Piaget and Vygotsky in XXI century: Discourse in early childhood education* (pp. 41–64). Springer International Publishing. https://doi.org/10.1007/978-3-031-05747-2_4

Watts, N., Adger, W. N., Agnolucci, P., Blackstock, J., Byass, P., Cai, W., & Costello, A. (2015). Health and climate change: Policy responses to protect public health. *The Lancet, 386*(10006), 1861–1914. https://doi.org/10.1016/S0140-6736(15)60854-6

Wdowinski, S., Bray, R., Kirtman, B. P., & Wu, Z. (2016). Increasing flooding hazard in coastal communities due to rising sea level: Case study of Miami Beach, Florida. *Ocean & Coastal Management, 126*, 1–8. https://doi.org/10.1016/j.ocecoaman.2016.03.002

Wing, O. E., Lehman, W., Bates, P. D., Sampson, C. C., Quinn, N., Smith, A. M., et al. (2022). Inequitable patterns of US flood risk in the Anthropocene. *Nature Climate Change, 12*(2), 156–162. https://doi.org/10.1038/s41558-021-01265-6

Zhang, W., Li, H., Li, Y., Liu, H., Chen, Y., & Ding, X. (2021). Application of deep learning algorithms in geotechnical engineering: A short critical review. *Artificial Intelligence Review, 1-41*, 5633. https://doi.org/10.1007/s10462-021-09967-1

Zhou, F., Endendijk, T., & Wouter Botzen, W. J. (2023). A review of the financial sector impacts of risks associated with climate change. *Annual Review of Resource Economics, 15*, 233. https://doi.org/10.1146/annurev-resource-101822-105702

Linking Environmental Performance and Financial Profitability of Companies under Greenhouse Gas Emissions Trading

Isabel Cristina Mañas-Álvarez, Federico Galán-Valdivieso, and María del Carmen Caba-Pérez

Abstract Given the current climate crisis, several international measures have been taken to combat air pollution, including the European greenhouse gas emissions trading scheme (EU-ETS). This tool is vital for those organizations forced to participate in the system (because of their sector of activity) since their production could be restrained by the emission limits, or their profitability could be reduced because of the acquisition of emission rights or investments in carbon-reducing policies or technologies. Through an OLS multiple linear regression, this paper analyzes the relationship between three measures of environmental performance and the financial performance of European companies, both belonging to the EuroStoxx600 index and participating in the EU-ETS, in the period 2018–2022. The results show a significant and negative relationship between the company's financial profitability and the level of CO_2 emissions, that is, low carbon emitters present higher levels of financial performance. This result implies that, although in the short term, it is possible that investing in carbon-reducing measures impacts firms' profitability, in the long term, these obstacles are compensated by higher returns for EU-ETS participants.

1 Introduction

The climate crisis is an unavoidable reality today. States at the global level must develop an effective response to the risks involved through cooperation. As a result, over the last few decades, they have developed various international regulations focused on achieving sustainable objectives designed to tackle climate change.

The first step in this initiative of international collaboration was taken on the Meadows Report, leading countries to meet at the United Nations Conference on the Human Environment, held in Stockholm from 5 to 16 June 1972. Since then,

I. C. Mañas-Álvarez (✉) · F. Galán-Valdivieso · M. del C. Caba-Pérez
Department of Economics and Business, University of Almeria, Almería, Spain
e-mail: ima263@ual.es; fgalan@ual.es; ccaba@ual.es

N. Naifar (ed.), *Climate Change and Finance*, Sustainable Finance,
https://doi.org/10.1007/978-3-031-56419-2_10

numerous assemblies and treaties emerged until the regulation that would give rise to the tool under study appeared: the system of greenhouse gas emission rights. The result of the Framework Convention on Climate Change (UNFCCC) of June 04, 1992, was the Kyoto Protocol, in which the foundations of this mechanism were established together with the Clean Development Mechanism and Joint Implementation. The purpose of this system was (and is) to reduce greenhouse gas emissions to 1990 levels for the period from 2008 to 2012 (subsequently extended for the periods 2013–2020 and 2021–2030). The beginning of this international strategy affected the traditional way of organization and production of companies, which began to worry about the limitations that would entail their profitability.

As is the case in such international treaties, each state must incorporate international law into its domestic law. The European Union transposed this rule through Directive (EU) 2003/87/EC of the European Parliament and of the Council of 13 October 2003 establishing a scheme for greenhouse gas emission allowance trading and amending Council Directive 96/61/EC of 25 October 2003. However, the European legislator, in addition to meeting the objectives, managed to incentivize companies to improve their environmental practices through an economic reward. In this way, economic performance is not compromised by environmental performance. The reason for this is its main feature of transferability, which allows companies that meet the minimum emissions to sell their surplus allowances to those that do not meet through the market (López Gordo, 2006). It is precisely this characteristic that has aroused interest among academics in the possible benefit it can bring to the economic-financial profitability of the company.

But the beginnings of this emission rights market in 2005 were not successful as it did not meet the proposed purpose. Companies were not incentivized to reduce their emissions due to low prices, caused by the over-allocation of securities, the 2008 crisis that implied a decrease in the level of production, and the free allocation method. However, from 2018 onward, they began to rise significantly due to the change in allocation method (now by auction), the withdrawal of surplus allowances from the market (carried over from years ago), and the change in the nature of these securities, becoming financial instruments (Sánchez García, 2019).

As an instrument that limits the production of companies, there was uncertainty, especially at the beginning, about the impact it would have on economic and financial profitability. Therefore, academics and professionals have studied since then the effects on organizations, thus the interest in the topic and also the literature have increased and, consequently, the diversity of the results. Furthermore, new questions arise when considering the participation in the EU-ETS system, since those firms forced to participate may face additional costs to reduce their carbon footprint, in carbon-reducing technologies and/or the acquisition of carbon emission allowances. In short, academics and professionals seek to know whether "it pays to be green" or not.

The aim of this chapter is to shed some light on the relevance of sustainability performance, analyzing the effect of three sustainability variables (the volume of CO_2 emissions, the efficiency of energy policies, and the score on ESG principles) on companies' profitability (those firms participating in the EU-ETS) through a multiple linear regression.

This work has the following structure: after the introduction, we present a brief description of the current state of emissions trading, and a review of the main existing literature on the relationship between environmental variables, such as carbon dioxide emissions or energy efficiency policies, with the financial performance of companies. After positing our hypotheses, we will proceed to the quantitative part of the work in which, by means of a multiple linear regression by ordinary least squares, the proposed model and the relationship between financial performance and independent economic and environmental variables. Finally, based on the results obtained, a series of conclusions will be drawn applicable to the practical development of the business activity.

2 Conceptual Framework

2.1 Developments in greenhouse gas emissions trading

The evolution of the greenhouse gas emissions trading in the European Union can be divided into four phases: 2005–2007, 2008–2012, 2013–2020, and 2021–2030. Due to its own regulation and other external causes, its development has not been homogeneous over the years (Garnacho, 2018).

In the first inaugural phase (2005–2007), the system of free allocation and the calculation based on historical issues led to an over-allocation of securities that resulted in prices close to zero, which clearly did not encourage the trading of these rights. This situation continued during the second phase (2008–2012) due to the boost in renewable energy policies and the surplus of emission allowances, coupled with the 2008 crisis, which resulted in a significant decrease in production. In those years, the impact of the emission allowance system on the operation and profitability of companies was analyzed by the researchers. They came to different conclusions as this tool was being developed. Based on data from the first phase, Commins et al. (2011) concluded that the greenhouse gas emissions trading scheme had a significantly negative impact on the capital performance ratios of European companies. In the same vein, Oberndorfer and Rennings (2007) saw that the competitiveness and creativity of companies were affected, but not with the same intensity that the previous study concluded. Likewise, Demailly and Quirion (2008) found slight losses of competitiveness in the steel sector of the European Union, losses considered potentially serious in the long term according to Zhang and Wei (2010). In contrast, there are several studies that conclude that the introduction of the greenhouse gas emission allowance system had a positive impact on the financial profitability of companies. Based on the first phase data, Chan et al. (2013) concluded that this scheme had a positive impact on the material costs and turnover of the energy industry in 10 European Union countries for the period

2001–2009. Its findings were supported by other studies that had a positive impact on business performance, such as Benz and Trück (2006), Veith et al. (2009), and Oberndorfer (2009), the latter revising its previous work in 2007. It is also to be expected that the allocation of allowances, while produced by auctioning, will lead to an increase in the cost of production of companies, reducing their profitability.

In the third phase, the system began to improve thanks to the change in the distribution system. First, the free allocation was mainly reserved for sectors and subsectors at risk of carbon leakage, taking as a reference the benchmarks and the historical level of activity of each sub-installation. As a result, the auction method was introduced as the main method of allocation, in contrast to the previous two phases, even withdrawing the electricity sector from free allocation. Therefore, the companies belonging to that activity have had to obtain all the necessary rights in this financial market to hedge their emissions and act within the legal framework, which entails a considerable increase in their costs. Secondly, the back-loading system (2014) and the Market Stability Reserve (2019) were created, the latter focused on withdrawing those rights whose price was at a low level that did not encourage the reduction of emissions. Thirdly, securities were redefined as financial instruments, so that trade would not only involve companies with surpluses of securities and loss-making companies but also other participants. Finally, another measure adopted was the acceleration of the decrease in the emissions ceiling from 1.74% to 2.2% from 2021. The aim of the new regulation is to raise prices so that companies not only feel controlled by greenhouse gas emission rights, but that, on their own, and in order to be more efficient, they seek to improve sustainability in their process through technological innovations. However, in 2020, with the international health crisis caused by COVID-19, the levels of activity of companies and their production levels dropped. As a result, the demand for the securities fell again and, therefore, their prices on the market, although not reaching the levels of previous years. From the academic perspective, studies continued on the relationship between emission allowances and business profitability. Clarkson et al. (2015), by measuring the level of pollution of a company in relation to its shortage of allowances, they extracted a negative relationship between the level of pollution and the market value, although this relationship is moderately presented for companies that have better environmental results compared to the rest. However, other studies have shown that the factor of allocation of verified allowances is decisive, since a higher proportion of verified allowances results in a negative effect on the company's turnover (Chan et al., 2013). In the following years, Fan et al. (2017) presented a study on the relationship between energy efficiency and financial performance of a sample of companies in China, where they demonstrated that low energy intensity can lead to better financial results. Makridou et al. (2019) analyze the company's profitability, concluding that profitability is positively associated with the proportion of verified allowances obtained and emissions allowances sales in the European market, both when examining all phases as a whole or each phase separately.

The fourth phase is now developing significantly. In September 2020, the European Commission presented a plan included in the "European Green Deal" to reduce emissions by 55%. This is why the "Fit for 55" package was created in July 2021. The allocation methods remain the same as in the third phase, with a nuance in the industrial sectors not exposed to leakage, for which the free allocation will be gradually reduced by 30% of what they would be entitled to until 2026 and reach 0% in 2030. At the end of 2020, lower prices resulting from the health crisis rose again, exceeding the €50/ton bracket in June 2021. From November 2021, they rose considerably due to a delay in the start of the auctions that same year and another mandatory reduction in gas emissions.

2.2 *Influence of environmental performance on business profitability*

Considering the relevance of the emissions trading system in European business activity and the unstoppable environmental regulations in the field of emissions, companies must devise strategies that adapt their environmental performance to this increasingly demanding legislation.

Some studies show the connection between economic growth, trade, financial profitability, energy use, renewable energy, and CO_2 emissions (Khan et al., 2019). According to Hassine and Harrathi (2017), renewable energy and trade have boosted the economy to overall levels, as well as achieved a decrease in air pollution. National economic growth (Sadeghieh, 2016), global market opening (Shahbaz et al., 2014; Al-Mulali & Ozturk, 2015), increased financial development (Shahbaz et al., 2013; Al-Mulali & Ozturk, 2015; Nasreen et al., 2017) and the progress of renewable energy (Al-Mulali & Ozturk, 2015; Khan et al., 2018) are factors that have led to CO_2 reductions. The reason behind the influence of these elements on the level of emissions is the long-term cointegration between the financial development index, renewable energy, and greenhouse gas emissions per capita, as Khan et al. (2019) show by analyzing these variables in 34 countries in Asia, Europe, and America, highlighting the bidirectionality found in the European case. That is why policymakers and environmental policymakers should focus on linking the pursuit of profitability with the promotion of renewable energy and clean technologies, which reduce the volume of emissions.

Therefore, for better protection of the environment, it is essential to invest in the innovation of new technologies and the correct application of environmental regulations. This leads companies to adapt their strategies to two possible alternatives: innovation, which would allow them to reduce their emissions and, thanks to emissions trading, sell surplus securities (Demailly & Quirion, 2008); or, as Anger and Oberndorfer (2008) point out, adopt the attitude of the survival principle by not modifying its strategy and investing in higher costs in order to remain at the minimum required by the legal framework.

2.2.1 Contribution to Carbon Dioxide Emissions

Some authors argue that the profits made by companies in various industrial sectors are due to higher final prices motivated by the need to cover costs related to emission allowances (Kara et al., 2008; Chen et al., 2008; Lise et al., 2010). However, those companies adhering to the emission rights regime have promoted and implemented greater energy efficiency as they have obtained greater profitability as a result (Aupperle et al., 1985). Change can reduce costs and combat potential risks, such as fluctuations in energy prices and energy shortages, which are increasingly common problems due to today's international conflicts. In addition, socially and environmentally conscious customers are willing to shoulder the burden and pay for more sustainable products (Telle, 2006). By decreasing emissions and improving profits, the company is in a more efficient situation in terms of profitability when it sells more and issues less or has to buy less or sell more rights. In this way, there is a "*win-win*" condition in the environmental policy as there are negative effects between the variables of CO_2 emissions and sales of the company.

Considering the importance of the volume of carbon dioxide emissions, the first hypothesis is posited:

H1: The volume of carbon dioxide emissions has a negative impact on the financial profitability of companies.

2.2.2 Environmental Performance

With the incipient concern developed in recent decades about the climate crisis and the connection between the energy factor and the economic and firms' financial performance, numerous analyses have been carried out on the impact that environmental performance has on organizations. Apart from offering non-monetary incentives, especially associated with renewable energy and environmental quality, such as improving the company's image from the consumer's perspective, it can also lead to economic incentives (Lin & Okoye, 2023). In fact, that is the approach of the European legislator when creating the European Union's system of greenhouse gas emission allowances.

However, the existing literature studying the relationship between economic and environmental performance shows contradictory results. Several studies show an increase in energy consumption in parallel with financial development as a result of increased production (Sadorsky, 2010; Chang, 2015; Çoban & Topcu, 2013; Islam et al., 2013; Ozturk & Acaravci, 2013; Sadorsky, 2011). On the other hand, there are studies that show a negative relationship between these two variables, with greater profitability at lower energy consumption, driven by innovation (Destek, 2018; Ouyang & Li, 2018; Shahbaz et al., 2016; Shahbaz et al., 2017; Topcu & Payne, 2017). Whatever the case, this causality between financial development and CO_2 emissions is bidirectional (Zhao & Yang, 2020).

As mentioned above, energy intensity is decisive as it is how financial development drives energy efficiency (Zakari et al., 2022). According to Fan et al. (2017), better financial outcomes can be achieved thanks to low energy intensity. However, it is a two-way support since the increase in the economic development of the company allows it to outsource energy and consequently improve its energy efficiency (Ahmed, 2017). This conclusion is supported by Chang (2015) who argues that profitability indicators can promote energy efficiency. However, the results also depend on the company's revenue and economic performance. According to a study by Zakari et al. (2022), sustainable innovation can push companies and institutions to gradually abandon the use of non-renewable energy. Therefore, the commitment to sustainable technologies has a positive effect on energy efficiency and, according to the above analyses, on profitability.

Furthermore, and due to its growing importance, studies have been carried out that analyze the effects of ESG (Environmental, Social, and Governance) reputation on companies. Zhang et al. (2023) and Yao et al. (2023) conclude that participation in carbon emissions trading has a positive impact on ESG performance; Duque-Grisales and Aguilera-Caracuel (2021) state that there is a negative relationship between the three dimensions of ESG *scoring* and financial profitability, highlighting that scores in the social sphere have a more significant negative impact on financial performance than governance and environmental scores; Dorfleitner et al. (2020) indicate that the inclusion of the Reuters controversy score can help build more efficient portfolios in the face of corporate scandals.

With all the information presented in this section, two other hypotheses are posited:

H2: Energy policy has a positive impact on the financial profitability of companies.

H3: The ESG Principles score positively influences the financial performance of companies.

3 Components and Data Analysis

The purpose of this work is to test the proposed model by means of an OLS multiple linear regression to verify the previous hypotheses and to verify the relationship between environmental performance and the economic-financial profitability of the company.

The data correspond to the companies listed in the European EuroStoxx600 index from 2018 to 2022, the last year for which data are available. The database used to obtain it was Thomson Reuters Refinitiv. With a sample of 600 companies and data between 2018 and 2022 (5 years), it totaled 3000 observations, removing those observations for which information was incomplete, finally resulting in 2204 company-year observations.

The dependent variable is the financial performance, proxied by the Return on Assets ratio, defined as the ratio between net profit and total assets. The choice of this variable responds to the necessity of focusing on the cost structure and avoiding the influence of the financial structure in the analysis. Based on the previous literature, the economic-financial and control variables that are most commonly used in the literature (Fan et al., 2017) will be included as control explanatory variables:

size, growth, risk, sector, and year. Regarding the environmental variables, the score of energy efficiency policies, ESG scores of environmental, social, and governance performance (as a whole), CO_2 emissions, and the participation of companies in the EU-ETS rights trading mechanism have been included.

The sustainability scores by the Thomson Reuters Refinitiv database were designed to measure a company's ESG performance on 10 main topics based on data provided by the company through various sources (sustainability reports, information on the web, etc.). These scores are available to more than 6000 companies worldwide, and are percentile-ranked scores, available in both scores (percentiles), and letter grades from A+ to C (LSEG, 2023). The energy efficiency policy item is also provided by Refinitiv, and it is a dummy variable that takes the value of 1 if the company has a policy to improve its energy efficiency, and 0 if it does not (Table 1).

With the proposed variables, the model to be tested is as follows:

$$FP_i = \beta_0 + \beta_1 \cdot YEAR_i + \beta_2 \cdot IND_i + \beta_3 \cdot SIZE_i + \beta_4 \cdot RISK_i + \beta_5 \cdot GRW_i + \beta_6 \cdot CO2_i + \beta_7 \cdot ESG_i + \beta_8 \cdot EEP_i$$

4 Results

A preliminary descriptive analysis shows that the dependent variable ROA does not follow a normal distribution, not meeting the essential requirements to allow to carry out the relevant analysis. Even when testing with different variables of economic and financial profitability such as ROE (Return On Equity), ROIC (Return

Table 1 Variables

Variable	Measurement	Calculation
Financial Performance (FP)	ROA	Net profit / Total assets
Year (YEAR)	Year	–
Sector (IND)	Sector	Industrial sector according to SIC (Standard Industrial Classification)
Size (SIZE)	Assets	Ln (assets)
Risk (RISK)	Debt ratio	Debt / net worth
Growth (GRW)	Δ sales	(sales t–sales t-1) / sales t-1
Contribution to CO_2 emissions (CO_2)	Estimated CO_2 emissions (Mt)	Ln (estimated CO_2 emissions)
ESG score (ESG)	ESG score	–
Energy efficiency policies (EEP)	Energy efficiency policies	0: The company does not have a policy to improve its energy efficiency 1: The company has a policy to improve its energy efficiency

On Invested Capital), or EBIT, the normality tests do not exhibit a normal behavior of the dependent variable. These results led the researchers to segment the sample according to the variable of participation in the greenhouse gas emission rights market.

Once the data has been segmented, the model has been retested. As a result, it has been found that those companies that do not belong to this system of emission rights show abnormal behavior in the dependent variable ROA; in contrast to those that do participate in this emissions trading, a segment for which the dependent variable ROA shows a more normal distribution of the data. This result is in line with the focus of the work, since this ratio highlights the profitability of a firm's main economic activity, and also eliminates the effect of the capital structure. Therefore, taking this segment of firms from the total sample, we analyze the proposed model by means of a multiple linear regression by ordinary least squares.

This subsample includes 139 companies with data from 2018 to 2022. Therefore, a total of 634 observations have been collected (of which 449 are valid).

1. Descriptive analysis

Table 2 shows the main statistics of the variables used, and Table 3 shows the correlations between the variables.

2. Inferential analysis

Once the general statistical parameters of the variables have been extracted, multiple linear regression by ordinary least squares is performed. According to the main parameters (see Table 4), the model has an adjusted R-squared of 15%, a percentage by which the behavior of the dependent variable is explained by the independent variables. Considering the number of explanatory variables, a R-squared of 13.5% is obtained. The Durbin-Watson (DW) test results in a 1.531.

Table 5 shows the coefficients of the model. Also, it can be verified that the values of the Variance Inflation Factor (VIF) are lower than the critical figure of 10, so it is considered that there are no problems of multicollinearity. Thus, the main requirements for accepting the model (normality, linearity, homoscedasticity, and absence of multicollinearity and autocorrelation) are met.

Table 2 Descriptive statistics ($n = 634$)

	Minimal	Maximum	Mean	S.D.
FP	−0,231,367	0,353,369	0,0370244	0,056877
YEAR	2018	2022	2019	1,411,010
IND	3	65	32,72	18,395,963
SIZE	20,196,354	29,044257	24,583,623	1,559,252
RISK	0	23,258,333	1,332,214	1,750,809
GRW	−2,085,929	1580,566,359	19,766,953	93,661,389
CO_2	4,276,666	19,0561988	13,877,156	2,879,375
ESG	0,273,146	0,9,550,250	0,778,007	0,108,046
EEP	0	1	0,639,694	0,114,368

Table 3 Bivariate correlations

	YEAR	IND	SIZE	RISK	GRW	CO_2	ESG	EEP
YEAR	1	–	–	–	–	–	–	–
IND	−0,079	1	–	–	–	–	–	–
SIZE	−0,054	0,190	1	–	–	–	–	–
RISK	0,090	0,045*	−0,230	1	–	–	–	–
GRW	−0,038*	−0,083	−0,104	0,035*	1	–	–	–
CO_2	−0,133	−0,120	0,031*	0,046*	−0,112	1	–	–
ESG	0,227	−0,120	−0,273	0,157	−0,022*	−0,147	1	–
EEP	−0,200	−0,028*	−0,028*	−0,124	0,089	−0,020*	−0,253	1

*The correlation is significant at the 0.05 level (two-sided)
**The correlation is significant at the 0.01 level (two-sided)

Table 4 Model summary (dependent variable, FP; $n = 449$)

Model	R	R-squared	Adjusted R-squared	Standard Estimation Error	Durbin-Watson
1	0,387	0,150	0,135	0,0481827	1531

In line with previous literature, the economic-financial control variables exert a great weight on the profitability of the company, measured in this case by the dependent variable ROA (Return On Assets). In this analysis, the most significant factors for the profitability of the companies in the sample are "size", measured by the normalized variable of total assets, and "risk" (RISK), measured by the debt ratio. By looking at the sign of their coefficients, it can be concluded that they negatively affect profitability. Therefore, the greater the size and risk, the lower the return; and vice versa.

Regarding the sustainability variables analyzed, it can be observed that the most significant variable is the "level of estimated CO2 emissions" (CO_2), whose significance level is 0.019. In addition, it maintains a negative relationship with the dependent variable since the sign of the coefficient is negative, i.e., the greater the volume of issuances, the lower the profitability; and vice versa.

5 Discussion, Conclusion, and Implications

Considering the results extracted from the analysis of the companies participating in greenhouse gas emissions trading, we can accept one of the hypotheses raised based on the previous literature. First, the first proposed hypothesis (H1: The volume of carbon dioxide emissions has a negative influence on the financial profitability of firms) can be accepted, since the corresponding independent variable (CO_2) has been found to be significant for the profitability-dependent variable and to have a negative relationship with it. Therefore, it is correct to assume that the economic and financial profitability of companies is affected by the volume of emissions. The lower the levels of emissions, the higher the profitability. This supports previous

Table 5 Collinearity coefficients and diagnostics

Model	Non-standardized coefficients		Standardized			Collinearity Statistics	
	B	Desv. Error	Beta	t	p-value	Tolerance	FIV
(constant)	2.501	4317		0,579	0,563		
YEAR	−0,001	0,002	−0,025	−0,542	0,588	0,911	1098
IND	0	0	0,068	1464	0,144	0,894	1119
SIZE	−0,004	0,002	−0,107	−2214	**0,027**	0,824	1213
RISK	−0,009	0,001	−0,325	−7063	**<0.001**	0,914	1094
GRW	−0	0	−0,045	−0,995	0,320	0,962	1040
CO_2	−0,002	0,001	−0,108	−2360	**0,019**	−0,081	−0,112
ESG	−0,005	0,024	−0,010	−0,198	0,843	−0,039	−0,009
EEP	0	0,020	0	0,011	0,992	0,871	1148

literature drawing conclusions about the benefit of better environmental performance relative to carbon dioxide emissions. That is why the best option for companies is to bet on technological innovation to reduce their emissions, thus opting for the innovation route (Demailly & Quirion, 2008) instead of the survival principle (Anger & Oberndorfer, 2008). In this way, companies can trade their surplus emission rights on the secondary market, taking advantage of those that have not opted for the same route and thus obtaining an economic incentive.

As for the second hypothesis (H2: Energy policy positively influences the financial profitability of companies) and third hypothesis (H3: ESG principles score positively influences companies' financial profitability), they cannot be accepted since the independent variables of energy policy (ESP) and ESG score have not been found to be significant. It is interesting that setting an energy policy does not seem to exert an influence on a firm's performance. Although there could be many reasons, perhaps the mere existence of an energy policy does not guarantee an effective reduction in energy consumption or, at least, the cost of energy consumption. The ESG score could not be significant due to their own construction, that is, it includes not only environmental aspects, but also social and governance, and this could distort the results.

However, it would not be appropriate to neglect these two aspects of companies that are becoming increasingly important, not only in decision-making about corporate strategy but also in that of the consumer himself. The growing concern for caring for the planet and the correct interaction between social groups means that the practices of companies are increasingly closely monitored. Likewise, an exhaustive regulatory block is currently being developed that begins to bind all companies (not just those voluntarily adhering to the legislation on the Code of Good Practice), such as the Corporate Sustainability Reporting Directive (CSRD) approved this year 2023 ("Directive (EU) 2022/2464 of the European Parliament and of the Council of 14 December 2022 amending Regulation (EU) No 537/2014, Directive 2004/109/EC, Directive 2006/43/EC and Directive 2013/34/EU as regards sustainability reporting by undertakings").

However, the results of this research could be complemented in future work with the use of other methodologies, such as panel data analysis or even non-parametric analyses, as they would provide more consolidated conclusions. Furthermore, profiling some of the selected variables (such as the ESG score) or adding new environmental variables will be advisable for future analysis, given the new regulations being generated in the climate framework (for example, the carbon border adjustment mechanism, linking all companies that import into the European Union).

References

Ahmed, K. (2017). Revisiting the role of financial development for energy-growth-trade nexus in BRICS economies. *Energy, 128*, 487–495.

Al-Mulali, U., & Ozturk, I. (2015). The effect of energy consumption, urbanization, trade openness, industrial output, and the political stability on the environmental degradation in the MENA (Middle East and north African) region. *Energy, 84*, 382–389.

Anger, N., & Oberndorfer, U. (2008). Firm performance and employment in the EU emissions trading scheme: An empirical assessment for Germany. *Energy Policy, 36*(1), 12–22.

Aupperle, K. E., Carroll, A. B., & Hatfield, J. D. (1985). An empirical examination of the relationship between corporate social responsibility and profitability. *Academy of Management Journal*, 446–463.

Benz, E., & Trück, S. (2006). CO2 emission allowances trading in Europe—Specifying a new class of assets. *Problems and Perspectives in Management*, 30–40.

Chan, H., Li, S., & Zhang, F. (2013). Firm competitiveness and the European Union emissions trading scheme. *Energy Policy, 63*, 1056–1064.

Chang, S. C. (2015). Effects of financial developments and income on energy consumption. *International Review of Economics & Finance, 35*, 28–44.

Chen, Y., Sijm, J., & Hobbs, B. F. (2008). Implications of CO2 emissions trading for short-run electricity market outcomes in Northwest Europe. *Journal of Regulatory Economics*, 251–281.

Clarkson, P. M., Li, Y., Pinnuck, M., & Richardson, G. D. (2015). The valuation relevance of greenhouse gas emissions under the European Union carbon emissions trading scheme. *European Accounting Review, 24*, 551–580.

Çoban, S., & Topcu, M. (2013). The nexus between financial development and energy consumption in the EU: A dynamic panel data analysis. *Energy Economics, 39*, 81–88.

Commins, N., Lyons, S., Schiffbauer, M., & Tol, R. S. (2011). Climate policy and corporate behavior. *The Energy Journal, 32*, 51–68.

Demailly, D., & Quirion, P. (2008). European emission trading scheme and competitiveness: A case study on the iron and steel industry. *Energy Economics, 30*(4), 2009–2027.

Destek, M. A. (2018). Financial development and energy consumption nexus in emerging economies. *Energy Sources, Part B: Economics, Planning, and Policy, 13*(1), 76–81.

Dorfleitner, G., Kreuzer, C., & Sparrer, C. (2020). ESG controversies and controversial ESG: About silent saints and small sinners. *Journal of Asset Management, 21*(5), 393–412.

Duque-Grisales, E., & Aguilera-Caracuel, J. (2021). Environmental, social and governance (ESG) scores and financial performance of multilatinas: Moderating effects of geographic international diversification and financial slack. *Journal of Business Ethics, 168*(2), 315–334.

Fan, L. W., Pan, S. J., Liu, G. Q., & Zhou, P. (2017). Does energy efficiency affect financial performance? Evidence from Chinese energy-intensive firms. *Journal of Cleaner Production, 151*, 53–59.

Garnacho, L. G. (2018). From the beginnings of emissions trading to today. *Working Paper, 2*(2).

Hassine, M. B., & Harrathi, N. (2017). The causal links between economic growth, renewable energy, financial development and foreign trade in gulf cooperation council countries. *International Journal of Energy Economics and Policy, 7*(2), 76–85.
Islam, F., Shahbaz, M., Ahmed, A. U., & Alam, M. M. (2013). Financial development and energy consumption nexus in Malaysia: A multivariate time series analysis. *Economic Modelling, 30*, 435–441.
Khan, M. T. I., Ali, Q., & Ashfaq, M. (2018). The nexus between greenhouse gas emission, electricity production, renewable energy and agriculture in Pakistan. *Renewable Energy, 118*, 437–451.
Khan, M. T. I., Yaseen, M. R., & Ali, Q. (2019). Nexus between financial development, tourism, renewable energy, and greenhouse gas emission in high-income countries: A continent-wise analysis. *Energy Economics, 83*, 293–310.
Kara, M., Syri, S., Lehtilä, A., Helynen, S., Kekkonen, V., Ruska, M., & Forsström, J. (2008). The impacts of EU CO2 emissions trading on electricity markets and electricity consumers in Finland. *Energy Economics*, 193–211.
Lin, B., & Okoye, J. O. (2023). Towards renewable energy generation and low greenhouse gas emission in high-income countries: Performance of financial development and governance. *Renewable Energy, 215*, 118931.
Lise, W., Sijm, J., & Hobbs, B. F. (2010). The impact of the EU ETS on prices, profits and emissions in the power sector: Simulation results with the COMPETES EU20 model. *Environmental and Resource Economics*, 23–44.
López Gordo, J. F. (2006). *The community environment and the Kyoto protocol: The harmonisation of energy taxation or a market on greenhouse gas emissions.*
LSEG (2023). *LSEG ESG scores*. Retrieved on 23 June 2023 from: https://www.lseg.com/en/data-analytics/sustainable-finance/esg-scores
Makridou, G., Doumpos, M., & Galariotis, E. (2019). The financial performance of firms participating in the EU emissions trading scheme. *Energy Policy, 251-252*, 254–258.
Nasreen, S., Anwar, S., & Ozturk, I. (2017). Financial stability, energy consumption and environmental quality: Evidence from south Asian economies. *Renewable and Sustainable Energy Reviews, 67*, 1105–1122.
Oberndorfer, U., & Rennings, K. (2007). Costs and competitiveness effects of the European Union emissions trading scheme. *European Environment, 17*, 1–17.
Oberndorfer, U. (2009). EU emission allowances and the stock market: Evidence from the electricity industry. *Ecological Economics, 68*, 1116–1126.
Ouyang, Y., & Li, P. (2018). On the nexus of financial development, economic growth, and energy consumption in China: New perspective from a GMM panel VAR approach. *Energy Economics, 71*, 238–252.
Ozturk, I., & Acaravci, A. (2013). The long-run and causal analysis of energy, growth, openness and financial development on carbon emissions in Turkey. *Energy Economics, 36*, 262–267.
Sadeghieh, M. (2016). *Financial development, CO2 emissions, fossil fuel consumption and economic growth: The case of Turkey* (Master's thesis, Eastern Mediterranean University (EMU)-Doğu Akdeniz Üniversitesi (DAÜ)).
Sadorsky, P. (2010). The impact of financial development on energy consumption in emerging economies. *Energy Policy, 38*(5), 2528–2535.
Sadorsky, P. (2011). Financial development and energy consumption in central and eastern European frontier economies. *Energy Policy, 39*(2), 999–1006.
Sánchez García, I. Á. (2019). The free allocation of emission allowances and the auctions of allowances in Spain. *Budget and Public Expenditure, 97.*
Shahbaz, M., Tiwari, A. K., & Nasir, M. (2013). The effects of financial development, economic growth, coal consumption and trade openness on CO2 emissions in South Africa. *Energy Policy, 61*, 1452–1459.

Shahbaz, M., Sbia, R., Hamdi, H., & Ozturk, I. (2014). Economic growth, electricity consumption, urbanization and environmental degradation relationship in United Arab Emirates. *Ecological Indicators, 45*, 622–631.

Shahbaz, M., Mallick, H., Mahalik, M. K., & Sadorsky, P. (2016). The role of globalization on the recent evolution of energy demand in India: Implications for sustainable development. *Energy Economics, 55*, 52–68.

Shahbaz, M., Van Hoang, T. H., Mahalik, M. K., & Roubaud, D. (2017). Energy consumption, financial development and economic growth in India: New evidence from a nonlinear and asymmetric analysis. *Energy Economics, 63*, 199–212.

Telle, K. (2006). “it pays to be green” - a premature conclusion? *Environmental and Resource Economics*, 195–220.

Topcu, M., & Payne, J. E. (2017). The financial development–energy consumption nexus revisited. *Energy Sources, Part B: Economics, Planning, and Policy, 12*(9), 822–830.

Veith, S., Werner, J. R., & Zimmermann, J. (2009). Capital market response to emission rights returns: Evidence from the European power sector. *Energy Economics, 31*, 605–613.

Yao, R., Fei, Y., Wang, Z., Yao, X., & Yang, S. (2023). The impact of China’s ETS on corporate green governance based on the perspective of corporate ESG performance. *International Journal of Environmental Research and Public Health, 20*(3), 2292.

Zakari, A., Khan, I., Tan, D., Alvarado, R., & Dagar, V. (2022). Energy efficiency and sustainable development goals (SDGs). *Energy, 239*, 122365.

Zhang, Y.-J., & Wei, Y.-M. (2010). An overview of current research on EU ETS: Evidence from its operating mechanism and economic effect. *Applied Energy, 87*, 1804–1814.

Zhang, Y., Zhang, Y., & Sun, Z. (2023). He impact of carbon emission trading policy on Enterprise ESG performance: Evidence from China. *Sustainability, 10*(15), 8279.

Zhao, B., & Yang, W. (2020). Does financial development influence CO2 emissions? *A Chinese province-level study. Energy, 200*, 117523.

Part III
Sector-Specific Impacts and Strategies

The Impact of Carbon Emission and Energy Factors on Climate Change in BRICS Countries

Vanitha S and N. Thangaiyarkarasi

Abstract The main causes of climate change and global warming are human activities and are primarily due to burning or combustion. Due to the development and improvement of the Industrial Revolution, the burning of fossil fuels has increased carbon emissions. It is one of the main challenges of climate change and global warming in the economic world. The main objective of this study is to examine climate change and its impact on carbon emission and energy factors in BRICS countries for the period of 20 years from 2003 to 2022. The study tests the relationship between carbon emission, GDP, and energy factors with the tools used for applied econometric analysis of Granger causality between all variables. Variables of Energy Use, Electric Power Consumption, Access to Electricity, and Renewable Energy Consumption are considered energy factors. The empirical results show that increasing GDP reduces carbon emissions to a low-carbon economy in the long run. The traditional relationship between per capita CO_2 emissions and per capita GDP. Both indicators appear to have increased from 1990 to 2008, after which CO_2 emissions began to decline in line with the global economic crisis, while GDP per capita continued to rise (Jambor and Balogh, Determinants of CO2 emission: A global evidence, 2018). As a policy implication, developed countries agree to provide finance to developing countries should increase the technology and infrastructure for adaptation and a new renewable energy-based economy.

1 Introduction

In the past decade, climate change has been one of the most essential issues affecting the economic activities in our world. Climate change is a long-term process, it can be done naturally, due to the solar magnetic activity cycle and human activities

V. S (✉) · N. Thangaiyarkarasi
Department of Commerce and Financial Studies, School of Business Studies, Bharathidasan University, Tiruchirappalli, India
e-mail: vanitha@bdu.ac.in

N. Naifar (ed.), *Climate Change and Finance*, Sustainable Finance,
https://doi.org/10.1007/978-3-031-56419-2_11

are the main source of climate change and global warming. Deforestation can create unwanted greenhouse gas emissions and release carbon dioxide into the atmosphere. Human activities are responsible for causing global warming, and climate change, it also includes droughts, fires, water scarcity, floods, and sea levels. The solution to climate change can provide economic activity while improving and protecting our environmental activities.

Our nations have some of the global agreements and frameworks to make progress, such as the Sustainable Development Goals (SDGs), the Paris Agreement, and the United Nations Framework Convention on Climate Change (UNFCCC). Its main challenges are to reduce emissions and modify to change climate impacts and financial requirements. When we human people shift from energy grids to renewables like wind and solar will reduce carbon emission and energetic climate change. But the developing countries have pledged to reduce net-zero emissions by 2050, half of the emission must be cut by 2030 to keep warming well below 1.5°C.

The population of BRICS countries—Brazil, Russia, India, China, and South Africa about 3.2 billion people, seems to be more important for preventing climate disasters. BRICS nations can account have more than half of the total emissions among 43 countries (By Ralph Torrie April 22, 2022). The BRICS Earth Index for 2019 is 95%, indicating that their emissions increased by approximately the amount required to remain on the path for their long-term net-zero commitments. China and India account for 82% of BRICS emissions. Both countries saw increases in all sectors in 2019 and need not expect their emissions to decline until after 2030. They are developing economies, their GDP per capita is 20% of European Union levels and 13% of G7 economies. It is nearly four times faster than in the G7 and EU economies, strengthening the strategic relationship between economic development, energy efficiency, and decarbonization.

Brazil has the suspicious distinction of reality the only major emitting country to increase its reduction emissions at the time of the pandemic. Unsustainable logging of the rainforest increased the country's net emissions by 9.5% in 2020, more than five times the reductions realized in all other sectors in 2019. Russia's economy ranked in the G20's top five in terms of greenhouse gas emissions in 2019, which was a result of its high reliance on oil and gas, poor energy efficiency, and aging infrastructure and technology. The development of solar photovoltaics and batteries is the focus of recent economic stimulus spending in India; however, this is hampered by ongoing investments in both new and old coal-fired electricity. The application of renewable resources, wind power, solar energy, and the process of electronic vehicles is improving faster in China than in any other country in the world. South Africa is a coal country, and the electric sector accounted for 41% of its greenhouse gas emissions in 2019, it is one of the best wind and solar regimes in the economy.

1.1 *The Transformation of the Energy Sector and Related Sustainability Responsibilities*

In accordance with their share of global energy production and consumption, the BRICS countries have a crucial role to play in the global energy agenda. The first BRICS Energy Ministerial was conducted in 2015 by the BRICS Economic Partnership policy that was established during the BRICS Ufa Summit, institutionalizing energy cooperation within BRICS. The BRICS presidents urged further discussion on the creation of a BRICS Energy Research Co-operation Platform (BRICS ERCP) in the Xiamen Declaration of 2017. As a consequence, the BRICS ERCP Platform was created, and this was acknowledged in the Johannesburg Summit Declaration of 2018. The BRICS ERCP is now a global forum for advancing energy-based sustainable development, exchanging cutting-edge energy technologies, increasing collaboration on educational initiatives, exchanging statistical data and plans for the development of national energy systems, and disseminating knowledge of best practices and regulatory frameworks in the energy sector. The platform also aims to encourage collaboration between the Academic Forum, the Program of Economic Co-operation, the New Development Bank, and the BRICS Science, Technology, and Innovative Initiative to promote BRICS energy cooperation.

In this study, an attempt has been made to analyze the impact of carbon emission and energy factors on climate change for the sample of BRICS countries using the variables of CO_2 emissions, GDP per capita, urban population, and energy factors.

The remainder of this study is structured as follows: Section 2 provides a literature review. Section 3 describes the theoretical analysis and hypothesis. The results are discussed in Sect. 4 and Sect. 5 provide the conclusion.

2 Review of Literature

In this study, an attempt has been made to briefly review the literature, which is the work already done by earlier researchers.

Muhammad Shahbaz et al. (2021), in the paper entitled, "Analyzing Nonlinear Impact of Economic Growth Drivers on CO2 Emissions: Designing an SDG Framework for India." India's situation is comparable to that of other Asian countries that are struggling to meet the Sustainable Development Goals (SDGs). This study looks at how per capita income, energy use, trade openness, and oil prices have an impact on CO_2 emissions between 1980 and 2019. The long-term impact of independent variable fluctuations on CO_2 emissions is established as an asymmetrical long-term and the results demonstrate that India's existing economic growth pattern, which depends on imported crude oil and fossil fuel-based energy consumption is environmentally unsustainable.

The paper entitled, “Electricity Consumption, Carbon Emissions and Economic Growth in Nigeria” by Godwin Effiong Akpan and Akpan (2012). The study investigated the long-term and causality relationship between power consumption, carbon emissions, and economic growth in Nigeria using a Multivariate Vector Error Correction (VECM) methodology. The lack of any connection between electricity and growth, in any direction, adds support to the Nigerian electricity dilemma. According to research, economic development over the long term is linked to rising carbon emissions, while rising power usage also raises carbon emissions. These imply that Nigeria's growth process is highly polluting, and the country's electricity consumption has significantly increased carbon emissions, as shown by the negative relationship between electricity consumption and emissions or the positive relationship.

Sahbi Farhani and Rejeb (2012) examined, “Energy Consumption, Economic Growth, and CO2 Emissions: Evidence from Panel Data for MENA Region.” Energy is essential to the growth of the economy. It plays a crucial role in sustainable development. The direction of causality between energy consumption (EC), economic growth (GDP), and CO_2 emissions has thus been the subject of numerous studies 15 MENA nations, including Algeria, Cyprus, Egypt, Iran, Israel, Jordan, Kuwait, Morocco, Oman, Saudi Arabia, Sudan, Syria, Tunisia, Turkey, and the United Arab Emirates, have annual statistics selected to span the years 1973–2008. Energy Consumption (EC) measured in kg of oil equivalent per capita, GDP per capita measured in constant 2000 US dollars, and CO_2 emissions measured in metric tons per capita are the variables considered in this study. The results of this study show that there is no short-term causal relationship between GDP and EC or between CO_2 emissions and EC. However, there is a unidirectional correlation in the long-term connecting GDP and CO_2 emissions to EC. The FMOLS and DOLS approaches are used in this study to assess the long-term association between the three variables.

The research paper entitled “The Role of economic growth and energy consumption on CO2 Emissions in E7 Countries,” by Doğan & Değer, (2018). This study uses panel data analysis to examine the relationships between carbon emissions, financial development, overall energy consumption, and economic growth in the E7 (Brazil, China, Indonesia, India, Mexico, Russia, and Turkey) countries between 1990 and 2014. when utilized Data from the BP World Energy Statistics database is used to calculate total energy consumption as the sum of energy from oil, natural gas, coal, and water. According to the panel data analysis, the author found, there is no consistent link between financial development and carbon emissions over the long term.

Lakhi Narayan Bharadwaj, (2020) in the paper entitled, “The Nexus Among Foreign Direct Investment, Economic Growth, And Carbon Emissions: Evidence from India.” The author used secondary data for India over the period from 1980 to 2017. The primary goal of the study is to examine how economic growth and foreign direct investment affect India's carbon emissions and analyze the nature of the link between the variables of Real GDP Per Capita, FDI Inflows, and CO_2 Emission in India. The parameters are Real GDP per capita (2010 US dollars), Energy use,

CO_2 emissions, and Net FDI inflow. The results of this analysis show that there is a substantial nonlinear positive association between GDP per capita and CO_2 emissions for both FDI inflows and FDI outflows. This investigation also supports the notion that there is pollution in this area.

Wang and Xu (2021), in an article entitled, "Internet Usage, Human Capital and Co2 Emissions: A global perspective." This study employs system GMM and a threshold regression model to empirically investigate the relationship between internet usage, human capital, and CO_2 emissions under various degrees of economic development using panel data from 70 countries from 1995 to 2018. The findings demonstrate that human capital and Internet use are key factors in the growth of a low-carbon economy and that human capital can control the negative effects of internet use on CO_2 emissions.

The study examined, "The impact of foreign direct investment, tourism electricity consumption, and economic development on CO2 emissions in Bangladesh," by Rahaman et al. (2022). The purpose of the study is to look into how economic growth, tourism, power use, and foreign direct investment (FDI) have affected CO_2 emissions in Bangladesh between 1990 and 2019. Environmental degradation is used as a substitute for CO_2 emissions, while GDP is used as a substitute for economic progress. This shows that the relationship between CO_2 emissions and economic development in Bangladesh is U-shaped inverted. The EKC postulate is validated as being true as a result. The short-term effects of tourism, GDP2, economic growth, and electricity consumption on CO_2 emissions are negligible.

3 Data and Methodology

The design of the study included the statement of the problem. Need of the study, Hypothesis of the study, the objective of the study, and methodology of the study.

3.1 Data Source

The study proposes to analyze the relationship between energy use or energy factors and carbon emissions, as well as how it affects the BRICS nations' economic development, is a complicated and urgent problem that necessitates careful consideration. Highlighting the effects of carbon emissions and energy consumption is essential for sustainable development as these emerging economies seek vigorous economic growth. Determine the degree to which rising carbon emissions affect economic growth trajectories, analyze the impact of energy factors like consumption, efficiency, and the adoption of renewable energy sources on economic growth, evaluate the policy ramifications and trade-offs related to balancing growth goals with emission reduction targets, and investigate the potential of technological advancements and innovation to mitigate these issues. Investigating these issues will give policymakers and stakeholders important insights into how to promote environmentally responsible and flourishing economies within the BRICS countries.

The Need for Study was to identify the role of energy factors, such as consumption patterns, energy efficiency, and renewable energy adoption, which can inform policy decisions for sustainable growth. Studying this relationship can help identify best practices and innovative solutions to balance economic growth with environmental sustainability. The impact of carbon emissions and energy factors on economic growth in BRICS countries is crucial for sustainable development. Climate change concerns and international commitments to reduce carbon emissions necessitate a comprehensive study of the relationship.

The objectives of the present study are proposed to examine the relationship between carbon emission and energy factors on economic growth in BRICS countries and to analyze the impact of energy factors and measures to reduce carbon dioxide emission in BRICS countries. The present study tested the following null hypothesis: H1: The impact of energy factors with carbon emission on economic growth in BRICS countries and H2: The energy factors can help to reduce carbon emission and to help sustainability development.

The BRICS Countries were selected as a sample for this study. The sample was selected for the BRICS countries with the websites of https://brics2021.gov.in/index. The sample selection of BRICS Countries—Brazil, Russia, India, China, and South Africa. The Study covers a period of 20 years from 2003 to 2022 and the Tools used for Analysis are Descriptive Statistics, Correlation, Unit Root Tests, and Granger Causality Tests. The main source of information for this study is secondary data. *Data are from the World Bank and* the data were collected from www.worldbankindicators.com. The sample variables are shown in the Table 1.

3.2 Econometric Model

The Granger causality test, initially developed in 1969, is a statistical hypothesis test for detecting if a one-time series helps predict another. Regressions generally show "mere" correlations, but according to Clive Granger, one way to test for causation in economics is to measure how well one-time series can predict future values

Table 1 Variables description

S. No.	Variables	Role	Description	Source
1	CO_2	Response variables	CO_2 emissions (metric tons per capita)	WDI
2	EU	Explanatory variables	Energy use	WDI
3	UP	Explanatory variables	Urban population	WDI
4	EPC	Control variables	Electric power consumption (kWh per capita)	WDI
5	GDP	Control variables	GDP (constant 2015 US$)	WDI
6	AE	Control variables	Access to electricity	WDI
7	REC	Control variables	Renewable energy consumption	WDI

Source: Data collected from www.worldindicators.com

using data from a different time series. The Granger causality test determines whether *X* forecasts *Y* rather than determining whether *X* causes *Y*.

A statistical theory known as Granger causality is utilized to establish the causal connection between two-time series variables. The Granger causality test determines whether a variable's previous values can offer additional useful information beyond those of its past values in predicting a different variable. The following is a representation of the Granger causality test formula: The null hypothesis (H0) for the Granger causality test asserts that *X* does not Granger cause *Y* given two variables, *X* and *Y*. In other words, it is difficult to forecast *Y* based on the historical values of *X*. According to the alternative theory (H1), *X* does Granger cause *Y*. Let the time series for *y* and *x* be stationary. Finding the appropriate lagged values of *y* to include in a univariate autoregression of *y* is the first step in testing the null hypothesis that *x* does not Granger cause *y*:

$$y_{t=} a_{0+} a_1 y_{t-1} + a_2 y_{t-2} + \ldots + a_m y_{t-m} + \text{error}_t$$

Next, lagged values of *x* are added to the autoregression:

$$y_{t=} a_{0+} a_1 y_{t-1} + a_2 y_{t-2} + \ldots + a_m y_{t-m} + b_p x_{t-p} + \ldots + b_q x_{t-q} + \text{error}_t$$

All lagged *x* values that are individually significant based on their t-statistics are kept in the regression, provided that they collectively add explanatory power to the regression based on an F-test (whose null hypothesis is that there is no explanatory power jointly added by the *x*s). In the notation of the aforementioned augmented regression, p is the lag length for which the lagged value of *x* is significant which is the smallest, and q is the longest.

If and only if no lag values of *x* are kept in the regression, the null hypothesis that *x* does not Granger because y is not rejected.

4 Results and Analysis

Table 2 revealed the results of the descriptive statistics. The mean value for the series shows that GDP has the highest value of 3.5252 when compared to AE, EPC, REC, UP, EU, and CO_2. The median of GDP has the highest value at 3.7960. The minimum value of CO_2 revealed at −0.0431, shows that negative one, and the maximum value of GDP revealed at 4.0488. Additionally, looking at the values for the standard deviation, variables of EPC and EU were higher than 1.000 to indicate that the values are closer to the mean values. All variables of skewness were negatively skewed in distribution.

Table 3 presents the results of correlation matrix. The variables of AE, EPC, UP, EU, and GDP in these countries are positively correlated and the other variables are negatively correlated. Here, the values of CO_2 and REC are negatively correlated. It indicates that one variable increases and the other variables are also increases. The

Table 2 Results of descriptive statistics

Variables	AE	EPC	REC	UP	EU	CO_2	GDP
Mean	1.7655	2.0344	1.0099	1.6734	1.9388	0.5377	3.5252
Median	1.9898	2.8680	1.0739	1.7949	2.7767	0.5660	3.7960
Maximum	2.0000	3.8207	1.6895	1.9411	3.7132	1.0750	4.0488
Minimum	0.0000	0.0000	0.0000	0.0000	0.0000	−0.0431	0.0000
Std. dev.	0.5936	1.6929	0.5709	0.4138	1.6114	0.4039	0.8699
Skewness	−2.6323	−0.3294	−0.5210	−3.2995	−0.3338	−0.0716	−3.3433
Kurtosis	8.0009	1.1992	2.0501	13.6926	1.2042	1.3670	13.8001
Jarque-Bera	219.6854	15.3204	8.2837	657.8305	15.2942	11.1973	672.3082
Probability	0.0000	0.0005	0.0159	0.0000	0.0005	0.0037	0.0000

Source: Data collected from www.worldindicators.com and Computed from E Views
Source: https://databank.worldbank.org/source/world-development-indicators
Note: *AE* denotes Access to Electricity, *EPC* denotes Electric Power Consumption, *REC* denotes Renewable Energy Consumption, *UP* denotes Urban Population, *EU* denotes Energy Use, CO_2 denotes Carbon Emission and *GDP* denotes Gross Domestic Product

Table 3 Results of correlation matrix

Variables	AE	EPC	REC	UP	EU	CO_2	GDP
AE	1						
EPC	0.3920	1					
REC	0.5692	0.3255	1				
UP	0.6430	0.2535	0.2884	1			
EU	0.3917	0.9994	0.3208	0.2481	1		
CO_2	0.4705	0.3449	−0.0886	0.4096	0.3441	1	
GDP	0.6432	0.2405	0.2520	0.9911	0.2352	0.4592	1

Source: Data collected from www.worldindicators.com and Computed from E Views

variable of CO_2 emission, compared to other variables AE, EPC, UP, EU, and GDP are positively correlated at 0.3255, 0.2535, 0.2481, 0.3441, and 0.2352, respectively.

Table 4 exhibits the results of the panel unit root test namely ADF (Augmented Dickey-Fuller) and (Phillips-Perron) Fisher Chi-Square statistics during the study period for sample GDP per capita, Carbon Emission, Energy Use, Urban Population, Electric Power Consumption, Access to Electricity, and Renewable Energy Consumption were used in the study. For analysis, the unit root test covering ADF test statistical probability and PP test statistical probability were used to understand the stationarity of the sample variable. It is to be noted that the statistical values of PP test the variables GDP in this case has the lowest *P*-value 0.0000, which is a strong indication that the null hypothesis is rejected. Similarly, the EU and AE variables also exhibit low p-values of 0.0029 and 0.0002, respectively, providing evidence for their stationarity.

Table 5 revealed the results of the Granger causality test shows the null hypothesis says that there is no Granger causation between the variables under investigation based on the findings. To evaluate the significance of the causation, the F-statistic and probability values are given. The test results show that there is no

Table 4 Results of unit root test

Variables	Augmented Dickey-Fuller Test		Phillips-Perron Test	
	t-Statistic	Prob.*	t-Statistic	Prob.*
GDP	−7.1033	0.000	−7.3489	0.000
CE	−2.7671	0.0667	−2.8473	0.0554
EU	−3.9196	0.0029	−3.364	0.0146
UP	−7.939	0.000	8.0465	0.000
EPC	3.8705	0.0034	−3.3567	0.0149
AE	6.8896	0.000	−4.7305	0.0002
REC	3.8197	0.0166	3.1403	0.0268

Source: Data collected from www.worldindicators.com and Computed from E Views

Table 5 Results of granger causality tests for Brazil

Null hypothesis:	Obs	F-Statistics	Prob.
GDP does not Granger Cause CO_2	18	0.0624	0.9397
CO2 does not Granger Cause GDP		3.0670	0.084
UP does not Granger Cause the EU	18	2.0088	0.1737
EU does not Granger Cause UP		0.4437	0.6510
EPC does not Granger Cause AE	18	0.7129	0.5084
AE does not Granger Cause EPC		1.3590	0.2911
REC does not Granger Cause GDP	18	2.1479	0.1594
GDP does not Granger Cause REC		0.8052	0.4697

Source: Data collected from www.worldindicators.com and Computed from E Views

causal connection between GDP and CO_2 emissions because there is a low F-statistic of 0.0624 and a high probability value of 0.9397. Similarly, the test's relatively low F-statistic 3.0670 and probability value of 0.084 shows that CO_2 emissions are not Granger's cause of GDP. According to the F-statistics and probability values, the tests performed for other pairs of variables, such as UP and EU, EPC and AE, and REC and GDP, similarly do not support any strong Granger cause and effect between the variables.

Table 6 presents the Granger Causality Tests. The probability value of CE at 0.0038, it is carbon emission can granger cause GDP since the p-value is less than 0.05. but GDP does not granger cause CE, so there is a unidirectional relationship between the variables. The F-statistic result of 9.1936 with a probability of 0.0038 shows that CE does have a substantial causal effect on GDP. Granger Cause is not UP with EU: According to the F-statistic value of 1.8440 with a probability of 0.1972, the EU is not significantly impacted by UP. The EU rejects the Granger Cause UP: According to the F-statistic value of 0.4197 and probability 0.6658, there is no clear causal relationship between the EU and UP. EPC fails to prove Granger Cause AE: The F-statistic value of 0.8208 with a probability of 0.4616 indicates that there is no clear causal relationship between AE and EPC. According to the F-statistic value of 1.3080 and likelihood of 0.3037, AE does not significantly affect EPC causally. Granger Cause in REC GDP. The F-statistic, which has a value of

Table 6 Results of granger causality tests for Russia

Null hypothesis	Obs	F-Statistic	Prob.
GDP does not Granger Cause CE	18	1.3135	0.3049
CE does not Granger Cause GDP		9.1936	0.0038
UP does not Granger Cause the EU	18	1.8440	0.1972
EU does not Granger Cause UP		0.4197	0.6658
EPC does not Granger Cause AE	18	0.8208	0.4616
AE does not Granger Cause EPC		1.3080	0.3037
REC does not Granger Cause GDP	18	106.015	9.0009
GDP does not Granger Cause REC		0.9250	0.4211

Source: Data collected from www.worldindicators.com and Computed from E Views

Table 7 Results of granger causality for India

Null hypothesis	Obs	F-Statistic	Prob.
GDP does not Granger Cause CE	18	0.5765	0.5767
CE does not Granger Cause GDP		1.9523	0.1845
UP does not Granger Cause the EU	18	3.4268	0.0638
EU does not Granger Cause UP		0.5770	0.5753
EPC does not Granger Cause AE	18	1.4657	0.2694
AE does not Granger Cause EPC		0.0435	0.9576
REC does not Granger Cause GDP	18	1.5721	0.2475
GDP does not Granger Cause REC		0.3814	0.6909

Source: Data collected from www.worldindicators.com and Computed from E Views

106.015 and a likelihood of 9.0009, there is a substantial causal relationship between REC and GDP. GDP does not cause inequality REC. The F-statistic value of 0.92503 and probability of 0.4211 shows that GDP does not significantly affect REC causally. Russia has a low level of energy efficiency, and also antiquated infrastructure and technology. In a BRICS country, Russia currently appears restricted to a significant and sustained economic fall, with an uncertain impact on the future of the nation's response to the effects of climate change (By Ralph Torrie April 22, 2022).

Table 7 shows the results of Granger causality tests for India. A probability of 0.5767 and an F-statistic of 0.5765 indicates that GDP does not granger cause directly with CE. The F-statistic and probability values of GDP, CE, UP, EU, EPC, AE, and REC all variables are greater than the significant level of 0.05. So, there is no relationship between the variables. When the probability values of urban population and energy use are 0.0638 and 0.5753, both variables do not granger cause relation, since the *p*-value is greater than 0.05 significant level. In India, the power sector is 80% dependent on energy sources, renewable energy, and both agricultural and industrial sectors.

Table 8 presents the Granger Causality Tests for China Country, if one variable does not Granger cause the other, the null hypothesis for each pair of variables is whether it occurs or not. The table gives the F-statistic, the number of observations, and the probability related to each test. At a significance level of 0.05, it is not

Table 8 Results of granger causality for China

Null hypothesis	Obs	F-Statistic	Prob.
GDP does not Granger Cause CE	18	1.3895	0.2865
CE does not Granger Cause GDP		2.9761	0.0892
UP does not Granger Cause the EU	18	1.8765	0.1954
EU does not Granger Cause UP		0.3147	0.7358
EPC does not Granger Cause AE	18	0.3652	0.7015
AE does not Granger Cause EPC		1.9759	0.1812
REC does not Granger Cause GDP	18	26.5685	0.6645
GDP does not Granger Cause REC		1.6599	0.2331

Source: Data collected from www.worldindicators.com and Computed from E Views

Table 9 Results of granger causality tests for South Africa

Null hypothesis	Obs	F-Statistic	Prob.
GDP does not Granger Cause CE	17	1.60893	0.2404
CE does not Granger Cause GDP		2.06427	0.1696
UP does not Granger Cause the EU	17	2.30034	0.1427
EU does not Granger Cause UP		0.61482	0.5569
EPC does not Granger Cause AE	17	0.76878	0.4835
AE does not Granger Cause EPC		2.04095	0.1695
REC does not Granger Cause GDP	17	0.42294	0.6645
GDP does not Granger Cause REC		1.64825	0.2331

Source: Data collected from www.worldindicators.com and Computed from E Views

possible to rule out the null hypothesis that GDP does not directly impact CE because the *p*-value is 0.2865. The *p*-value is lower at 0.0892. The probability value of variables is not granger because of their other variables, so there is no relationship between all variables. The value of Carbon Emissions (CE) is revealed at 0.0892 was a lower value than other variables. So, it is the world's leader in the use of renewable energy and the electrification of cars, but it is also the country with the fastest-growing use of coal-fired power.

Table 9 reveals the results of Granger Causality Tests for South Africa Country. With the *p*-value of 0.2404, the test results for the relationship between GDP and CE show that CE is not Granger caused by GDP, indicating that changes in GDP have little impact on CE. A *p*-value of 0.1696 indicates that the test between CE and GDP does not demonstrate that CE causes GDP. The test indicates that, for the relationship between UP and EU. EU does not Granger cause UP, and vice versa, both with *p*-values greater than 0.05 significant level. Renewable energy resources exist in South Africa, it provides one of the world's most significant solar regimes. The Renewable Energy Independent Power Producer Procurement Programme will help to improve and reduce carbon emissions.

Further research may prefer to add various variables to our analyses with various countries or regions. The variables like CO_2 emissions from other sectors excluding residential buildings and commercial and public services (% of total fuel combustion), CO_2 intensity (kg per kg of oil equivalent energy use), etc. used to analyze further studies, that will help to develop emerging energy technologies.

5 Conclusion

The study concluded that increased investment in energy infrastructure is the key factor in reducing the low-carbon economy for BRICS nations as a whole. In addition to increasing power efficiency, this will increase the capacity for producing electricity to keep up with demand. As a result, there will be an increase in power generation and a decrease in CO_2 emissions. Individually designed strategies will be generally welcomed in light of the overarching objectives of enhancing and sustainable economic growth and development, energy security, and climate change prevention.

The NDC commitments made by the government are:

- Brazil promises to cut its emissions of greenhouse gases by 37% in 2025 compared to 2005 and by 43% in 2030 compared to that year. The Brazilian government declared its intention to achieve climate neutrality (net-zero emissions) in 2050 during the Summit of Leaders on Climate, moving the target forward by 10 years from the revision of the Nationally Determined Contribution that was announced in 2020.
- Russia made its first Nationally Determined Contribution (NDC) is expected to be sustainable and balanced, and it is specifically planned to reduce greenhouse gas emissions by 70% by 2030 compared to the level of 1990, taking into account the maximum absorption capacity of forests and other ecosystems.
- India committed to achieving the goals by the year 2030 in its Nationally Determined Contributions (NDCs). To decrease the GDP's emissions intensity by 33%–45% below 2005 levels by 2030 and to increase the share of non-fossil fuels to 40% of installed electric power capacity by 2030, with assistance from technology transfer and low-cost international financing, including from the Green Climate Fund (GCF).
- China continues to work toward implementing the Paris Agreement and has made significant contributions to its adoption. China wants to reach carbon neutrality before 2060 and reach a peak in carbon dioxide emissions by 2030.
- In October 2015, South Africa submitted its first NDC to the UNFCCC, pledging to limit its country's greenhouse gas emissions to a range between 389 Mt. CO_2-eq and 2050.

This chapter analyzed a Granger causality test, the results revealed that all the variables in the BRICS country had positive values only. It suggests that the relationship between their variables is interrelated to each other. Today, our economic world adopts modern innovation and new technologies like people using digitalization, electrification for all power sectors, industrialization, and renewables from coal, oil, and gas to renewables sectors like solar, wind, and biofuels. The BRICS countries establishing the conceptual framework of new business models and technologies that will be required to sustain the gradual process of decarbonization in the next decades have been strengthened by government action. This will help to achieve the targets, goals, and objectives of each member nation, who are taking responsibility for NDC commitments and to reach net-zero emissions.

References

Akpan, G. E., & Akpan, U. F. (2012). Electricity consumption, carbon emissions, and economic growth in Nigeria. *International Journal of Energy Economics and Policy, 2*(4), 292–306.

Bharadwaj, L. N. (2020). The nexus among foreign direct investment, economic growth and carbon emissions: Evidence from India.

Doğan, B., & Değer, O. (2018). The role of economic growth and energy consumption on CO 2 emissions in E7 countries. *Theoretical & Applied Economics, 25*(2).

Farhani, S., & Rejeb, J. B. (2012). Energy consumption, economic growth, and CO2 emissions: Evidence from panel data for MENA region. *International Journal of Energy Economics and Policy, 2*(2), 71–81.

https://www.corporateknights.com/rankings/earth-index/2022-earth-index/earth-index-brics-nations/

Jambor, A., & Balogh, J. M. (2018, November). *Determinants of CO2 emission: A global evidence.*

Rahaman, M. A., Hossain, M. A., & Chen, S. (2022). The impact of foreign direct investment, tourism, electricity consumption, and economic development on CO2 emissions in Bangladesh. *Environmental Science and Pollution Research, 29*(25), 37344–37358.

Ralph Torrie April 22, 2022. https://www.corporateknights.com/rankings/earth-index/2022-earth-index/earth-index-brics-nations/

Shahbaz, M., Sharma, R., Sinha, A., & Jiao, Z. (2021). Analyzing nonlinear impact of economic growth drivers on CO2 emissions: Designing an SDG framework for India. *Energy Policy, 148*, 111965.

Wang, J., & Xu, Y. (2021). Internet usage, human capital and CO2 emissions: A global perspective. *Sustainability, 13*(15), 8268.

A Nexus Between Climate Change Risk and Financial Performance of Agricultural Businesses in Developing Economy

Muhammad Sheeraz, Nadeem Iqbal, and Muhammad Sajid

Abstract Considering the increasing systemic risks, it is crucial to analyze climate-related concerns' impact on agricultural firms' financial performance, especially in developing countries. The research sought to examine the possible influence of climate change risks on the financial viability of publicly listed agricultural companies in Vietnam. Previous research has mostly used statistical and benchmarking methods, focusing on retrospective questionnaire surveys and corporate environmental reports. The methods used resulted in linear, probabilistic, and static understandings of how climate change affects company operations and economic performance. Nevertheless, they fail to adequately capture the intricate long-term impacts of climate change on the financial performance of corporations. Our research aims to enhance the existing body of knowledge by constructing a dynamic model that investigates the changing connections among climate change threats, financial performance, and operational processes. The project aims to develop a comprehensive model that will improve the understanding of climate change and its impact on company performance for managers and academics. Our model is based on corporate climate change management, systems thinking, and system dynamics. It leverages Stella software as its fundamental framework. This comprehensive approach will provide useful insights into the complex relationship between climate change and the financial performance of agribusiness enterprises in Vietnam.

M. Sheeraz
Department of Agricultural Business and Marketing, FAS&T, Bahauddin Zakariya University, Multan, Pakistan

N. Iqbal (✉)
Faculty of Management and Communication Studies, National Skills University, Islamabad, Pakistan

M. Sajid
Department of Plant Pathology, FAS&T, Bahauddin Zakariya University, Multan, Pakistan

N. Naifar (ed.), *Climate Change and Finance*, Sustainable Finance,
https://doi.org/10.1007/978-3-031-56419-2_12

1 Introduction

Climate change has a substantial impact on several elements, greatly impeding the sustainable development of countries. The effect includes the intensification of temperatures, changes in patterns of rainfall, the rise in sea levels, the intrusion of saltwater, and a higher likelihood of severe weather events including extreme droughts and floods (IPCC, 2022). Moreover, the increasing concentration of carbon dioxide and other greenhouse gas emissions will only worsen this worldwide problem in the forthcoming millennium (Bhaktikul et al., 2021). Given the widespread recognition of climate change as one of the most urgent issues of the century, policymakers are confronted with an urgent need to implement significant measures. Individuals living in economically poor nations are more susceptible to the consequences of natural calamities and climate change as a result of restricted resources, insufficient infrastructure, and unstable government systems. The intrinsic susceptibility of these individuals, arising from their restricted ability to adjust to the difficulties presented by climate change, might lead to intensified inequities in health and diminished availability of crucial resources like food and clean water. As a result, these constraints impede endeavors focused on alleviating poverty.

Climate-related hazards, including natural calamities and unfavorable meteorological circumstances, have a significant impact on the financial outcomes of organizations. The IPCC (2022) highlights many sectors that are susceptible to climate change, such as agricultural activities, aquifers and water reserves, shoreline areas, aquatic ecosystems, medical systems, and power generation. Within these many industries, the agricultural sector has significant negative impacts as a result of climate change. These include rising production expenses, shrinking areas of cultivable land, lower crop and animal outputs, and degraded product quality. Prior research has examined the consequences of climate change in a number of countries, including both high- and low-income ones (Cadman & Sarker, 2022). Natural catastrophes, due to their amplified magnitude and severity, impose substantial economic consequences (Rahman et al., 2022). In addition, scientists have investigated the economic implications of long-term climate changes, namely the effects of increasing temperatures and less precipitation on agricultural productivity (Ali et al., 2023). Nevertheless, previous studies have mostly focused on climate change and its association with the yield of agricultural crop production (Singhania, 2021). Climate change offers several threats to the structure and productivity of industrial processes, which in turn affects firms' financial measures, including revenue, costs, profitability, and financial efficiency. It is important to realize that climate change is now firmly established as a significant risk factor that influences investment choices. This recognition is based on the understanding that financial markets play a crucial role in allowing informed capital allocation decisions by effectively pricing risks, as highlighted by TCFD (2017). This heightened awareness is not limited to financial stakeholders but also encompasses strategic investors. Therefore, it is necessary to carefully examine firm records in order to analyze the impact of climate change on financial performance and provide guidance for investment plans.

The frequency of climate change risk assessments has surged in recent years. In 2010, recognizing the critical need for improved reporting of climate-related financial information, the Financial Stability Board established the Task Force on Climate-Related Financial Disclosures (TCFD) (Dou & Istraživanja, 2022). Under the auspices of TCFD guidelines, publicly traded firms have the mandate to divulge climate change concerns. Despite the paramount significance of this requirement, research into the impact of climate risks on the financial performance of agricultural enterprises, particularly those listed on public markets, remains scarce (Kasperzak et al., 2023). The study fills this gap by meticulously examining how climate change exerts its influence on the financial outcomes of publicly traded agricultural enterprises. In doing so, our research significantly contributes to the existing body of knowledge in this domain.

The study is first in a nature to investigate the effect of climate change on agriculture in Vietnam, a country where agriculture holds high importance in driving economic development (Wang et al., 2021). Vietnam, characterized by high population density in susceptible coastal and low-lying delta regions, grapples persistently with the threat of natural disasters (Williams & Love, 2022). As is common in other developing nations, Vietnam faces substantial challenges arising from natural disasters and climate change in its pursuit of goals such as eradicating hunger, alleviating poverty, and achieving the Sustainable Development Goals (SDGs). The agricultural sector in Asia remains highly vulnerable to the changing climate, a trend underscored by the IPCC (2022), and Vietnam is no exception (Umar & Safi, 2023). Forecasts suggest that Vietnam's GDP could experience a decline ranging from 0.7% to 2.4% by 2050, primarily attributable to direct impacts on crop output and indirect effects on irrigation, water availability, and potential evapotranspiration, as reported by the World Bank in 2010. It is crucial to emphasize that these consequences stem from both natural disasters and climate change. An important peril is the incursion of saltwater, which is expected to impact 1 million hectares, or nearly 70% of cultivated land in Vietnam's coastal districts by 2030. This is primarily caused by an upsurge in the delta of the Mekong River. The UNDP noted in 2007 that this delta is one of the three most vulnerable in the world, along with the Nile (Egypt) and the Ganges (Bangladesh and India). Therefore, Vietnam serves as a compelling case study.

In order to investigate the climate change consequences in publicly traded agribusinesses located in Southeast Asia (Vietnam), we have selected key climate variables as proxies. These variables include temperature, wind, rainfall, and humidity, all of which significantly influence crop growth and agricultural productivity. Crops are highly sensitive to fluctuations in these climate parameters, a fact supported by research on the JADEE climate system (Sadiq et al., 2022). Temperature shocks, for instance, have been shown to exert detrimental effects on crop output, total factor productivity, and labor productivity. Similarly, the influence of precipitation on crops is a critical factor in farming (Tamasiga et al., 2022). The increasing precipitation levels are a prominent marker of climate change, primarily driven by the rising average global temperature. This elevation in temperature leads to longer rain cycles, as warmer air can retain more moisture (Nocca et al., 2023). Consequently, these altered rain patterns have the potential to result in both severe flooding and drought scenarios (Ho & Shimada, 2019). Moreover, humidity, closely tied to temperature and precipitation, also plays a significant role in the context of climate change. Lastly, it is worth noting that wind patterns can impact plant physiological processes (Alam et al., 2022). These selected climate variables collectively contribute to our understanding of how climate change affects the agricultural sector in Vietnam.

In order to fill this knowledge vacuum, our research will create a dynamic model to analyze the interplay between operational processes, financial performance, and climate change concerns in publicly listed Vietnamese agricultural enterprises. Research on climate change risk and financial performance in public-listed agriculture enterprises in developing nations is crucial. The climate change impact on poor nations' farming practices highlights the necessity for immediate mitigation initiatives (Saeed et al., 2023). Climate change risk reporting is insufficient, indicating a gap in understanding and mitigating climate-related hazards (Abhayawansa & Adams, 2021). High weather unpredictability and climate hazards have hurt firms in very climate-vulnerable regions, suggesting a relationship between climate risk and financial performance (Bae et al., 2021).

Additionally, carbon emissions and their direct influence on climate change have been highlighted as negative economic impacts on enterprises (Makan & Kabra, 2021). Climate risk also leads to unfavorable corporate financing terms for bank loans, such as higher interest paid and more covenant constraints, which may affect financial performance (Huang et al., 2022). The research emphasizes the necessity to understand how climate change affects the financial industry and financial risks (Krasnova et al., 2021). Supporting the transition to a sustainable economic model may help the finance sector and combat climate change (Mattera & Soto, 2022).

The literature highlights the need for more research in order to comprehensively comprehend the correlation between climate change risk and financial performance in publicly traded agribusiness companies in developing countries. The research topic is of significant importance because of the gaps in risk reporting, the adverse economic impacts of climate change, and the potential financial business solutions. We examine the effects of climate change on the financial performance of different companies by using OLS regression and subsample analysis. We examined 77 distinct companies that were listed on both the HSX and the HNX between 2010 and 2022. Wind was the only climate change component that adversely affected business financial performance. While temperature and precipitation have beneficial impacts, humidity yielded discernible influence.

The subsequent sections of the chapters are structured in the following manner: The literature review is given in Sect. 2. Section 3 introduces the assumptions. The research methodologies are extensively discussed in Chapter "Investor Attitudes and Preferences Toward Climate Change and Sustainable Investments: A Bibliometric Analysis". The details are elucidated in Sect. 5. Section 6 is dedicated to the dissemination and discussion of our research results. The concluding portion includes the closing remarks and any relevant ramifications.

2 Literature Review

2.1 Agricultural Climate Change Sensitivity

On a global scale, both governments and businesses harbor significant concerns regarding the looming threat of climate change. Alarming observations indicate that climate change predictions have materialized more rapidly than anticipated (UNISDR, 2019). It is increasingly evident that human-induced climate change is exerting its influence, manifesting in continuing alterations in average heat and rainfall norms, and an upsurge in the frequency of severe climatic events (Leitold et al., 2021). These extreme climate events have the potential to significantly impact agricultural, forestry, and fisheries productivity, encompassing factors such as heightened flooding, prolonged flood seasons, sea-level rise-induced inundation, and alterations in salinity levels (Thuc Huong GIANG et al., 2021). Although there are exceptions in certain regions and for specific crops, the prevailing trend indicates a negative effect of climate change on global agriculture. Natural catastrophes, which constitute a substantial 35% of the variance in GDP and agricultural output, underscore the economic repercussions (Mngumi et al., 2022). Wheat and maize, for instance, may encounter productivity challenges attributable to climate change by 2030, especially in situations with significant greenhouse gas emissions, this might occur. Rising temperatures, shifts in precipitation patterns, and impacts on irrigation water availability all bear direct or indirect consequences for crop yields. In addition to agriculture, other industries like fishing and aquaculture are highly dependent on water access and the availability of coastal resources, face heightened vulnerability to climate change. Elevated temperatures can lead to the depletion of aquatic plants and organic compound enrichment in water bodies, causing nocturnal oxygen levels to plummet and impeding the growth of farmed organisms. Furthermore, climate change has the potential to modify the occurrence and severity of forest disruptions, including problems such as pest invasions, the spread of non-native species, wildfires, and storms. As a result, this might have a significant impact on several businesses and their overall economic performance. This chapter explores the many and complex effects of climate change on the performance of businesses.

Globally, governments and businesses are highly worried about the threat posed by climate change. Predictions about climate change have come to pass far more quickly than expected (UNISDR, 2019). There is growing evidence that human-induced climate change is affecting long-term patterns of temperature and precipitation, as well as increasing the occurrence of extreme weather events. Severe climatic events, such as intensified floods, prolonged flood seasons, rising sea levels causing inundation, and changes in salt intrusion, may have a substantial influence on the productivity of agriculture, forestry, and fisheries. Despite climate change, certain regions and crops could do well, but this is not the norm. Climate change may have a negative effect on world agriculture. Natural catastrophes account for 35% of the variation in GDP and agricultural output. According to the high greenhouse gas emission scenario, the productivity of wheat and maize might be impacted by climate change as early as 2030. Rising temperatures, changing precipitation patterns, and effects on irrigation water availability all have an impact on crop yields, either directly or indirectly. Industries like fishing and aquaculture, which rely significantly on water access and coastal resources, are especially susceptible to the impacts of climate change. When temperatures rise, aquatic plants die off or the water becomes fertilized with organic compounds, which causes oxygen levels to drop substantially overnight. The growth of farmed organisms is hampered by low oxygen levels. Climate change may also have an impact on the frequency and intensity of forest disturbances such as insect infestations, invasive species, wildfires, and storms climate change's implications on business performance.

In the last 20 years, the Mekong Delta in Vietnam has become more susceptible to the impacts of climate change (Leitold et al., 2021). Remarkably, this delta serves as the source of 90% of the nation's rice exports and contributes to 54% of the nation's total rice production, underlining its critical role in Vietnam's agricultural landscape (Leitold et al., 2021). However, the rising sea levels have inflicted substantial damage upon the Mekong Delta. Saltwater intrusion, coupled with the looming threat of floods driven by relative sea level rise, has significantly diminished the available arable land for rice cultivation. The ramifications of climate change have compelled a pressing need to reduce the extent of land under cultivation. The nation has recently grappled with prolonged and severe heatwaves and droughts, serving as stark evidence of the escalating severity of climate change.

The Ministry of Agriculture and Rural Development estimated that more than 14,000 agricultural enterprises in 2021 will be hosted by Vietnam. An observable pattern in the agricultural commodities value chain is the growing participation of major companies in allocating funds toward agricultural and rural sectors Furthermore, it is essential to recognize the substantial economic significance of Vietnam's agriculture, forestry, and aquaculture sectors, with a combined export value of $41.2 billion in 2020, as reported by the British Chamber of Commerce in Vietnam (NGUYEN, 2020). Despite this economic vitality, climate change poses a multitude of threats to these industries (Leitold et al., 2021). Climate change affects various aspects of these businesses, including their human capital, revenues, costs, profitability, and overall financial performance.

2.2 *How Climate Change Endangers the Development of Agricultural Businesses?*

Two distinct approaches are commonly employed for assessing climate risks. The first approach involves risk analysis centered on evaluating the likelihood of specific climate hazards, with a primary focus on the nature of the hazard itself rather than its consequences (Leitold et al., 2021). The second approach adopts a broader perspective, utilizing risk analysis to assess the probability of surpassing predetermined criteria associated with desirable or undesirable future states, including critical thresholds. These criteria-driven assessments enable informed decision-making and proactive measures. In the context of climate change's impact on various sectors, both approaches find utility. Despite a wealth of research investigating the repercussions of climate change on various industries, relatively few studies have delved into the direct influence of climate risk on company performance.

Furthermore, it was shown that firms operating in areas with more severe weather conditions tended to hold larger amounts of cash reserves in order to have better financial adaptability. These companies also kept a larger part of long-term debt and provided cash dividends. Additional research examined the effects of climate change on the financial performance of listed mining firms in China, specifically focusing on five different climate risks (Van Khuc et al., 2018). A further study analyzed how climate change influences bank performance in high-income nations by using forest land, CO_2 emissions, and temperature change as indices of climate change (Shahzad et al., 2022). In another research, the Germanwatch Climate Risk Index was used to examine the relationship between climate risk and enterprises' involvement in accrual-based and real profits management across 64 nations (Nasir et al., 2022). The study revealed a positive influence of climate risk on firms' participation in these management practices. Ultimately, some other authors collected information from 71 nations to clarify the manner in which climate sensitivity intensifies the expenses associated with repaying debts (Sun & Drakeman, 2022). This impact occurs both directly and indirectly, since it restricts the ability to get credit. These studies provide insight into the complex correlation between climate risk and company success.

In various national contexts, researchers have explored the financial performance of agribusinesses, often with a focus on specific economic aspects or emerging concerns. For instance, Hai and Minh-Tu (2023) delved into the financial performance of young agribusinesses in India, while Zhu et al. (2022)) investigated Nigerian-listed enterprises. However, studies bridging the gap between climate change and agricultural profitability remain relatively uncommon. Tsai et al. (2022)) demonstrated the favorable and significant impact of the El Niño Southern Oscillation on stock returns within the US food and agriculture sectors, although empirical findings have occasionally faced scrutiny. Notably, Palmer et al. (2019)) quantified the impact of yearly rainfall and temperature variations on agricultural earnings, ultimately determining that climate change has the potential to increase annual profits for agricultural land in the United States by 4%. In a distinct regional focus, Ito and Tourism (2023) employed spatial micro-econometrics to assess climate change's effects on wine companies in Moldavia, Romania, revealing localized impacts on vineyard revenue, suggesting benefits for grape farmers amidst change.

Recent years have witnessed an upsurge in academic investigations into the potential impact of global warming on Vietnam's agricultural sector. Several authors have applied Ricardian analysis to examine how climate change is affecting farmers and agricultural productivity in Vietnam (Zhu et al., 2022). Notably, there have been studies focusing on specific regions or provinces. For example, Raimi et al. (2021) analyzed the methods used by farmers to adapt to drought and the intrusion of saline. Liu et al. (2023a) examined household adaptation in the Vietnamese Mekong

River Delta region. Chinhamu (2022) evaluated the economic impacts of climate change on agriculture in the Northwestern area. These investigations often include climate and household data. Notably, there remains a notable gap in the examination of how climate change might impact the financial outcomes of Vietnam's agro-forestry-fishery industries. Our findings stand poised to capture the attention of managers, decision makers, and investors in these pivotal and climate-vulnerable sectors, given their substantial significance within the Vietnamese economic landscape.

Climate change poses a formidable challenge to businesses, particularly those heavily reliant on environmental conditions for their operations, as underscored by the Intergovernmental Panel on Climate Change (IPCC, 2022). Its disruptive impact extends across multiple dimensions, including the loss of tangible assets, alterations in capital structures, escalated operational costs, adoption of novel production techniques and systems, business interruptions, diminished profitability, and an array of other unfavorable consequences, all of which exert a detrimental influence on enterprises' financial performance, a phenomenon substantiated by (Shin & Choi, 2019). The manifestations of climate change encompass shifts in average temperatures, precipitation patterns, humidity levels, and wind speeds, collectively creating a complex landscape of challenges for businesses to navigate.

Anticipate more precipitation when humidity levels rise throughout spring, leading to a heightened risk of soil erosion and a decrease in available workdays during the growth season due to waterlogging. Increasing humidity leads to the formation of water vapor, fog, and mildew. The economy has significant negative impacts when excessive levels of humidity lead to product spoiling, rancidity, and mould development, all of which result in a deterioration of product quality. The mean annual humidity in the lowlands is around 80%, but it may reach up to 86% in the highlands (MONRE, 2021). Working in hot, humid conditions may have a number of negative health impacts, including dehydration, headaches, strain, and overall discomfort (Ehlers et al., 2021). These outcomes have a significant impact on outdoor workers' productivity and motivation, which in turn has an impact on corporate performance. Farmers and agricultural workers may suffer from the effects of humidity, which may have an adverse effect on a company's productivity. On the other hand, dampness could enhance agricultural productivity. In cotton fields, humidity is positively correlated with both blossom and boll yields (Peradilan, 2020).

The frequency of very rainy days has increased, especially in central Vietnam, and this tendency is seen in the majority of climatic zones (MONRE, 2021). Extreme floods brought on by rainstorms may wipe out whole agricultural fields. Flooding poses a continuous risk to small and medium-sized enterprises (SMEs), especially in developing countries such as Vietnam where the flood protection infrastructure may be incomplete (Wu et al., 2020).

Along with the risk of flooding, irregular precipitation patterns often aggravate protracted droughts that are bad for the aquaculture sector. Rainfall alters the local radiation environment and its impact on plant life in addition to changing the soil's moisture content (Shao et al., 2022) drought is a climate change-related disaster that is most likely to have a large impact on economic activity. Raimi et al. (2021))

assessed the economic damages resulting from floods in Turkey between 2010 and 2022 and found that the agricultural sector was the most severely impacted. When Alessandro et al. (2023)) examined the impact of the devastating floods that occurred in the Veneto region in 2010 on company performance, they discovered that 2 years later, the increase in value generated by the affected businesses was 7% greater than it had been before the disaster. According to Chinhamu (2022), more precipitation in the winter and spring will reduce agricultural income, whereas increased precipitation in the summer and fall will increase it.

Wind damage to forests, farms, and city trees may have serious financial repercussions. In many plant ecosystems, wind has a substantial impact on successional phases and plant regeneration, according to Ehlers et al. (2021)) study. Strong winds may affect agricultural enterprises since they can cause short-term damage to trees. The study conducted by Hudecova and Rajcaniova in 2023 also highlights the occurrence of branch stripping and abrasion of leaves and fruits, which can be exacerbated by the presence of sand or strong winds. Hurricanes or windstorms with powerful winds can impede economic growth and lead to financial losses (Bossman et al., 2023).

Thermometer: Global warming is a significant concern that has a profound impact on several aspects of business and individuals' daily routines. The global mean surface temperature saw a rise of 0.999 degrees Celsius over the first two decades of the twenty-first century (2001–2020) and the same period in the preceding century (1850–1900) (IPCC, 2021). Between 1958 and 2018, Vietnam's annual average temperature increased by 0.89 °C, and between 1986 and 2018, it increased by 0.74 °C nationwide (Gao & Zhang, 2023).

According to the United Nations Food and Agriculture Organization (2011), there are a number of ways that temperature affects farming. As a consequence of fewer freezing days in the north, growing seasons, seed choices, and crop types may need to adjust to changing climatic circumstances and commercial realities. However, a rise in the number of days with temperatures over 258 °C will be beneficial for tropical areas. Higher temperatures would accelerate crop development and shorten the plant's growth cycle. Owing to the heightened demand from the agricultural industry, the likelihood of encountering severe droughts and water scarcity for irrigation purposes is significantly amplified. Climate change may lead to an increase in the reproduction and spread of pests that prefer hotter climes and erratic rainfall patterns. These pests include caterpillars that devour rice and cut ears, black cutworms, bark-boring beetles, and fungus, among others. If evapotranspiration increased and agricultural water shortage intensified, the area covered by hygrophytes would decrease. Animal reproduction and development will be negatively impacted by the reduced feed supply. The adaptation of different animal breeds to the new, higher temperatures will also be difficult (UN Food and Agriculture Organization, 2011).

The impact of temperature on a company's bottom line has been investigated in a number of earlier empirical investigations. While (Nguyen & Nguyen, 2020) discovered that higher temperatures hurt the bottom lines of Chinese manufacturing, Anton (2021) found the converse to be true for companies in the oil and gas sectors in Europe. Climate change-related droughts and a lack of water for irrigation and human use pose a danger to agricultural productivity. Many agricultural districts will be compelled to cease operations since there will be less water available due to increasing temperatures (Dong et al., 2022). Temperature may reduce the nutritional value of crops, even though it is just one of several elements that impact harvest performance (Bugshan et al., 2023). The key source of supply for companies in the agricultural industry, raw materials is decreasing as a direct consequence of the rising global temperature, this is also accompanied by changes in rainfall patterns, the occurrence and length of periods of extreme heat, and periods of prolonged dryness (Nguyen & Nguyen, 2020). As a result of pain, fatigue, and cognitive impairment, high temperatures can have a negative impact on worker performance and productivity, especially in climate-exposed sectors like agriculture (Liu et al., 2023a). Consequently, we can infer that due to the pain, fatigue, and cognitive impairment, high temperatures can have a negative impact on worker performance and productivity, especially in climate-exposed sectors like agriculture (Liu et al., 2023a).

3 Methodology

We investigate how climate risk impacts firms' bottom lines using the same models as Sun and Drakeman (2022).

$$\begin{aligned} \textit{Financial performance}_{it} = {} & \beta_0 \textit{Wind}_{it} + \beta_1 \textit{Rain}_{it} + \beta_2 \textit{Humid}_{it} + \beta_3 \textit{Temp}_{it} \\ & + \beta_4 \textit{Sun}_{it} + \beta_x \textit{Controls}_{it} + e_{it} \end{aligned} \tag{1}$$

Three indicators are used to assess the financial performance of an organization: return on assets (ROA), return on equity (ROE), and return on sales (EPS), as shown in Table 1. Whereas, Table 2 exhibits the climate change variables represented by wind, precipitation, humidity, temperature, and solar radiation. Controlling variables include Debtit, Growthit, Sizeit, Holderit, Firmtypeit, GDPit, Inflationit, agriexportit, and Humanit.

The criteria used to evaluate a company's financial health include return on assets, return on equity, and return on stock (Li et al., 2022). The concept of Climate change risks and business performance using Return on Assets (ROA) is profit for the year divided by average total assets. In order to calculate return on equity, which is a measure of financial success, divide net income by shareholders' equity. A tool

Table 1 Financial performance and metrics

Financial performance	Metrics
ROA	Return on Assets
ROE	Return on Equity
EPS	Earnings per Share

Table 2 Climate change, metrics, and data sources

Climate change	Metrics	Data source
Temp	Average temperature (2010–2022)	NASA Power
Wind	Wind index (2010–2022)	Wind surface pressure index
Rain	Rainfall index (2010–2022)	Total rainfall at monitoring stations
Humid	Humidity index (2010–2022)	Specific humidity at 2 meters

Table 3 Control variables and metrics

Control variables	Metrics/Calculation
Debtas	Leverage: Total liabilities/Total assets
Growth	Firm growth: (Total assets at year end - Total assets at year beginning) / Total assets at year end
Size	Firm size: Natural logarithm of total assets at year end
Holder	Shareholding ratio of controlling shareholders: Number of shares held by the largest shareholder/Total number of shares
GDP	Economic development: Per capita GDP
Human	Human capital: Percentage of laborers aged 15 years and overworking in the economy
Inflation	Inflation rate (2010–2022)
Agri. Export	Calculated agricultural export (2010–2022)

for evaluating a company's profitability in relation to its sales volume is the Return on Sales (EPS) ratio. As substitutes for independent variables, temperature, precipitation, wind speed, and relative humidity are employed. Average surface temperatures are used to calculate global temperatures, as indicated by Dong et al. (2022)). The reference point for measuring wind is surface wind speed. By examining the long-term average, the amount of rain may be approximated (Liu et al., 2023b). The reference point for measuring humidity is the relative humidity at 2 m.

Table 3 presents the initial set of control variables, which consist of the capital structure (Debts), firm size (Size), shareholding percentage of controlling shareholders (Holder), ownership structure (Firm type), and growth (Growth). Debt finance, as per agency theory, serves as a governance mechanism that may mitigate the negative effects on performance caused by the agency costs associated with equity capital. One way to define leverage is the entire debt divided by the total assets. The corporation could do better financially if the state buys a bigger interest. Depending on the size of the organization, financial outcomes might either go better or worse. Larger businesses are seen as being more varied, well-run, and technologically savvy. Because of economies of scale, bigger companies can be able to take greater risks. However, organizational structure inside larger businesses makes managerial inefficiencies more likely, which raises agency costs. The pace at which a company's total assets grow serves as a gauge of its growth (Peradilan & 2020, 2020).

The market share, level of competition, and financial performance of an organization all increase with the pace of expansion. However, research by Zahoor et al. (2022)) suggests that a company's performance may be harmed by its expansion. In order to approximate a company's real size, the natural logarithm of its total assets is utilized. The proportion of shares owned by the largest shareholder determines the ratio of a controlling shareholder. Whether a firm is governed by the government or by a private company may have an impact on how well it performs. The additional group of control variables composed of the Human capital (Human) and economic development (GDP) indicators for each province. States with stronger human capital levels tend to have more successful businesses. When provincial GDP growth rates are higher than average, business.

Financial data is gathered from Vietstock (Nguyen & Nguyen, 2020). The macroeconomic information at the provincial level, pertaining to the registered industrial enterprises, is obtained from the website of the General Information Office of Vietnam (Jiang et al., 2021). A typical economic metric is the gross domestic product (GDP) per person. The percentage of the working age population that has finished formal education or training programs is the standard method for calculating human capital.

4 Results

Table 4 shows the yearly averages for return on assets (ROA), return on equity (ROE), and return on sales (EPS) for the sampled enterprises. It can be observed that the Return on Equity, Return on Shares, and Return on Assets yielded highest value of 72%, 91% and 168%, respectively. This is true because the sample includes companies from a variety of geographical areas, from the colder north to the warmer south. According to Debt the average debt-to-assets ratio is 41%.

The agricultural sector in Vietnam exhibits a wide range of firm sizes, as indicated by the standard deviation of 83.3%. There is a chance that this, with a standard variation of 47.53 million VND, will affect profitable firms. Table 5 displays the Pearson connection between the facility's weather. EPS is greater than ROE and ROA in ratio.

Table 5 depicted that among climate change indicators, only precipitation has a significant inverse association with reactive oxygen species (EPS), but temperature has a high positive link with EPS. Non-financial indicators have no bearing on output in monetary terms. Leverage has a negative relationship with all performance metrics for the financial sector. As a business grows and expands, most financial health indicators get better. The percentage of ownership has an adverse relationship with both ROA and ROE. EPS and GDP have a positive association, but not ROA. Human capital and ROI are related only in a good way. Investments in human capital have a positive correlation to financial performance. The OLS regression results of the impact of climate change on firms' financial performance are shown in Table 6. Rain is a significant favorable factor, according to ROA, when examining

Table 4 Descriptive statistics

Variable	N	Mean	Sd	Min	Median	Max
ROA	382	0.09	0.07	−0.198	0.08	0.72
ROE	384	0.15	0.58	−0.956	0.12	0.91
EPS	387	0.09	0.13	−0.877	0.07	1.68
Wind_F	388	99.67	2.37	91.18	100.39	101.05
Rain_F	387	1582.52	415.85	350.88	1548.16	3210.19
Humid_F	389	16.32	1.08	13.49	16.17	18.65
Temp_F	391	26.65	2.43	20.32	27.81	30.79
CO_2	394	27.98	0.87	21.98	29.13	30.98
Wind_H	393	99.81	2.32	91.18	100.43	101.05
Rain_H	395	1543.19	424.78	350.86	1489.59	3210.17
Humid_H	397	16.35	1.09	13.47	16.16	18.49
Temp_H	399	26.93	2.46	20.83	28.71	30.79
Debt	401	0.46	0.19	0.98	0.63	0.59
Growth	403	0.115	0.24	−0.53	0.07	1.76
Size	405	17.21	8.34	4.65	14.78	31.47
Holder	406	0.41	0.21	0.03	0.46	0.98
GDP	407	83.78	47.57	25.47	75.67	300.39
Human	409	0.23	0.06	0.01	0.18	0.41
Inflation	410	0.32	0.57	0.03	0.19	0.41
Agri export	412	0.63	0.64	0.09	0.05	0.02

how the weather affects a company's financial results. The rate of return (EPS) and temperature have a positive correlation. Vietnam frequently experiences humid tropical monsoon weather. The warmth and plenty of water allow trees to stay green and flourish all year long.

We use robustness testing for panel regressions. The findings of the Hausman test indicate that random regression is the preferred approach. The results of the stochastic regressions are shown in Table 8. Climate change measurements made at the company's plant revealed that Humidity alone had a positive influence on ROA and that Temperature also had a positive impact on EPS. This ecosystem enables the cultivation of rice and vegetables with the possibility of 2–3 annual harvests. Additionally, it sustains a wide variety of plant and animal species. Ample rainfall is beneficial for wet-rice cultivation, intensive farming to enhance crop productivity, and the expansion of agricultural and animal varieties. According to the majority of models, precipitation and temperature have a significant beneficial influence. This finding aligns with previous research indicating that elevated temperatures have a positive effect on rice production. However, a recent study conducted (Raimi et al., 2021) contradicts this notion by suggesting that although warmer regions caused by global warming may lead to reduced yields, colder areas may actually see an increase in rice yields. There is an inverse relationship between leveraging loans and achieving financial success in the agricultural, forestry, and fishing sectors. This discovery aligns with previous research conducted (Lin et al., 2022). An

Table 5 Correlation analysis

	ROA	ROE	EPS	Wind	Rain	Humid	Temp	Debt	Growth	Size	Holder	GDP	Human	Inflation
ROE	0.23													
EPS	0.58*	0.44*												
Wind	−0.05	−0.05	0.07											
Rain	−0.09	−0.04	−0.17*	−0.17*										
Humid	0.04	−0.04	0.05	0.46*	−0.18*									
Temp	0.05	−0.09	0.17*	0.60*	−0.57*	0.50*								
Deb	−0.34*	−0.12*	−0.36*	−0.71*	0.23*	−0.18*	−0.61*							
Growth	0.14	0.14*	0.12*	−0.04	−0.04	−0.05	0.02	0.19*						
Size	0.14*	0.11*	0.14*	−0.19*	−0.24*	0.10*	0.04	−0.91*	0.06					
Holder	−0.16*	−0.17*	−0.07	0.08	−0.15*	0.08	0.13*	0.01	−0.61*	0.09				
GDP	−0.06	−0.17*	0.19*	0.15*	−0.18*	0.09	0.16*	−0.14*	−0.09	0.25*	0.19*			
Human	0.02	−0.07	0.11*	0.09	−0.15*	0.03	0.16	−0.26*	−0.19*	0.51*	0.09	0.57*		
Inflation	0.66	0.55	0.65	0.58	0.58	0.54	0.54	0.58	0.46	0.44	0.47	0.98	0.87	
Agri export	0.66	0.55	0.87	0.86	0.74	0.74	0.54	0.75	0.07	0.07	0.05	0.04	0.03	0.54

Table 6 OLS regression analysis results

Variables	ROE	ROE	ROA	ROA	EPS	EPS
Wind	−0.002	−0.003	−0.005	−0.007	−0.009	−0.011
	(−0.651)	(−0.765)	(−1.087)	(−1.488)	(−0.939)	(−0.987)
Rain	0.43	0.656	0.986	0.002*	0.998	0.003
	−0.295	−1.188	−0.462	−1.757	−1.228	−1.319
Humid	−0.041	−0.700	−0.080	0.980	−0.061	−0.009
	(−0.623)	(−0.074)	(−0.016)	−0.738	(−0.907)	(−0.668)
Temp	0.983	0.015	0.984	0.987	0.015*	0.081**
	−0.397	−0.755	−0.587	−1.157	−1.976	−2.349
Debt	−0.152**	−0.132*	−0.178***	−0.167***	−0.285***	−0.258***
	(−1.999)	(−1.961)	(−9.109)	(−8.769)	(−8.673)	(−8.684)
Growth	0.094**	0.099**	0.055***	0.095***	0.099***	0.059***
	−2.436	−2.459	−2.976	−2.995	−3.142	−3.091
Size	0.080**	0.090**	0.876	0.998	0.768	0.987
	−2.156	−2.189	−1.676	−1.649	−1.429	−1.676
Holder	−0.048	−0.047	−0.028	−0.092	−0.009	−0.0165
	(−1.186)	(−1.189)	(−1.098)	(−1.078)	(−0.085)	(−0.280)
GDP	−0.009***	−0.008***	−0.007***	−0.060***	0.005**	0.004**
	(−3.269)	(−3.088)	(−3.845)	(−3.566)	−2.259	−2.368
Human	0.046	0.054	0.068	0.069	−0.066	−0.078
	−0.5.657	−0.618	−1.165	−1.237	(−0.628)	(−0.809)
Inflation	0.543	0.654	0.321	0.432	0.897	0.574
	−0.867	−0.867	−0.432	−0.676	−0.987	−0.543
Agri export	0.6897	0.675	0.765	0.576	0.543	
	−0.756	−0.6544	−0.654	−0.654	−0.876	−0.674
Intercept	0.465	0.332	0.358*	0.258	0.175	0.285
	−1.454	−0.995	0.334	−1.671	−0.464	−0.876
Year control	Yes	Yes	Yes	Yes	Yes	Yes
N	387	388	386	387	389	390
Adj. R-sq	0.045	0.046	0.169	0.176	0.179	0.187

Note(s): The table displays the results of ordinary least squares (OLS) regressions, which examine the relationship between corporate financial performance indicators (ROE, ROA, and EPS), and environmental elements (Wind, Rain, Humidity, and Temperature). All regressions include control variables for leverage (debts), growth (growth), firm size (size), holding shareholder ratio (holder), economic development (GDP), human capital (human), and year. The t-statistics corresponding to the regression coefficients are shown alongside the coefficients. The significance levels for 10%, 5%, and 1% are denoted as *, **, and ***, respectively

organization's financial consequences are closely correlated with its growth rate (Growth) and size (Size). The results are consistent with the findings of Pokharel and Featherstone (2021). There is no correlation between the percentage of controlling stockholders and financial performance.

Macroeconomic factors (GDP) have a positive impact on EPS but a significant negative impact on ROA and ROE. Vietnam has historically been an agricultural country, but recent industrialization and modernization efforts have reduced the importance of agriculture to the overall economy. The standards for workforce education are also lax in the fishery, forestry, and agriculture sectors. The performance of enterprises and the caliber of human resources do not directly correlate, according to research by Dang et al. (2021)). Agriculture, forestry, and aquaculture are examples of direct industries. Indirect industries include manufacturing of rubber, medicines, food and drinks, timber-based goods, paper goods, compounds, and fertilizers. The outcomes of the subsample analysis are shown in Table 7. Models 4, 8, and 12 show how wind has a negative impact on the return on equity, return on

Table 7 Climate change risks and firm performance

Variable	Direct industry	Indirect industry	Direct industry	Indirect industry	Direct industry	Indirect industry
	ROE	ROE	ROA	ROA	EPS	EPS
Wind	0	−0.01	−0.002	−0.001	0.32	−0.002
	−0.038	(−1.39)	(−0.198)	(−1.99)	−0.89	(−1.29)
Rain	0.003	−0.002	0.096	0.076	7	0.021
	−0.698	(−0.094)	−1.22	−0.258	−0.5175	−1.49
Humid	−0.02	0.014	−0.0181	0.013	−0.006	−0.016
	(−1.48)	−0.948	(−1.14)	−1.25	(−0.506)	(−0.987)
Temp	0.014	0.12	0.005	0.007	0.005	0.071*
	−0.397	−0.476	−0.187	−0.855	−0.087	−1.96
CO_2	0.564	0.574	0.749	0.797	0.942	0.990
	(−0.685)	(−0.568)	(−0.685)	(−0.57)	(−0.685)	(−0.568)
Debt	−0.112	−0.15***	−0.147***	−0.280***	−0.283***	−0.243***
	(−1.27)	(−2.78)	(−5.67)	(−6.94)	(−8.09)	(−5.18)
Growth	0.086	0.114	0.054***	0.087*	0.098***	0.172***
	−1.65	−1.58	−2.71	−1.97	−2.81	−2.87
Size	0.01***	0.012	0.001***	0.032	0.050***	0.076
	−3.16	−0.712	−3.13	−0.443	−2.63	−0.672
Holder	0.027	−0.087	0.013	−0.052	0.14***	−0.082
	−0.388	(−1.47)	−0.598	(−1.42)	−4.438	(−1.28)
GDP	−0.06***	−0.007	−0.009***	−0.007*	0.764	0.654
	(−4.30)	(−0.868)	(−4.58)	(−1.68)	−1.25	−0.982
Human	0.144	−0.023	0.176*	−0.018	−0.036	−0.017
	−1.65	(−0.185)	−1.79	(−0.108)	(−0.287)	(−0.089)
Inflation	0.987	0.974	0.457	0.743	0.987	0.689
	(−1.32)	(−0.685)	(−0.543)	(−0.564)	(−0.564)	(−0.642)
Agri export	0.547	0.543	1.33	1.54	2.19	2.40
	(−0.987)	(−0.876)	(−0.879)	(−0.56)	(−0.578)	(−0.997)
Intercept	0.187	0.836*	0.177	0.478	−0.293	0.493
	−0.343	−1.77	−0.627	−1.53	(−0.675)	−1.41
N	177	213	179	218	180	215
Adj. R-sq	0.049	0.058	0.229	0.176	0.286	0.127

assets, and return on stock of indirect industries. In both Models (7 and 12), rain is shown to be beneficial for ROA and EPS in the corresponding industries.

In our robustness testing, we take regional aspects into consideration. Vietnam's northern, center, and southern regions all have significantly different climates from one another. In the interest of clarity, the results table is excluded. Research indicates that wind adversely affects several financial performance criteria, but rain and warmth have a beneficial influence on all of them.

Table 8 displays the outcomes of random regressions examining the impact of temperature, humidity, wind, and rain on company profitability. All regressions use covariates for the variables leverage (Debts), growth (Growth), firm size (Size), holding company concentration (Holder), GDP gross domestic product (GDP), and human capital (Human). The regressions of the company's financial performance on climate are shown by models (1), (3), (5), and (7). The effects of climate change on a company's financial performance are shown in models (2), (4), (6), and (8). The t-statistics for the regression coefficient are shown inside brackets. The symbols *,

Table 8 Random regression results

Variable	ROE	ROE	ROA	ROA	EPS	EPS
Wind	−0.007	−0.009	−0.008	−0.003	−0.002	−0.001
	(−0.399)	(−0.471)	(−0.864)	(−1.804)	(−0.480)	(−0.890)
Rain	−0.001	0.76	0.98	0.72	0.89	0.99
	(−0.031)	−0.699	−0.78	−1.559	−0.967	−1.089
Humid	−0.001	0.778	0.018	0.081*	0.87	0.98
	(−0.043)	−0.394	−1.256	−1.717	−0.193	−0.205
Temp	−0.001	0.876	−0.009	0.876	0.081	0.091*
	(−0.097)	−0.247	(−0.098)	−0.347	−1.258	−1.689
Debt	−0.199***	−0.169***	−0.251***	−0.272***	−0.292***	−0.232***
	(−3.68)	(−3.63)	(−7.65)	(−7.89)	(−5.08)	(−5.28)
Growth	0.04***	0.029***	0.075***	0.095***	0.078***	0.078***
	−3.19	−3.176	−3.893	−3.871	−3.184	−3.191
Size	0.987	0.787	0.567	0.965	0.765	0.876
	−0.998	−0.995	−0.467	−0.487	−0.689	−0.796
Holder	−0.001	−0.002	−0.005	−0.004	0.9	−0.006
	(−0.004)	(−0.073)	(−0.126)	(−0.173)	−0.068	(−0.093)
GDP	−0.001*	−0.002*	−0.005**	−0.003*	0.007*	0.009*
	(−1.95)	(−1.77)	(−1.97)	(−1.79)	−1.89	−1.95
Human	0.086	0.068	0.088	0.077	−0.072	−0.094
	−0.741	−0.659	−1.37	−1.271	(−0.284)	(−0.39)
Inflation	0.463	0.546	0.473	0.657	0.547	0.479
	−0.574	−0.362	−0.576	−0.574	−0.574	−0.548
Agri export	0.547	0.325	0.574	0.685	0.689	0.569
	−0.463	−0.473	−0.326	−0.796	−0.658	−0.548
Intercept	0.522	0.336	0.359	0.385	0.191	0.282
	−0.915	−0.648	−1.126	−1.073	−0.284	−0.501

, and * represent statistical significance at the 1%, 5%, and 10% levels, respectively.

A multiple regression study examining the possible impact of climate change on companies found that rainfall and temperature, together with other climate change factors, had a notable positive effect on the financial performance of agricultural listed organizations. However, other indicators do not show the same influence. Another significant discovery was that while severe weather events may lead to floods and other forms of destruction, they really have a beneficial effect on the financial well-being of businesses. These judgements are further bolstered by doing subsample analysis to provide a comprehensive comparison between direct and indirect industries.

This study contributes to the growing body of information on the impact of climate change on economic production in Vietnam's agriculture sector by doing exploratory research, which has not been previously undertaken. GSO supplied the data for the local region, while Viet Stock Finance, a prominent data platform for Vietnamese listed enterprises, gave the financial data. NASA Power contributed the worldwide climatic statistics data for this research. The study's findings have significant value for various agricultural stakeholders, such as academics, businesses, practitioners, and policymakers, especially in developing and emerging nations. These findings provide a deeper understanding of the mechanism through which climate change risks impact firm performance. They also allow for an evaluation of the extent to which these risks affect firm performance and facilitate the development of strategies and policies for managing these risks.

Note(s): The results of random regressions looking at how temperature, humidity, wind, and rain affect business profitability are shown in the following table. All regressions include controls for the variables leverage (Debts), growth (Growth), business size (Size), holding company concentration (Holder), GDP gross domestic product(GDP), and human capital (Human). Models (1), (3), (5), and (7) show the regressions of the company's financial performance on climate change. The impacts of climate change on a company's bottom line are depicted in models (2), (4), (6), and (8). T-statistics for the regression coefficient are shown in brackets. Significance at the 90%, 95% and 99% level is *, **, *** respectively.

By openly and transparently addressing the risks and possible consequences associated with climate change, together with the measures taken to mitigate those risks, the firm may instill more confidence in investors about their investment choices. These sectors, including agriculture, food, and forestry, are among the non-financial organizations that get special advice from TCFD for the disclosure of climate change-related risks. Agricultural companies should develop their own strategies to tackle the risks linked to climate change, as well as the obstacles and difficulties in achieving sustainable agribusiness (Bhaktikul et al., 2021), while considering the expenses related to contract farming (Nguyen-Anh et al., 2022) and insurance (Dang et al., 2021).

If more data is made available to investors, they will be better equipped to communicate with businesses about the longevity of their plans and capital investments (TCFD, 2017). Even though their nation might not require disclosures, agriculture enterprises in emerging and developing nations should abide by the TCFD recommendations. This is due to the fact that the agriculture, food, and forest industries are among the non-financial organizations that have additional guidance from TCFD on disclosing risks related to climate change. Agriculture companies should create their own action plans to address risks associated with climate change, as well as the barriers and challenges to sustainable agribusiness when taking into account the costs of contract farming, insurance.

5 Conclusion

The objective of this research was to analyze the influence of climate change on the financial performance of agricultural companies in Vietnam. These prominent meteorological phenomena, including wind, rainfall, temperature, and humidity, have seen substantial recent alterations. Consequently, they have been selected to ascertain the potential impact of climate change on firms' financial outcomes.

There is room for policymakers and authorities at both the federal and provincial levels to enhance their efforts in facilitating the connection between agribusinesses and implementing effective measures to address the impacts of climate change. This may be achieved by implementing more effective rules and regulations, fostering novel and inventive ideas, or establishing social infrastructure. An inherent limitation of this study is the relatively limited sample size, mostly attributed to the prevalence of privately owned agribusiness firms in Vietnam.

Our research affects government, regulatory, and agricultural policy. Vietnam's agricultural business faces systemic threats from climate change, therefore authorities must implement resilience and sustainability measures. Standards must compel firms to disclose climate-related risks, mitigation, and adaptation. Agriculture employing climate-resilient techniques and technologies may get government subsidies, incentives, or tax breaks. Promoting green technologies and practices may help the environment and economy. Policymakers should prioritize knowledge transfer and capacity building. Climate change information may help farmers and enterprises make informed decisions and implement effective adaptation programs. Insurance and risk management authorities should engage with the insurance sector to develop innovative solutions to climate-related risks faced by agricultural businesses. It gives a monetary buffer for bad weather. Policymakers should finance research and development to promote climate-resilient crop types and new farming practices. Supporting agricultural science may help it adapt to climate change. Collaboration between public and private entities is vital. Joint initiatives may assist governments and agricultural companies in managing climate concerns with comprehensive legislation, data sharing platforms, and early warning systems. Policymakers should encourage enterprises to diversify their sourcing and build local markets to create climate-resilient supply chains. This helps agriculture adapt to climate change. Education and training in climate-smart agriculture may be advantageous. Policymakers, educators, and industry organizations should educate farmers and agricultural professionals in sustainable ways. Climate resilience gains should be spread equally throughout the agricultural community via inclusive and equitable policies. Policymakers should consider smallholder farmers and underprivileged populations.

Future research should widen its scope to include privately held companies in order to better understand how climate change is affecting agricultural enterprises. The study's scope could perhaps be expanded to cover more industries.

References

Abhayawansa, S., & Adams, C. (2021). Towards a conceptual framework for non-financial reporting inclusive of pandemic and climate risk reporting. *Meditari Accountancy Research, 30*(3), 710–738. https://doi.org/10.1108/medar-11-2020-1097

Alam, A., Du, A. M., Rahman, M., Yazdifar, H., & Abbasi, K. (2022). SMEs respond to climate change: Evidence from developing countries. *Technological Forecasting and Social Change, 185*, 122087. https://doi.org/10.1016/j.techfore.2022.122087

Alessandro, A., Maski, G., & Pangestuty, F. W. (2023). The implementation of sustainable finance: A case study in Bank performance. *E Journal. Aibpmjournals.Com, 6*(1), 27–37. https://doi.org/10.32535/ijafap.v6i1.2074

Ali, M., Seraj, M., Türüç, F., Tursoy, T., & Raza, A. (2023). Do banking sector development, economic growth, and clean energy consumption scale up green finance investment for a sustainable environment in South Asia: Evidence for newly developed RALS co-integration. *Environmental Science and Pollution Research, 30*(25), 67891–67906. https://doi.org/10.1007/S11356-023-27023-Z

Anton, S. G. (2021). The impact of temperature increases on firm profitability. Empirical evidence from the European energy and gas sector. *Applied Energy, 295*, 117051.

Bae, S., Masud, M., Rashid, M., & Kim, J. (2021). Determinants of climate financing and the moderating effect of politics: Evidence from Bangladesh. *Sustainability Accounting Management and Policy Journal, 13*(1), 247–272. https://doi.org/10.1108/sampj-04-2019-0157

Bhaktikul, K., Aroonsrimorakot, S., Laiphrakpam, M., & Paisantanakij, W. (2021). Toward a low-carbon tourism for sustainable development: A study based on a royal project for highland community development in Chiang Rai, Thailand. *Environment, Development and Sustainability, 23*(7), 10743–10762. https://doi.org/10.1007/S10668-020-01083-4

Bossman, A., Gubareva, M., & Teplova, T. (2023). EU sectoral stocks amid geopolitical risk, market sentiment, and crude oil implied volatility: An asymmetric analysis of the Russia-Ukraine tensions. *Resources Policy, 82*, 103515. https://doi.org/10.1016/j.resourpol.2023.103515

Bugshan, A., Bakry, W., & Li, Y. (2023). Oil price volatility and firm profitability: An empirical analysis of Shariah-compliant and non-Shariah-compliant firms. *International Journal of Emerging Markets, 18*(5), 1147–1167. https://doi.org/10.1108/IJOEM-10-2020-1288

Cadman, T., & Sarker, T. (2022). De Gruyter handbook of sustainable development and finance. *De Gruyter Handbook of Sustainable Development and Finance, 1–682*. https://doi.org/10.1515/9783110733488/PDF

Chinhamu, K. (2022). Value-at-risk estimation of precious metal returns using long memory GARCH models with heavy-tailed distribution. *Researchgate.Net, 11*(1), 89. https://doi.org/10.18576/jsap/110107

Dang, T. C., Nguyen, T. B., Nguyen, T. Y., Trinh, T. H., & Banh, T. T. (2021). Factors affecting the profitability of listed commercial banks in Vietnam: Does agriculture finance matter? *AgBioforum, 23*(1), 32–41. https://agbioforum.org/menuscript/index.php/agb/article/view/35

Dong, F., Zhu, J., Li, Y., Chen, Y., Gao, Y., Hu, M., Qin, C., & Sun, J. (2022). How green technology innovation affects carbon emission efficiency: Evidence from developed countries proposing carbon neutrality targets. *Environmental Science and Pollution Research, 29*(24), 35780–35799. https://doi.org/10.1007/S11356-022-18581-9

Dou, D., & Istraživanja, L. L.-E. R.-E. (2022). Does sustainable financial inclusion and energy efficiency ensure green environment? Evidence from BRICS countries. *Taylor & Francis, 35*(1), 5599–5614. https://doi.org/10.1080/1331677X.2022.2032785

Ehlers, T., Gao, D., & Packer, F. (2021). A taxonomy of sustainable finance taxonomies. *SSRN Electronic Journal.* https://doi.org/10.2139/ssrn.3945635

Gao, Y., & Zhang, J. (2023). Investigating financialization perspective of oil prices, green bonds, and stock market movement in COVID-19: Empirical study of E7 economies. *Environmental Science and Pollution Research., 30*, 64111. https://doi.org/10.1007/S11356-023-26808-6

Hai, N., & Minh-Tu, L. (2023). Survival of SMEs: Business environment, financial performance, bank credit, and accounting errors. *Taylor & Francis.*, 1. https://doi.org/10.1080/17520843.2023.2238993

Ho, T. T., & Shimada, K. (2019). The effects of climate smart agriculture and climate change adaptation on the technical efficiency of rice farming—An empirical study in the mekong delta of Vietnam. *Agriculture (Switzerland), 9*(5). https://doi.org/10.3390/agriculture9050099

Huang, H., Kerstein, J., Wang, C., & Wu, F. (2022). Firm climate risk, risk management, and bank loan financing. *Strategic Management Journal, 43*(13), 2849–2880. https://doi.org/10.1002/smj.3437

Hudecova, K., & Rajcaniova, M. (2023). Geopolitical risk and energy market. *Peace Economics, Peace Science and Public Policy., 29*, 171. https://doi.org/10.1515/PEPS-2022-0033/HTML

IPCC. (2022). Climate change 2022: Impacts, adaptation, and vulnerability. In H.-O. Portner, D. C. Roberts, M. Tignor, E. S. Poloczanska, K. Mintenbeck, A. Alegrıa, M. Craig, S. Langsdorf, S. Loschke, V. Moller, A. Okem, & B. Rama (Eds.), *Contribution of working group II to the sixth assessment report of the intergovernmental panel on climate change* (p. 3056). Cambridge University Press. https://doi.org/10.1017/9781009325844

Ito, E., & Tourism, J. H. (2023). An evidence-base for reducing the CO2 emissions of national mega sports events: Application of the three-hub model to the Japan 2019 Rugby world cup. *Taylor & Francis.*, 1. https://doi.org/10.1080/09669582.2023.2177301

Jiang, Q., Khattak, S. I., & Rahman, Z. U. (2021). Measuring the simultaneous effects of electricity consumption and production on carbon dioxide emissions (CO2e) in China: New evidence from an EKC-based assessment. *Energy, 229*, 120616. https://doi.org/10.1016/j.energy.2021.120616

Kasperzak, R., Kureljusic, M., & Reisch, L. (2023). Accounting for carbon emissions—Current state of Sustainability reporting practice under the GHG protocol. *Mdpi.Com., 15*. https://doi.org/10.3390/su15020994

Krasnova, I., Prymostka, L., & Lavreniuk, V. (2021). Climate change risks in financial business. *The Problems of Economy, 3*(49), 140–146. https://doi.org/10.32983/2222-0712-2021-3-140-146

Leitold, R., Garschagen, M., Tran, V., & Revilla Diez, J. (2021). Flood risk reduction and climate change adaptation of manufacturing firms: Global knowledge gaps and lessons from Ho Chi Minh City. *International Journal of Disaster Risk Reduction, 61*, 102351. https://doi.org/10.1016/j.ijdrr.2021.102351

Li, Q., Sharif, A., Razzaq, A., & Yu, Y. (2022). Do climate technology, financialization, and sustainable finance impede environmental challenges? Evidence from G10 economies. *Technological Forecasting and Social Change, 185*, 122095. https://doi.org/10.1016/j.techfore.2022.122095

Lin, C. Y., Chau, K. Y., Tran, T. K., Sadiq, M., Van, L., & Hien Phan, T. T. (2022). Development of renewable energy resources by green finance, volatility and risk: Empirical evidence from China. *Renewable Energy, 201*, 821–831. https://doi.org/10.1016/j.renene.2022.10.086

Liu, F., Umair, M., & Gao, J. (2023a). Assessing oil price volatility co-movement with stock market volatility through quantile regression approach. *Resources Policy, 81*, 103375. https://doi.org/10.1016/j.resourpol.2023.103375

Liu, F., Xu, J., & Ai, C. (2023b). Heterogeneous impacts of oil prices on China's stock market: Based on a new decomposition method. *Energy, 268*, 126644. https://doi.org/10.1016/j.energy.2023.126644
Makan, L., & Kabra, K. (2021). Carbon emission reduction and financial performance in an emerging market: Empirical study of indian firms. *Indonesian Journal of Sustainability Accounting and Management, 5*(1), 23–32. https://doi.org/10.28992/ijsam.v5i1.292
Mattera, M., & Soto, F. (2022). Dodging the bullet: Overcoming the financial impact of Ukraine armed conflict with sustainable business strategies and environmental approaches. *The Journal of Risk Finance, 24*(1), 122–142. https://doi.org/10.1108/jrf-04-2022-0092
Mngumi, F., Shaorong, S., Shair, F., & Waqas, M. (2022). Does green finance mitigate the effects of climate variability: Role of renewable energy investment and infrastructure. *Environmental Science and Pollution Research, 29*(39), 59287–59299. https://doi.org/10.1007/S11356-022-19839-Y
MONRE. (2021). Climate change scenarios for Vietnam. Publishing House of Natural Resources, Environment and Cartography.
Nasir, M. H., Wen, J., Nassani, A. A., Haffar, M., Igharo, A. E., Musibau, H. O., & Waqas, M. (2022). Energy security and energy poverty in emerging economies: A step towards sustainable energy efficiency. *Frontiers in Energy Research, 10*. https://doi.org/10.3389/FENRG.2022.834614/FULL
Nguyen, P.-H. J.-F. H.-P. V.-T. T.-K. (2020). Assessing the unemployment problem using a Grey MCDM model under COVID-19 impacts: A case analysis from Vietnam. *The Journal of Asian Finance, Economics and Business, 7*(12), 53–62. https://doi.org/10.13106/JAFEB.2020.VOL7.NO12.053
Nguyen, T. N. L., & Nguyen, V. C. (2020). The determinants of profitability in listed enterprises: A study from vietnamese stock exchange. *Journal of Asian Finance, Economics and Business, 7*(1), 47–58. https://doi.org/10.13106/jafeb.2020.vol7.no1.47
Nguyen-Anh, T., Hoang-Duc, C., Nguyen-Thi-Thuy, L., Vu-Tien, V., Nguyen-Dinh, U., & To-The, N. (2022). Do intangible assets stimulate firm performance? Empirical evidence from Vietnamese agriculture, forestry and fishery small- and medium-sized enterprises. *Journal of Innovation and Knowledge, 7*(3), 100194. https://doi.org/10.1016/j.jik.2022.100194
Nocca, F., Bosone, M., De Toro, P., & Fusco Girard, L. (2023). Towards the human circular Tourism: Recommendations, actions, and multidimensional indicators for the tourist category. *Sustainability, 15*(3), 1845. https://doi.org/10.3390/SU15031845
Palmer, M., Nguyen, C. V., Mitra, S., Mont, D., & Groce, N. E. (2019). Long-lasting consequences of war on disability. *Journal of Peace Research., 56*, 860. https://doi.org/10.1177/0022343319846545
Peradilan, I. A.-J. H.. (2020). Consumer protection and fintech companies in Indonesia: Innovations and challenges of the financial services authority. *Jurnalhukumdanperadilan.Org*, 9(1), 132–153. https://doi.org/10.25216/JHP.9.1.2020.132-153.
Pokharel, K. P., & Featherstone, A. M. (2021). Examining the productivity growth of agricultural cooperatives: The biennial malmquist index approach. *Journal of Co-operative Organization and Management, 9*(2), 100148.
Rahman, S., Hossain Moral, I., Hassan, M., Shakhawat Hossain, G., Perveen, R., & Circle, S. B. (2022). A systematic review of green finance in the banking industry: Perspectives from a developing country. *Aimspress.Com, 4*(3), 347–363.
Raimi, L., Olowo, R., & Shokunbi, M. (2021). A comparative discourse of sustainable finance options for agribusiness transformation in Nigeria and Brunei: Implications for entrepreneurship and enterprise development. *World Journal of Science, Technology and Sustainable Development, 18*(4), 325–350. https://doi.org/10.1108/WJSTSD-05-2021-0051/FULL/HTML
Sadiq, M., Nonthapot, S., Mohamad, S., Chee Keong, O., Ehsanullah, S., & Iqbal, N. (2022). Does green finance matter for sustainable entrepreneurship and environmental corporate social responsibility during COVID-19? *China Finance Review International, 12*(2), 317–333. https://doi.org/10.1108/CFRI-02-2021-0038

Saeed, R., Ali, W., & Nadeem, A. (2023). Weather induced risks, mitigation strategies and farmers' willingness to participate in flood insurance scheme in Punjab, Pakistan. *Journal of Economic Impact, 5*(1), 15–21. https://doi.org/10.52223/jei5012302

Shahzad, U., Madaleno, M., Dagar, V., Ghosh, S., Do, B., & Gan, ˇ. (2022). Exploring the role of export product quality and economic complexity for economic progress of developed economies: Does institutional quality matter? *Elsevier., 62*, 40. https://doi.org/10.1016/j.strueco.2022.04.003

Shao, H., Cheng, J., Wang, Y., (2022). *Can digital finance promote comprehensive carbon emission performance?* Evidence from Chinese cities. *Mdpi.Com.* https://doi.org/10.3390/ijerph191610255.

Shin, Y. J., & Choi, Y. (2019). Feasibility of the fintech industry as an innovation platform for sustainable economic growth in Korea. *Sustainability (Switzerland), 11*(19). https://doi.org/10.3390/su11195351

Singhania, M. (2021). Institutional framework of ESG disclosures: Comparative analysis of developed and developing countries. *Taylor & Francis., 13*, 516. https://doi.org/10.1080/20430795.2021.1964810

Sun, Y. Y., & Drakeman, D. (2022). The double-edged sword of wine tourism: The economic and environmental impacts of wine tourism in Australia. *Journal of Sustainable Tourism, 30*(4), 932–949. https://doi.org/10.1080/09669582.2021.1903018

Tamasiga, P., Onyeaka, H., Ouassou, E., & houssin. (2022). Unlocking the green economy in African countries: An integrated framework of FinTech as an enabler of the transition to Sustainability. *Energies, 15*(22). https://doi.org/10.3390/en15228658

TCFD. (2017). Recommendations of the task force on climate related financial disclosures, Report, available at: https://www.fsb-tcfd.org/recommendations/

Thuc Huong Giang, N., Minh Hanh, T., Thu Hien, P., Trong Trinh, N., Thi Khanh Huyen, N., & Huyen Trang, V. (2021). *The impacts of climate change risks on financial performance: Evidence from listed manufacturing firms in Vietnam.* Atlantis-Press.Com. https://www.atlantis-press.com/proceedings/icech-21/125965442

Tsai, F.-S., Cabrilo, S., Chou, H.-H., Hu, F., & Tang, A. D. (2022). Open innovation and SME performance: The roles of reverse knowledge sharing and stakeholder relationships. *Journal of Business Research, 148*, 433–443. https://doi.org/10.1016/J.JBUSRES.2022.03.051

Umar, M., & Safi, A. (2023). Do green finance and innovation matter for environmental protection? A case of OECD economies. *Energy Economics, 119*, 106560. https://doi.org/10.1016/j.eneco.2023.106560

UNISDR. (2019). Global assessment report on disaster risk reduction, united nations international strategy for disaster reduction secretariat, Geneva, available at: https://digitallibrary.un.org/record/3825375

Van Khuc, Q., Tran, B. Q., Meyfroidt, P., & Paschke, M. W. (2018). Drivers of deforestation and forest degradation in Vietnam: An exploratory analysis at the national level. *Forest Policy and Economics., 90*, 128. https://doi.org/10.1016/j.forpol.2018.02.004

Wang, X., Huang, J., Xiang, Z., & Huang, J. (2021). Nexus between green finance, energy efficiency, and carbon emission: Covid-19 implications from BRICS countries. *Frontiers in Energy Research, 9*. https://doi.org/10.3389/fenrg.2021.786659

Williams, J., & Love, W. (2022). Low-carbon research and teaching in geography: Pathways and perspectives. *Professional Geographer, 74*(1), 41–51. https://doi.org/10.1080/00330124.2021.1977156

Wu, C. F., Huang, S. C., Chang, T., Chiou, C. C., & Hsueh, H. P. (2020). The nexus of financial development and economic growth across major Asian economies: Evidence from bootstrap ARDL testing and machine learning approach. *Journal of Computational and Applied Mathematics, 372*, 112660. https://doi.org/10.1016/j.cam.2019.112660

Zahoor, Z., Khan, I., & Hou, F. (2022). Clean energy investment and financial development as determinants of environment and sustainable economic growth: Evidence from China. *Environmental Science and Pollution Research, 29*(11), 16006–16016. https://doi.org/10.1007/S11356-021-16832-9

Zhu, Y., Bashir, S., & Marie, M. (2022). Assessing the relationship between poverty and economic growth: Does sustainable development goal can be achieved? *Environmental Science and Pollution Research, 29*(19), 27613–27623. https://doi.org/10.1007/S11356-021-18240-5

Effects and Opportunities of Climate Change on a Company's Goodwill

Amir Ahmad Dar, Akshat Jain, and Mehak Malhotra

Abstract Goodwill refers to reputation. In business and accounting, it refers to the intangible asset representing the excess value of a company's tangible and identifiable assets over its liabilities. It is essentially the value of a company's reputation, brand, customer loyalty, and other intangible factors that contribute to its success and ability to generate earnings above the value of its net assets. There are various methods for estimating the value of the Goodwill of a firm. The methods are the super profit, annuity, capitalization, etc. The main parameters used to estimate the value of goodwill as per the above models are normal profit, super profit, interest rate, number of purchased years, etc., which are all factors that impact the value of the goodwill. But some other factors impact it. In this chapter, we will discuss the effects and opportunities of climate change on a company's reputation. The effects of climate change on goodwill will vary depending on the industry and region, and the steps taken by the company to manage climate-related risks it is vital to emphasize. A real-world example of how climate change can affect a company's goodwill and reputation is the case of BP (British Petroleum) and the Deepwater Horizon oil spill in 2010 was explained in this study.

1 Introduction

The concept of goodwill finds its origins in the history of accounting and business customs. Initially, goodwill was introduced as an accounting term to represent the value of a business that surpassed the total of its tangible assets. This notion

A. A. Dar (✉)
School of Chemical Engineering and Physical Sciences, Lovely Professional University, Phagwara, Punjab, India

A. Jain · M. Malhotra
Department of Statistics and Data Analytics, Lovely Professional University, Phagwara, Punjab, India

N. Naifar (ed.), *Climate Change and Finance*, Sustainable Finance,
https://doi.org/10.1007/978-3-031-56419-2_13

originated from medieval business practices and has undergone continuous evolution throughout the centuries (Li & Sloan, 2017). During medieval times, business dealings relied heavily on trust and personal connections. Merchants and traders built reputations for being fair, honest, and dependable, which greatly influenced their business prosperity. These intangible attributes were deemed valuable assets linked to the business and were regarded as integral components of its overall worth.

In the initial stages of accounting systems, goodwill was not explicitly identified as an independent asset. Nonetheless, certain early accounting records acknowledged the business's value beyond its tangible assets, demonstrating an awareness of intangible elements that played a crucial role in the company's achievements. Luca Pacioli's introduction of double-entry bookkeeping during the fifteenth century was a crucial advancement in accounting methods. This system facilitated a more methodical recording of business transactions, encompassing the acknowledgment of both assets and liabilities. Although goodwill was not directly addressed, this era led to an improved comprehension of intangible value (Smith, 2008).

During the seventeenth and eighteenth centuries, the prominence of goodwill increased with the emergence of joint stock companies. These companies permitted numerous investors to partake in business ventures, making a company's reputation and established relationships essential in attracting investors and customers. As a result, the concept of goodwill gained greater significance during this period. As business practices and accounting principles progressed, it became clear that standardization in financial reporting was necessary. In the early twentieth century, multiple accounting bodies and standard-setting organizations took the initiative to develop accounting principles, which included guidelines for handling goodwill in financial statements.

The official recognition of goodwill as an intangible asset in the United States came about through the enactment of the Securities Act of 1933 and the Securities Exchange Act of 1934. These acts set forth the regulations for financial reporting by publicly traded companies, which included the inclusion of goodwill in accounting standards. Goodwill has been extensively studied in the fields of accounting, finance, and business valuation. Researchers have delved into numerous facets of goodwill, encompassing its definition, evaluation, impairment, and influence on financial reporting.

The literature has concentrated on offering a distinct and all-encompassing definition of goodwill, along with understanding its position as an intangible asset. Researchers have explored the conceptual framework of goodwill, examining its characteristics, criteria for recognition, and appropriate accounting treatment (McGee & Preobragenskaya, 2006; Collett et al., 2000; Vashisht & Kukreja, 2020).

.

Examining whether the case companies elaborate on important climate change indicators that (may) change the numbers included in their Financial Statements, with regard to the value of production assets, reserves for future cash flows, and income streams (TCFD, 2017). In particular, the institutional structure of financial accounting rules raises the question of how much it encourages businesses to disclose the ways in which climate change affects their operations and the worth of their production assets (Scholten et al., 2020). Climate change can affect a firm's goodwill both directly and indirectly. Goodwill represents the intangible value of a company's reputation, brand, and customer loyalty, all of which can exert a significant impact on its overall success and financial performance. In this chapter, various external factors are discussed about how climate change impacts the reputation of the firm. How a firm's response to climate change can have a big impact on its goodwill? A company's reputation can be improved by taking proactive steps, being environmentally conscious, and being dedicated to sustainability; on the other hand, indifference or detrimental effects on the environment can cause goodwill to erode. In this study, the authors explained an actual instance of how climate change might impact a company's goodwill and reputation which was the 2010 Deepwater Horizon oil spill involving British Petroleum (BP). Even though a drilling rig explosion was the main cause of the catastrophe, BP's response and the environmental damage had a big influence on the company's reputation (Smith et al., 2011; Shultz et al., 2015).

2 Literature Review

One of the biggest issues facing humanity and a topic of widespread concern in the international community is climate change (Gaffney & Steffen, 2017). In certain areas, climate change creates an invisible catastrophe and has an impact on agriculture, health, finance, and other fields (Akbari et al., 2020). The unpredictability of climate change is its most notable aspect. As a result, it is exceedingly challenging to assess how climate change will affect the world economy. Climate change-related natural disasters have severely damaged global financial assets, resulting in losses of up to US $24 trillion (Dietz et al., 2016). An economy's lifeblood is its finance sector. Globally, academics are closely monitoring how climate change affects financial stability. It is challenging to assess the economic effects of climate change because it is impossible to forecast with precision the frequency and severity of natural disasters brought on by climate change (Zhang et al., 2018). The progressive increase in global temperatures has led to a rise in the economic losses resulting from natural disasters including hurricanes, droughts, and floods. This has resulted in heightened financial risks for the impacted areas. Furthermore, as greenhouse gas concentrations in the atmosphere rise, the likelihood of climate disasters will rise as well. These events might harm the economy indirectly by upsetting global supply chains, which would then undermine financial stability (Stern & Taylor, 2007).

Goodwill is an intangible asset in economics that stands for the worth of a business's reputation, brand, and customer relationship. It is regarded as a part of the total worth, or market capitalization, of a corporation. When a business purchases another business for more money than the fair market value of its net assets, goodwill is reported on the balance sheet of the acquiring business.

The valuation of goodwill has garnered significant attention in accounting, finance, and business valuation literature. Researchers and professionals have delved into diverse methodologies to assess the worth of goodwill in various scenarios, including mergers and acquisitions, business combinations, and overall business valuations. Within the literature, some prominent themes encompass:

The Purchase Price Allocation (PPA) method is extensively employed both in practical applications and research. Various studies have addressed the difficulties and intricacies associated with distributing the purchase price among identifiable assets and goodwill, particularly concerning intangible assets (Lynn & Lynn, 2020).

The Excess Earnings Method (EEM) has gained popularity as a preferred approach for assessing goodwill in situations where there is no acquisition involved, but the value of goodwill requires determination. Researchers have thoroughly explored the implementation of this method across different industries and contexts. Certain research studies have expanded the EEM by incorporating multiple future periods to calculate excess earnings and ascertain the present value of goodwill (Pratt, 1985).

Researchers have investigated the correlation between market capitalization and goodwill to gain insights into how market perception and investor sentiment influence the valuation of goodwill. Studies have examined the utilization of comparable transactions as a method to approximate the value of goodwill by drawing insights from previous acquisitions and business combinations. Numerous valuation models and methods have been put forward to assess the value of goodwill, encompassing discounted cash flow (DCF) models, excess earnings models, and residual income models.

In general, the literature on goodwill valuation methods underscores the significance of precisely estimating and disclosing goodwill as an intangible asset. Researchers persist in investigating ways to enhance the valuation process, tackle challenges, and harmonize accounting practices with the genuine economic value of goodwill. As accounting standards progress and novel valuation techniques emerge, the literature is expected to further expand and develop.

For calculating the value of the goodwill using different methods, different parameters are used such as capital employed, number of purchased years, normal profit, and capitalized average profits. Every element influences the goodwill valuation in some way. For instance, Dar et al. (2023) employed an experiment design to determine which parameter has the most influence on goodwill valuation using the super profit approach. It was discovered that the most important component influencing the goodwill value was the capital employed (Dar et al., 2023). A similar approach can be used by researchers to determine which factors are most important for the goodwill value of the other models. Still, there are some more factors that affect the company's reputation. In this research, we will go over the different approaches, how to calculate the value of goodwill using specific techniques, and the external elements that affect it.

3 Goodwill

Goodwill in the accounting and business context is an intangible asset reflecting the surplus value of a company's identifiable and tangible assets compared to its liabilities. It signifies the worth of a company's reputation, brand, customer loyalty, and other intangible aspects that play a vital role in its success and capacity to generate profits beyond its net asset value.

Goodwill is commonly generated during an acquisition when a company is purchased at a price surpassing the net book value of its identifiable assets (like buildings, equipment, and inventory) minus its liabilities. The surplus amount paid is assigned to the worth of intangible assets, encompassing the company's brand recognition, customer base, skilled workforce, patents, and other elements that enhance the overall value of the business beyond the total of its individual parts (Boennen & Glaum, 2014).

Following are some key points about goodwill:

Intangible Asset: Goodwill lacks physical presence but is recognized as an intangible asset on a company's financial statement.

Purchased not separately: In contrast to other tangible assets or patents, goodwill cannot be bought or sold independently. It only emerges through the acquisition of an entire business.

Amortization: Previously, companies used to spread the cost of goodwill over a defined period, often up to 40 years, through a process called amortization. However, accounting practices have advanced, and nowadays, most companies do not amortize goodwill. Instead, they assess it annually for any potential impairment.

Impairment Testing: Companies are required to evaluate the value of goodwill at least once a year to determine if it may be impaired. If the value of the reporting unit (a specific segment of the company) linked to the goodwill drops below its recorded value, the company must recognize an impairment loss.

Influence on financial statements: The presence of goodwill as an intangible asset on a company's balance sheet can lead to an expense recorded on the income statement in the event of goodwill impairment, potentially reducing the company's reported earnings.

Importance for Mergers and Acquisitions: Goodwill plays a vital role in mergers and acquisitions, representing the premium paid for acquiring another company. A substantial amount of goodwill can indicate that the acquiring company paid a significant premium above the fair value of the acquired company.

4 Methods of Goodwill

Methods of goodwill pertain to the various approaches employed for calculating or valuing goodwill in the fields of accounting and finance. Goodwill, an intangible asset, emerges when a company is acquired at a price exceeding the fair market value of its identifiable net assets (assets minus liabilities):

1. Purchase Price Allocation (PPA) Method: The most prevalent approach for calculating goodwill in a business combination when a company is acquired is the PPA method (Paugam et al., 2015). The following steps to calculate the goodwill of a firm are:

   (a) Identify Fair Market Value: To identify the fair market value, the assessment of all assets and liabilities of the acquired company is conducted as of the

acquisition date. This encompasses tangible assets (such as property, equipment, and inventory) and identifiable intangible assets (like patents, and trademarks).
   (b) Calculate Net Assets: To determine the net assets of the acquired company, subtract the fair market value of its liabilities from the fair market value of its assets.
   (c) Determine Purchase Price: Afterward, the acquiring company compares the purchase price paid to acquire the target company with the net assets calculated in step (b). If the purchase price surpasses the net assets, the surplus amount is classified as goodwill.
   (d) Record Goodwill: The determined goodwill is recorded as an intangible asset on the balance sheet of the acquiring company.

2. Excess Earnings Method (EEM): The EEM is an alternative technique used to assess the value of goodwill, particularly in business valuation situations where no actual acquisition occurs (Harper, 1985). This method involves the following steps:

   (a) Calculate Normal Rate of Return: Determine the normal rate of return expected from the net tangible assets of the business. This is the rate of return that investors would typically expect given the risk associated with the business.
       Identify the normal rate of return expected from the net tangible assets of the business. This rate represents the return that investors would typically expect, considering the associated business risks.
   (b) Estimate Excess Earnings: Compute the actual earnings of the business and then deduct the expected normal rate of return from step (a). This computation will yield the excess earnings generated by the business.
   (c) Determine Capitalized Value: To estimate the value of the business's intangible assets, including goodwill, capitalize the excess earnings by dividing it by the rate of return determined in step (a).

4.1 Market Capitalization Method

The capitalization method is a goodwill valuation technique where the value of goodwill is determined by subtracting the actual capital employed from the capitalization value of average profits, considering the normal rate of return.

There are two ways of calculating profits in the capitalization method and these methods are:

1. Average Profit Method
2. Super Profits Method

4.1.1 Average Profit Method

The capitalization of the average profit method involves calculating the goodwill value by deducting the actual capital employed from the capitalized value of the average profits (Zadorozhnyi et al., 2018), considering the normal rate of return.

$$Goodwill = Normal\ Capital \quad Actual\ Capital\ Employed\ or$$

$$Goodwill = Capitalized\ Average\ Profits \quad Actual\ Capital\ Employed$$

where,

$$\begin{aligned} &Capitalized\ Average\ Profits\ or\ Normal\ Capital \\ &= Average\ estimated\ profits\ x100\ /\ Normal\ Rate\ of\ Return \end{aligned}$$

$$\begin{aligned} &Actual\ capital\ employed\ or\ Net\ Asset\ of\ Business \\ &= Total\ Assets\begin{pmatrix} excluding\ goodwill,non \\ -traded\ investments\ and\ fictitious\ assets \end{pmatrix} \\ &\text{-}Outsiders\ liabilities \end{aligned}$$

Steps involve estimating the goodwill with the Average profit method

Step 1: Measure estimated average profits.
Step 2: Calculate the capitalized average profits.
Step 3: Calculate the value of Actual capital employed or net assets of the business.
Step 4: Calculate goodwill by subtracting the actual capital employed from the capitalized average profit.

4.1.2 Super Profit

Super profit represents the surplus of expected future profits over normal profit and serves as a means to determine the additional earnings generated by the business (Colborne & Hall, 1993). Goodwill is calculated by multiplying the value of super profits by a specific factor, which corresponds to the number of years of purchase.

"The following steps are involved in the calculation of goodwill using the super profit method:

1. Calculate the total capital of the business. It will be the sum total of all the net current and fixed assets along with the shareholders' equity.
2. Determine the normal profit by multiplying the total capital employed with the normal rate of return.
3. Calculate the average estimated profit or average manageable profit.
4. Calculate super profit by subtracting the value of normal profit from the average estimated profit to determine the super profit.
5. Multiply the super profit by the number of years of purchase to determine the goodwill.

It can be expressed in the formula as follows":

$$Normal\ Profit = Capital\ Employed\ x\left(Normal\ Rate\ of\ Return/100\right)$$

$$Super\ Profit = Average\ estimated\ profit\quad Normal\ Profit$$

$$Goodwill = Super\ Profit\ x\ No.of\ years\ of\ purchase$$

It is essential to acknowledge that the methods used for valuing goodwill can differ based on the unique circumstances and the purpose of the valuation. Furthermore, accounting standards and regulations may influence the method chosen for financial reporting, particularly in the context of business combinations and mergers and acquisitions. Professional appraisers, accountants, and financial analysts often utilize a combination of these methods and consider multiple factors to arrive at a reasonable and well-supported estimate of goodwill.

The valuation of goodwill is influenced by various parameters, including average profit, super profit, interest rate, capitalized value, and others. Each of these factors has a certain impact on goodwill. By varying a single variable while holding others constant, it becomes possible to ascertain the positive or negative effect of each parameter on goodwill. This approach helps in identifying the specific factors that affect goodwill positively or negatively (Penman, 2013; Schadewitz, 2017; Palepu et al., 2013).

4.2 Advantages of Goodwill

Goodwill can bestow several advantages upon a business, playing a crucial role in its sustained success and financial performance. Some of the key advantages of goodwill include:

1. Customer Loyalty and Trust: Businesses with a favorable reputation and solid goodwill tend to gain more trust from their customers. This trust can result in higher customer retention rates and enhanced customer loyalty, leading to repeated business and word-of-mouth referrals.
2. Improved Reputation and Brand Value: A robust brand image and positive reputation can draw in more customers, cultivate customer loyalty, and set the business apart from its competitors. Goodwill embodies the intangible worth of a company's reputation and brand recognition.
3. Competitive Advantage: With a respected brand and a positive reputation, a company can distinguish itself in the market and attract new customers. Goodwill can serve as a valuable competitive edge, particularly in industries where products or services are comparable.
4. Pricing Power: Businesses with goodwill and a robust brand may have greater pricing flexibility for their products or services. Customers might be willing to pay a premium for a renowned and trusted brand, enabling the company to sustain healthy profit margins.

5. Employee Recruitment and Morale: Employees often take pride in working for a well-regarded company, leading to higher productivity and lower turnover rates. Moreover, a strong brand can facilitate the attraction of top talent when recruiting new employees. A company with a favorable reputation and goodwill can significantly influence employee morale and satisfaction.
6. Attraction for Investors and Partnerships: Investors show more interest in a company with a robust brand and positive reputation, and other businesses may seek partnerships to capitalize on the goodwill of a well-established brand. Goodwill can enhance a company's appeal to potential investors, partners, or collaborators.
7. Acquisitions and Mergers: Goodwill can play a crucial role in the valuation of a company during mergers and acquisitions. A strong brand and positive reputation can lead to a higher valuation, offering potential financial advantages to shareholders in case of an acquisition.
8. Risk Mitigation: A positive reputation can act as a buffer, mitigating the impact of adverse situations and preserving the trust of customers and stakeholders. During times of crisis or unfavorable events, companies with robust goodwill are often more resilient.

It is important to note that while goodwill offers many advantages, it is an intangible asset that is not directly quantifiable and can be challenging to manage and protect. Companies must continuously work to maintain and enhance their goodwill by prioritizing customer satisfaction, ethical practices, and responsible corporate behavior (Gomes, 1988).

4.3 *Disadvantages*

While goodwill can provide notable benefits to a business, it is not without its drawbacks and potential risks. Several significant disadvantages of goodwill encompass:

1. Unpredictable Nature and Intangible: Goodwill, being an intangible asset, lacks physical presence and precise quantification. This makes it challenging to measure its true value accurately. Moreover, shifts in customer perceptions, market conditions, or adverse incidents can unpredictably influence the worth of goodwill.
2. Price Premium: In some cases, the purchase price of a company may include a significant premium for goodwill. If the business fails to deliver the expected results or synergies after the acquisition, the goodwill may not be realized, leading to financial losses for the acquiring company.
3. Prone to Impairment: Goodwill is susceptible to regular impairment assessments. Any deterioration in the company's financial performance, alterations in market conditions, or adverse incidents impacting the business could necessitate the reduction of goodwill value, leading to impairment charges that may affect the company's financial statements and overall profitability.

4. Susceptibility to Reputation Risk: Goodwill is intricately linked to a company's reputation and brand perception. Adverse publicity, customer discontent, or unethical conduct can swiftly erode goodwill, resulting in customer attrition and a decline in market share.
5. Exposure to Market Volatility: The value of goodwill can be influenced by shifts in market conditions, industry trends, or economic fluctuations. Economic downturns or disruptions within the industry may result in reduced customer spending and a decline in the valuation of goodwill.
6. Limited Transferability: Goodwill is typically non-transferable as an independent asset. It remains tied to the company and cannot be sold or transferred separately from the business. This lack of transferability poses challenges in realizing the value of goodwill in isolation.
7. Challenges in Valuation: Assessing the value of goodwill is an intricate task that often involves employing diverse estimation methods. Variations in opinions and assumptions used during the valuation process can lead to disparities in the determined value of goodwill.
8. Influence on Financial Ratios: As an intangible asset listed on the company's balance sheet, goodwill can have an impact on financial ratios, such as return on assets and return on equity. The presence of substantial goodwill may influence perceptions of the company's financial stability and performance.
9. Growth Constraints through Acquisitions: Overreliance on acquiring companies to build goodwill may impose limitations on growth. Companies might become overly dependent on acquisitions, which can be risky if the integration process proves challenging or if suitable acquisition targets are scarce.
10. Complexity in Regulatory and Accounting Compliance: Accounting for goodwill requires strict adherence to specific accounting standards and regulations, which can be intricate and may necessitate the engagement of external auditors or valuation experts.

In conclusion, while goodwill offers significant advantages to a company, it also poses challenges and risks that necessitate careful consideration and effective management. This ensures the safeguarding of a company's reputation, financial performance, and sustained long-term success (Wen & Moehrle, 2016).

5 Factors That Impact on Goodwill as per Super Profit Method

The valuation of goodwill depends on different parameters such as average profit, super profit, interest rate, and capitalized value, among others. Each of these factors influences goodwill to some extent. To determine the positive or negative impact of each parameter on goodwill, one can vary a single variable while keeping other variables constant. This way it allows us to identify which factor has a positive or negative effect on goodwill. But Dar et al. (2023) used a design of experiment where it can vary each parameter at a time and found which parameter impacted more on the goodwill value (Dar et al., 2023).

The super profit method used the four parameters that impact the valuation of the goodwill. The parameters are the Number of purchase years (N), capital employed (CA), Average Profit (NP), and nominal rate (r). Dar et al. (2023) investigate the effect of all the factors using the design of an experiment with some statistical tools like analysis of variance and analysis of mean. The study uses the design of an experimental approach to carry out the tests and optimize the parameters for goodwill values. Interestingly, research is the first to use the DOE to determine the ideal set of input parameters. The Taguchi technique DOE, analysis of variance (ANOVA), and analysis of mean (ANOM) were used to investigate the effects of input parameters. MINITAB 18 software was utilized for the analysis. With an emphasis on how variables affect a desired outcome, ANOM enables to identification of the most efficient combinations of variables at various levels. Ranking the input components according to their impact on the corresponding outputs is also helpful. The traders can utilize regression analysis to examine the behavior of elements influencing goodwill valuation. The percentage contribution of each of these factors to the change in goodwill value is determined using ANOVA. The findings show that the following order applies to a particular dataset's goodwill value factors: capital employed (CA), nominal rate (r), average profit (AP), and purchase period (N) (Dar et al., 2023).

6 Factors, Challenges, and Opportunities of Goodwill

6.1 Factors that Impact Goodwill

Numerous elements can influence the value and resilience of goodwill in a business. These factors encompass both internal and external aspects and hold significant importance in shaping the company's reputation, brand perception, and customer loyalty (d'Arcy & Tarca, 2018). Here are several pivotal factors that can impact goodwill:

1. Customer Satisfaction: The degree of customer satisfaction and the caliber of products or services offered by the company can directly impact goodwill. Contented customers are more inclined to share positive word-of-mouth, contributing to a favorable brand image.
2. Brand Reputation: The strength and positivity of a brand's reputation, established through consistent efforts, can profoundly influence goodwill. A highly regarded brand is more apt to attract customers, foster loyalty, and set the business apart from its competitors.
3. Ethical Business Conduct: Companies exhibiting robust ethical practices and a commitment to corporate social responsibility tend to be perceived more positively by customers and stakeholders, thereby positively influencing goodwill.

4. Product or Service Excellence: The caliber and dependability of the products or services provided by the company are vital in fostering and sustaining goodwill. Ensuring consistent delivery of high-quality offerings can instill customer trust and loyalty.
5. Customer Support and Service Excellence: Offering exceptional customer service and support can result in heightened customer satisfaction and loyalty, contributing to the development of positive goodwill.
6. Flexibility and Innovation: Companies that prioritize innovation and adaptability to meet evolving market demands are better positioned to outperform competitors and sustain positive goodwill.
7. Financial Strength: The financial stability and performance of a company can influence goodwill, as a financially successful company is typically perceived as more dependable and trustworthy.
8. Marketing and Promotional Initiatives: Successful marketing and advertising endeavors can facilitate brand recognition, reinforce positive brand associations, and bolster goodwill.
9. Effective Corporate Governance: The efficiency of corporate governance practices and transparent management decisions can shape the perception of a company and, consequently, impact its goodwill.
10. Industry and Market Environment: External elements, including economic conditions, industry trends, and competitive landscape, can likewise influence goodwill. A company's adeptness in navigating and prospering amidst challenging environments can bolster its reputation and goodwill.
11. Influence of Social Media and Online Reputation: In the age of widespread social media and online reviews, a company's online presence and customer feedback can exert a substantial influence on its goodwill.
12. Emphasis on Environmental and Social Responsibility: Companies that exhibit dedication to environmental sustainability and social responsibility can garner positive recognition and goodwill from environmentally conscious consumers.

It is crucial to acknowledge that goodwill is not dependent on any single factor; instead, it stems from the collective perception of diverse internal and external elements. Preserving and augmenting goodwill demands a proactive approach from businesses to foster and sustain positive relationships with customers, employees, investors, and the wider community.

6.2 *Opportunities for Goodwill During Climate Change*

Companies can leverage climate change to improve their reputation and cultivate favorable connections with stakeholders. By adopting sustainable practices and taking proactive steps to tackle climate issues, they can create various chances that contribute to fostering goodwill (Mee et al., 2008; Gallego-Álvarez et al., 2011).

1. Enhanced Brand Reputation: Companies that adopt environmentally responsible practices and sustainability initiatives can build a strong reputation as socially conscious and environmentally friendly organizations. This positive brand image can enhance goodwill and attract environmentally conscious consumers.
2. Increased Customer Loyalty: In a growing trend, consumers are actively seeking out businesses that place a premium on sustainability. By providing products and services that are environmentally friendly, companies can nurture customer loyalty and bolster their positive reputation.
3. Investor Attraction: The issue of climate change is progressively becoming a pivotal aspect for investors. Companies equipped with strong sustainability strategies and effective climate risk management are apt to draw the attention of socially conscious investors, leading to a favorable influence on their reputation.
4. Employee Engagement: Highlighting sustainability and a commitment to environmental accountability can elevate employee satisfaction and involvement. Staff members are inclined to take pride in being part of a company that takes active steps to combat climate change, thus fostering a constructive work atmosphere and augmenting a positive reputation.
5. Innovation Opportunities: Tackling the hurdles posed by climate change can stimulate innovation within the organization. Businesses that allocate resources to develop green technologies and sustainable resolutions can set themselves apart in the market, garnering favorable recognition and enhancing their reputation.
6. Cost Savings and Efficiency: Enforcing energy-efficient measures and curbing wastage can result in cost reductions and heightened operational effectiveness. Exhibiting conscientious resource utilization can have a favorable impact on how stakeholders perceive the company, thereby elevating its reputation.
7. Positive Media Coverage: Companies that proactively address climate change and champion sustainability might draw favorable media coverage and public interest. This advantageous visibility has the potential to elevate brand standing and cultivate a positive reputation.
8. Regulatory Compliance: By actively adhering to environmental regulations and demonstrating a commitment to environmental responsibility, companies can avoid reputational risks and potential legal issues, contributing to goodwill.
9. Partnerships and Collaborations: Firms that uphold robust sustainability measures have the potential to attract partners and collaborators who share similar values. This can foster advantageous alliances that magnify the company's positive influence on climate change and reputation.
10. Access to New Markets: Occasionally, companies that give precedence to sustainability and environmental conscientiousness could open doors to novel markets or customer segments that hold eco-friendly products and services in high regard.

To recap, climate change offers substantial chances for companies to cultivate positive sentiment and reinforce their connections with stakeholders. By harmonizing their tactics with sustainability objectives, advocating environmental conscientiousness, and proactively tackling climate-related obstacles, businesses can elevate their standing, draw devoted customers and investors, and cultivate a favorable image among staff and the public. Making sustainability a foundational principle can stimulate ingenuity, economize expenses, and set them apart in the market, ultimately resulting in enhanced goodwill and enduring prosperity.

6.3 Climate Change Impacts on Goodwill

The goodwill of a company can be substantially influenced by climate change, both through direct and indirect means. Goodwill represents the intangible worth of a company's reputation, brand, and loyalty. As climate change takes on greater urgency as a global concern, it has the potential to impact different facets of a company's activities, its image, and how stakeholders perceive it. Consequently, this can bring about shifts in the company's goodwill. Here are several manners in which climate change can affect a company's goodwill:

1. Environmental Accountability and Image: With the emergence of climate change concerns, consumers, investors, and other stakeholders are attaching increased significance to a company's ecological responsibility and its efforts to maintain sustainable practices. Companies that take proactive steps to combat climate change, integrate sustainable strategies, and minimize their carbon impact are likely to improve their standing and positive reputation, consequently enhancing their goodwill.
2. Perception in the Public Eye and Societal Consciousness: Occurrences linked to climate change, like extreme weather incidents, ecological disasters, or environmental mishaps, can lead to unfavorable media attention and public disapproval. Companies perceived as contributing to climate change or failing to address its impacts adequately might experience harm to their reputation and a decrease in their goodwill.
3. Regulatory Risks: Governments globally are imposing more stringent environmental rules to combat climate change. Companies that fail to adhere to these regulations or are sluggish in adapting to them could encounter legal consequences and a less favorable perception from stakeholders, influencing their goodwill.
4. Resilience of Supply Networks: Climate change can interrupt supply networks due to elements such as severe weather occurrences, shortages of resources, or disruptions in transportation. Companies that proficiently handle these risks and guarantee the durability of their supply networks might elevate their goodwill by showcasing adaptability and conscientious business methods.
5. Shifting Consumer Desires: Consumer inclinations are evolving toward products and services that are environmentally mindful. Companies providing choices that are sustainable and eco-friendly might secure a competitive edge, attracting

a larger customer base and cultivating strong brand loyalty. This, in turn, bolsters goodwill.

6. Perception among Investors: Climate change is acknowledged as a financial concern. Investors are now factoring a company's approach to climate-related matters into their investment choices. Companies that adeptly tackle the risks linked to climate change and incorporate sustainability into their strategies could draw the interest of more socially conscious investors, thereby beneficially influencing their goodwill.
7. Employee Engagement: Employees are increasingly concerned about their employers' commitment to sustainability and climate action. Companies that prioritize environmental responsibility and implement green initiatives may improve employee morale, attract top talent, and foster a positive workplace culture, ultimately influencing goodwill.
8. Legal Actions and Legal Vulnerabilities: Matters connected to climate change can result in legal conflicts, including lawsuits associated with environmental harm or responsibilities. Companies implicated in such legal disputes could experience adverse publicity and potential financial setbacks, which can influence their goodwill.

In conclusion, the impact of climate change on a company's goodwill is extensive. Firms that adopt sustainable strategies, confront climate risks, and demonstrate environmental responsibility are apt to elevate their standing, secure customer allegiance, draw in investors, and uphold constructive interactions with employees and the wider community. Conversely, companies that disregard climate change concerns could encounter detrimental effects on both their reputation and finances, ultimately diminishing their goodwill. Hence, comprehending and effectively addressing the influence of climate change on goodwill is progressively vital for companies aiming to flourish within a more environmentally aware and socially accountable business environment.

6.4 Challenges of Goodwill During Climate Change

Amidst the context of climate change, businesses might confront numerous challenges impacting their goodwill. These challenges stem from the growing significance of environmental accountability and sustainability from the perspective of stakeholders. Here are some of the primary difficulties that companies could potentially face in terms of goodwill during the course of climate change:

1. Risk to Reputation: Events associated with climate change, like environmental mishaps, instances of pollution, or an inability to fulfil sustainability pledges, can result in harm to reputation. Unfavorable media exposure and public opinion can undermine goodwill, particularly when stakeholders perceive the company as insufficiently responsive to climate concerns.

2. Changing Consumer Choices: Consumers are progressively factoring a company's environmental actions into their buying choices. The inability to embrace sustainable methods or offer environmentally friendly products could lead to diminished customer allegiance and adverse repercussions on goodwill.
3. Investor Perspective: Investors are growing more aware of the risks and prospects linked to climate change. Businesses that fail to sufficiently reveal or handle climate-related risks could encounter diminished investor trust and a decreased valuation, which would impact goodwill.
4. Transparency in Communication: Stakeholders anticipate clear reporting regarding a company's actions and risks associated with climate change. Insufficient disclosure and inadequate communication about climate initiatives might create an impression of opacity, consequently influencing goodwill.
5. Greenwashing Concerns: Companies may face skepticism and scrutiny if they engage in "greenwashing," where they overstate or falsely claim their environmental efforts. Such practices can lead to a loss of trust and credibility, negatively impacting goodwill.

7 Case Study and Implications

British Petroleum (BP) and the Deepwater Horizon oil spill in 2010 provide a real-world example of how climate change can affect a company's goodwill and public reputation. BP's actions and the environmental effects of the catastrophe had a significant impact on the company's reputation and public perception, even though the explosion of a drilling rig was the main initiator of the crisis (Smith et al., 2011; Shultz et al., 2015).

Historically, one of the biggest environmental disasters is the Deepwater Horizon oil leak. Millions of barrels of oil were spilt into the Gulf of Mexico, severely damaging marine life, coastal ecosystems, and the local economies in the areas that were impacted. This tragedy served as a reminder of the inherent environmental risks associated with oil and gas exploration and production activities. Public awareness of the environmental harm increased as a result of the incident's widespread media coverage. Pictures of contaminated beaches and wildlife covered in oil gave the public a very bad impression of BP. The business received harsh criticism for its part in the mishap (Smith et al., 2011; Shultz et al., 2015).

A lot of people thought that BP handled the disaster inadequately in the beginning. The corporation faced criticism for downplaying the extent of the spill, miscalculating its effects on the environment, and not having a solid containment and clean-up plan. These elements played a part in BP's goodwill and reputation continuing to decline. The business also faced severe legal consequences, such as large penalties and demands to pay restitution. The financial consequences were significant, involving billions of dollars in costs associated with remediation efforts, court settlements, and a decrease in the company's stock price. As a result, investors and shareholders suffered large financial losses (Smith et al., 2011; Shultz et al., 2015).

The business BP, which had previously positioned itself as ecologically conscious, suffered a serious setback to its standing as a sustainable and ethical energy supplier. The incident questioned the company's commitment to environmental management and brought attention to the risks inherent in the oil and gas sector. Years after the event, BP was still struggling to deal with the long-term effects. Although the business made changes to its environmental and safety policies, the Deepwater Horizon leak is still a significant event in its corporate history. The public's perception of BP and its operations is still influenced by this incident.

This case illustrates how a major environmental event can have a profound effect on a company's goodwill and reputation even if it is not directly related to climate change. Companies that are perceived as being careless with the environment or unprepared to deal with environmental problems risk a significant decline in support and confidence from the public. As the public's awareness of climate change grows, businesses are under more scrutiny for their environmental policies and their capacity to handle related issues.

The implications of environmental incidents on business strategy and policymaking underscore the importance of integrating sustainability into core business practices. Businesses that place a high priority on environmental responsibility and take proactive measures to address climate change issues will be better able to maintain their competitive edge over time and build strong relationships with stakeholders. The establishment of regulatory frameworks that support and encourage ethical corporate activities is one way that policymakers help. Businesses, governments, and other stakeholders must work together in order to effectively address the difficulties that climate change presents.

8 Conclusion

Today, goodwill is recognized as a significant intangible asset in accounting, arising from business combinations and acquisitions when the purchase price exceeds the fair value of net assets acquired. It is an essential component of a company's overall valuation and can have a significant impact on financial statements and shareholder perception. Goodwill represents the intangible value of a company's reputation, brand, and other factors that contribute to its success and distinguish it from its competitors. It is an essential component of a company's overall valuation and financial health and can have a significant impact on its financial statements and shareholder perception.

In conclusion, climate change can significantly influence a firm's goodwill through its environmental responsibility, compliance with regulations, market perception, ability to manage physical risks, supply chain resilience, and investor perception. Companies that proactively address climate change challenges and integrate sustainable practices into their business strategies are more likely to enhance their goodwill and build long-term value with stakeholders. To effectively tackle these hurdles, businesses need to embrace proactive and open strategies toward climate change and sustainability. Showcasing authentic dedication to environmental accountability, harmonizing corporate strategies with climate objectives, and involving stakeholders in sustainable actions can contribute to the preservation and augmentation of goodwill in the face of climate change. It is imperative for companies to incorporate climate change concerns into risk management and strategic decision-making to ensure resilience and sustain favorable connections with customers, investors, employees, and the broader public.

References

Akbari, M., Alamdarlo, H. N., & Mosavi, S. H. (2020). The effects of climate change and groundwater salinity on farmers' income risk. *Ecological Indicators, 110*, 105893.

Boennen, S., & Glaum, M. (2014). Goodwill accounting: A review of the literature. *Available at SSRN, 2462516.*

Colborne, A., & Hall, P. C. (1993). The profit method of valuation. *Journal of Property Valuation and Investment, 11*(1), 43–49.

Collett, P. H., Hrasky, S., & Tearney, M. G. (2000). Accounting for goodwill: The debate and the future. *Journal of Business Ethics, 23*(1), 95–105.

Dar, A. A., Jain, A., Malhotra, M., Afzal, S., & Khan, M. S. (2023). Effect of factors on company's goodwill. *Journal of Global Entrepreneurship Research, 13*(1), 18.

d'Arcy, A., & Tarca, A. (2018). Reviewing IFRS goodwill accounting research: Implementation effects and cross-country differences. *The International Journal of Accounting, 53*(3), 203–226.

Dietz, S., Bowen, A., Dixon, C., & Gradwell, P. (2016). 'Climate value at risk' of global financial assets. *Nature Climate Change, 6*(7), 676–679.

Gaffney, O., & Steffen, W. (2017). The Anthropocene equation. *The Anthropocene Review, 4*(1), 53–61.

Gallego-Álvarez, I., Rodríguez-Domínguez, L., & García-Sánchez, I. M. (2011). Study of some explanatory factors in the opportunities arising from climate change. *Journal of Cleaner Production, 19*(9–10), 912–926.

Gomes, G. M. (1988). Excess earnings, competitive advantage and goodwill value. *Journal of Small Business Management, 26*(3), 22–31.

Harper, J. R. (1985). Excess earnings: Redefining the professional goodwill doctrine. *Pac. LJ, 17*, 165.

Li, K. K., & Sloan, R. G. (2017). Has goodwill accounting gone bad? *Review of Accounting Studies, 22*, 964–1003.

Lynn, S., & Lynn, S. (2020). *Intangibles valuation, purchase Price allocation (PPA) and goodwill impairment* (pp. 205–249). A Short Course Based on IFRS.

McGee, R. W., & Preobragenskaya, G. G. (2006). *Accounting and financial system reform in a transition economy: A case study of Russia*. Springer Science & Business Media.

Mee, L. D., Dublin, H. T., & Eberhard, A. A. (2008). Evaluating the global environment facility: A goodwill gesture or a serious attempt to deliver global benefits? *Global Environmental Change, 18*(4), 800–810.

Palepu, K. G., Healy, P. M., & Peek, E. (2013). *Business analysis and valuation: Using financial statements*. South-Western College Pub.

Paugam, L., Astolfi, P., & Ramond, O. (2015). Accounting for business combinations: Do purchase price allocations matter? *Journal of Accounting and Public Policy, 34*(4), 362–391.

Penman, S. (2013). *Accounting for value*. Columbia University Press.

Pratt, S. P. (1985). The excess earnings method. *Business Valuation Review, 4*(3), 4–12.

Schadewitz, H. J. (2017). *Goodwill Bilanzierung nach IFRS: Eine kritische Analyse unter besonderer Berücksichtigung des Goodwill Impairment Tests*. Springer.
Scholten, R., Lambooy, T., Renes, R., & Bartels, W. (2020). "the impact of climate change in the valuation of production assets via the IFRS framework: An exploratory qualitative comparative case study approach" *accounting. Economics, and Law: A Convivium, 10*(2), 20180032. https://doi.org/10.1515/ael-2018-0032
Shultz, J. M., Walsh, L., Garfin, D. R., Wilson, F. E., & Neria, Y. (2015). The 2010 Deepwater horizon oil spill: The trauma signature of an ecological disaster. *The Journal of Behavioral Health Services & Research, 42*, 58–76.
Smith, F. (2008). The influence of Amatino Manucci and Luca Pacioli. *BShM Bulletin, 23*(3), 143–156.
Smith, L. C., Smith, M., & Ashcroft, P. (2011). Analysis of environmental and economic damages from British Petroleum's Deepwater horizon oil spill. *Albany Law Review, 74*(1), 563–585.
Stern, N., & Taylor, C. (2007). Climate change: Risk, ethics, and the Stern review. *Science, 317*(5835), 203–204.
TCFD: Task Force on Climate-related Financial Disclosures. (2017). *Recommendations of the task force on climate-related financial disclosures* [online]. Retrieved June 3, 2018 from: https://www.fsb-tcfd.org/wp-content/uploads/2017/06/FINAL-TCFD-Report-062817.pdf
Vashisht, P., & Kukreja, P. (2020). Determinants of goodwill impairment in Indian companies. *International Journal of Management Studies, 7*(1), 119–130.
Wen, H., & Moehrle, S. R. (2016). Accounting for goodwill: An academic literature review and analysis to inform the debate. *Research in Accounting Regulation, 28*(1), 11–21.
Zadorozhnyi, Z. M., Sudyn, Y., & Muravskyi, V. (2018). *Goodwill assessment in enterprise management: Innovative approaches using computer and communication technologies*.
Zhang, H. L., Chang-Xin, L. I. U., & Zheng, W. A. N. G. (2018). Progress in damage function of integrated assessment models for climate change. *Advances in Climate Change Research, 14*(1), 40.

Harnessing the Potential of Green Cryptocurrencies: A Path Toward Climate Change Mitigation

Nicola Del Sarto, Elena Scali, and Roberto Barontini

Abstract The rise of cryptocurrencies has introduced a paradigm shift in global finance but has concurrently raised critical environmental concerns linked to energy-intensive cryptocurrency mining. This chapter explores the concept of "Green Cryptocurrencies" as a promising solution to alleviate the ecological consequences associated with cryptocurrency ecosystems. Beginning with analyzing cryptocurrency mining's environmental implications, including its health and climate impacts, we underscore the importance of transitioning from proof-of-work (PoW) to more sustainable consensus mechanisms. "Green Cryptocurrencies" encompass a range of initiatives and strategies designed to reduce carbon emissions within the cryptocurrency ecosystem. We provide a comprehensive literature review on cryptocurrency environmental impacts and the development of green alternatives. Our statistical analysis reveals the relationship between electricity costs and the adoption of green cryptocurrencies. Furthermore, we detail the methodology for selecting green cryptocurrencies as case studies and report their characteristics. In particular, we focus on 13 green cryptocurrencies. Notably, we highlight that the adoption of these green cryptocurrencies can contribute to CO_2 emission reduction.

1 Introduction

The advent of cryptocurrencies has ushered in a new era of financial innovation and digital transactions. It has disrupted traditional monetary systems, reshaping the landscape of global finance. However, this revolutionary development has not been without its controversies, particularly regarding the environmental impact of

N. Del Sarto (✉)
University of Florence, Florence, Italy
e-mail: nicola.delsarto@unifi.it

E. Scali · R. Barontini
Scuola Superiore Sant'Anna, Pisa, Italy

N. Naifar (ed.), *Climate Change and Finance*, Sustainable Finance,
https://doi.org/10.1007/978-3-031-56419-2_14

cryptocurrency mining. The mining process relies heavily on energy-intensive processes, raising concerns about its ecological footprint. As concerns about climate change and environmental sustainability intensify, it becomes increasingly imperative to address the environmental impact of cryptocurrencies.

From 1860 to 2014, the temperature rose by ~0.9°C due to a steady increase in atmospheric greenhouse gas (GHG) concentrations. In fact, over the same years, humanity has emitted around 584 Gt of CO_2 from fossil fuel use, industrial processes, and land-use change (Mora et al., 2018). Due to their high carbon footprint, cryptocurrencies are also being linked to global warming and the amount of electricity used to mine Bitcoin has been the subject of intense debate (de Vries, 2018). Bitcoin, the most well-known type of cryptocurrency, was the first decentralized cryptocurrency, created in 2009 by an individual known as Satoshi Nakamoto. The popularity of Bitcoin has led to the development of more cryptocurrencies. Consequently, the yearly power consumption associated with global cryptocurrency mining has significantly increased over time (Alzoubi & Mishra, 2023). According to the University of Cambridge, Bitcoin's greenhouse gas emissions increased dramatically (from 2.62 $MtCO_2e$ in 2014 to 48.41 $MtCO_2e$ in 2022).

Goodkind et al. (2020) expanded their investigation beyond energy to estimate the monetary impact of cryptocurrency mining. The study assessed costs from air pollution and health issues related to mining Bitcoin, Ethereum, Litecoin, and Monero. Findings showed that for every $1 of Bitcoin, $0.49 in damages occurred in the USA and $0.37 in China. This emphasizes the urgency of addressing the environmental impact of cryptocurrency mining.

Criticism primarily targets the proof-of-work consensus mechanism, abbreviated as PoW. Specifically, this consensus mechanism consumes a significant amount of energy because of the competitiveness involved in generating new blocks.

However, these critiques are often extended to all blockchains, although non-PoW blockchains require less energy and have a lower carbon footprint (Rieger et al., 2022). Inevitably, this raises concerns about the adoption of blockchain technology (Sedlmeir et al., 2020).

This chapter explores "Green Cryptocurrencies" as a promising avenue for mitigating the environmental consequences associated with cryptocurrency mining and blockchain technology. In a time where global agreement on the need for climate action is increasing, combining digital finance with ecological responsibility creates a significant opportunity. "Green Cryptocurrencies" include various efforts, strategies, and technologies aiming to use blockchain and cryptocurrency systems to minimize their carbon footprint and contribute to broader climate change mitigation.

In this chapter, we delve into the complex landscape of cryptocurrency mining, analyzing its environmental impact and challenges for sustainable development. We explore innovative solutions promoting eco-friendly practices in the cryptocurrency ecosystem. Through case studies, theoretical frameworks, and practical insights, we aim to provide a nuanced understanding of the evolving relationship between cryptocurrencies and climate change. We also provide a statistical analysis of the impact of the cost of electricity on the variation in the amount of green cryptocurrency.

In our exploration of Green Cryptocurrencies, we describe the environmental challenges within the crypto sphere while presenting a forward-looking perspective on how these technologies can be harnessed as a force for positive change. By delving into the evolving field of sustainable blockchain and cryptocurrency practices, we aim to provide researchers, policymakers, and industry stakeholders with the knowledge and resources necessary to navigate the complex terrain of climate-conscious digital finance.

In the upcoming chapters, we will explore the principles of Green Cryptocurrencies, examine innovative projects driving sustainable blockchain technology, and discuss regulatory frameworks influencing the future of environmentally friendly digital finance. Our journey aims to harness the potential of Green Cryptocurrencies for climate change mitigation and a sustainable digital future.

Our chapter contributes to the existing literature by providing an extensive review of the energy consumption of PoW cryptocurrencies. Additionally, we compare PoW consensus mechanisms with energy-efficient alternatives. Specifically, we have chosen certain green cryptocurrencies as case studies, thereby contributing to the limited literature on green cryptocurrencies (Alzoubi & Mishra, 2023). Furthermore, our chapter addresses the question of how blockchain can contribute to sustainability (Calandra et al., 2023). Lastly, we conduct a statistical analysis that reveals the negative impact of electricity costs on the fluctuation in the number of brown cryptocurrencies, while it does not affect the green ones.

In Sect. 2, we provide a literature review on the environmental impact of cryptocurrencies and the development of green cryptocurrencies. Then, in Sect. 3, we describe the methodology used to select some green cryptocurrencies as a case study and the data related to the statistical analysis. In Sect. 4, we focus on analyzing the characteristics of the previously selected green cryptocurrencies and report the results of the analysis. Finally, we conclude in Sect. 5.

2 Background

2.1 Cryptocurrencies and Environmental Impact

Cryptocurrency mining, a pivotal aspect of digital finance, has gained attention in recent years for its notable environmental impact. Several studies have investigated the environmental and social consequences linked to the energy-intensive process of cryptocurrency mining.

Indeed, while some academics have highlighted the benefits of cryptocurrencies (Rogojanu & Badea, 2015; Badea & Mungiu-Pupăzan, 2021), others have questioned them, particularly regarding their energy consumption. For example, Jones et al. (2022) underline that the blockchain technology used by Bitcoin (the PoW) is energy intensive and, simultaneously, the production of Bitcoin has increased climate damage over time. In particular, Bitcoin caused a total of $12 billion in global climate damage between 2016 and 2021, without considering the damage caused by other cryptocurrencies using the same consensus mechanism (or using different consensus mechanisms). The following paragraphs will review the literature on the environmental impact of cryptocurrencies.

Krause and Tolaymat (2018) explain the reasons why PoW cryptocurrencies consume significant energy. This is because miners engage in a competitive process to create new blocks, striving to quickly provide a unique identifier (hash) for transactions on the blockchain and earn new currencies. To stay competitive, miners utilize highly sophisticated computer equipment demanding substantial electricity, as success depends on the power of the computer (Saleh, 2020). The efficiency of a machine relies on the amount of power needed to perform a set of calculations, which will lead to a future where only the most efficient machines will remain. Additionally, agents seek maximum profit, driving them to seek the cheapest electricity (de Vries, 2019). Furthermore, mining profitability is expected to be linked to the mining location (e.g., Norway has a surplus of renewable energy and is a country with political and legal stability; Campbell, 2020). Summarizing, the profitability of mining Bitcoin depends on three elements: hardware efficiency, electricity cost, and the climate of the mining site.

Li et al. (2019) and Sedlmeir et al. (2020) highlight that determining the exact amount of energy consumed by blockchains is challenging. Li et al. (2019) point out that the number of cryptocurrencies is uncertain, the cryptocurrencies are untraceable, and the condition of the miners is unknown, while Sedlmeir et al. (2020) state that the number of contributors, the characteristics of their hardware, and the effort they put into mining are unknown.

However, numerous studies have focused on the energy consumption of cryptocurrencies, particularly Bitcoin. What stands out in the literature is the increasing energy consumption over time, primarily driven by the widespread use of Bitcoins. Specifically, annual energy consumption is expected to rise from 4 TWh in 2017 to approximately 100 TWh in 2023. Furthermore, the energy usage in Bitcoin mining is projected to reach its peak in China in 2024, amounting to 296.59 TWh, as reported by Jiang et al. (2021). Table 1 presents energy consumption estimates from some of these studies. It is important to note that the variety of results depends on the calculation models adopted by the authors. However, notwithstanding these estimates are certainly variables, they help us to understand the amount of energy consumption of cryptocurrencies.

Gallersdörfer et al. (2020) differ from the previous studies because, not only calculate the energy consumption for Bitcoin, but also provide evidence for the top 20 cryptocurrencies by market capitalization at the end of March 2020[1]. They found that Bitcoin represents approximately 66% of the total energy consumption. Assuming Bitcoin is an independent country, Digiconomist (2023) would rank it 34th in the world for energy consumption.

On the contrary, Sedlmeir et al. (2020) underline that "increasing energy efficiency of mining hardware would not reduce a PoW blockchain's energy requirements in the long term." In fact, the difficulty of cryptographic puzzles is periodically adjusted to the total computing power of the network to keep their solution time constant, ensuring the security of the network. As participants with better hardware can solve more puzzles for the same energy cost, other participants will follow suit, leading to an overall increase in computing power. This necessitates an increase in the difficulty of

[1] In order of market capitalization are: Bitcoin, Ethereum, Bitcoin Cash, Bitcoin SV, Litecoin, Monero, Dash, Ethereum C, Zcash, DogeCoin, Bitcoin Gold, Decred, RavenCoin, MonaCoin, Bytom, SiaCoin, DigiByte, Horizen, Komodo, and Bytecoin.

Table 1 Papers on the energy consumption of cryptocurrencies

Source	Year of estimation	Energy consumption/ Year	Cryptocurrencies considered
Bevand (2017)	2017	4.12 - 4.73 TWh	Bitcoin
Stoll et al. (2019)	2018	45.8 TWh	Bitcoin
Zade et al. (2019)	2018	3,852 MW (annual average power)	Bitcoin
		991 MW(annual average power)	Ethereum
Bondarev (2020)	2017	5.027 TWh	Bitcoin
	2019	55.27 TWh	Bitcoin
De Vries (2020)	2019	87.1 TWh	Bitcoin
Sedlmeir et al. (2020)	2020	60 - 125 TWh	Bitcoin
Jiang et al. (2021)	2024	296.59 TWh	Bitcoin
Bitcoin Energy Consumption Index	2019	73.12 TWh	Bitcoin
	2023	109.48 TWh	Bitcoin
Cambridge Bitcoin Energy Consumption Index	2019	74.54 TWh	Bitcoin
	2023	148.85 TWh	Bitcoin

Table 2 Papers on the CO_2 emissions of cryptocurrencies

Source	CO_2 emissions	Cryptocurrencies considered
McCook (2018)	63 $MtCO_2$ per year	Bitcoin
Krause and Tolaymat (2018)	3–15 $MtCO_2$ in the period: 01/01/2016 and 30/06/2016	Bitcoin, Ethereum, Litecoin and Monero
Stoll et al. (2019)	22.0–22.9 $MtCO_2$ per year	Bitcoin
Köhler and Pizzol (2019)	17.29 $MtCO_2$ in 2018	Bitcoin

the puzzle to ensure it continues to be solved in a constant amount of time. In summary, the energy requirements of PoW blockchains will not decrease in the long term.

At the same time, it is also interesting to consider research that calculates the CO_2 emissions generated by cryptocurrencies. Table 2 sums up the results of some of these studies (which are mainly focused on Bitcoin). Also in this case we need to consider that the methods for calculating the carbon footprint may differ.

McCook (2018) demonstrates that gold mining is more harmful than Bitcoin mining in terms of $MtCO_2$ produced. Nevertheless, the emissions from Bitcoin are incredibly high. Krause and Tolaymat (2018) highlight that the mining of the four cryptocurrencies that they considered, utilize a considerable portion of global energy production. In addition, this consumption will increase. Stoll et al. (2019) determine an amount of carbon emission comparable with that produced by Kansas City. Finally, Köhler and Pizzol (2019) find the carbon emission to be 17.29 $MtCO_2$.

Considering the environmental impact of energy consumption, we should consider different issues, starting with e-waste, i.e., electronic or electrical equipment that has been discarded (Wendl et al., 2023). According to de Vries (2019), the annual production of e-waste is estimated at 10,948 tonnes, an amount comparable to the e-waste generated by Luxembourg. In this context, de Vries and Stoll (2021) describe a methodology for calculating Bitcoin's e-waste, and show that the entire Bitcoin network produces 30.7 metric kilotons of equipment per year.

The environmental impact of cryptocurrencies can also relate to economic aspects such as trading activity (Wendl et al., 2023). Indeed, the higher the price of Bitcoin, the greater the motivation to mine, which increases energy consumption (Truby, 2018; de Vries, 2021). Furthermore, Huynh et al. (2022) demonstrate the existence of a bidirectional relationship between Bitcoin returns (and volumes) and energy consumption.

De Vries (2020) stresses that the higher profitability of mining activities might induce the use of cheaper and less energy-efficient equipment, which causes more pollution. Conversely, during the decline phase, miners have fewer alternatives to choose from, as only the most energy-efficient equipment remains profitable. In fact, during this phase, participants may decide to exit the market, thereby reducing energy demand.

Schinckus et al. (2020) illustrate a direct association between cryptocurrency trading activity and energy consumption, both in the short and in the long run, without distinguishing between cryptocurrencies employing the PoW consensus mechanism and those utilizing alternative mechanisms. Erdogan et al. (2022) observe that the demand for cryptocurrencies, as indicated by cryptocurrency prices, has a causal impact on environmental degradation. Specifically, they discover that increased demand for Bitcoin and decreased demand for Ripple and Ethereum result in heightened environmental pollution. Simultaneously, Schinckus et al. (2022) identify a positive correlation between hashrate and energy requirements; however, they clarify that this correlation is not attributable to trading volumes.

Summarizing, these studies indicate an upward trend in electricity consumption in cryptocurrency mining, particularly concerning Bitcoin. Moreover, cryptocurrency demand and trading activity have a negative environmental impact. These results have obviously caught the attention not only of researchers but also of international bodies such as the European Commission (de Vries, 2020) because the use of cryptocurrencies contributes to climate change and affects human health, creating negative social externalities (Goodkind et al., 2020).

2.2 *Innovations in Green Cryptocurrency Development*

Some researchers have wondered whether blockchain technologies could help solve the climate crisis. According to Chen (2018), to answer this question, we should examine the application of the blockchain technology in the energy and carbon markets. In particular, according to UNFCCC, blockchain can be useful in specific ways[2]: enhance the carbon trading system, ensuring transparency and the validity of transactions; enable the development of peer-to-peer renewable energy trading platforms; facilitate financial transactions to support climate action with transparent

[2] https://unfccc.int/news/how-blockchain-technology-could-boost-climate-action

funding allocation to projects; and, finally, provide greater transparency on greenhouse gas emission. In this work, our focus is on the green cryptocurrencies.

The PoW consensus mechanism stands out as the primary driver of energy consumption in cryptocurrencies. It requires significant energy due to the competitive process of creating new blocks. Miners compete to quickly provide a unique identifier for transactions, necessitating sophisticated and energy-intensive computer equipment. This competition favors the use of powerful machines, and efficiency is crucial for success. In essence, PoW cryptocurrency energy consumption is driven by the need for computational power in a competitive environment.

Researchers have proposed different solutions to reduce this energy use: develop more efficient hardware; reduce the difficulty of puzzles; or utilize renewable energy. Nevertheless, each solution has its own limitations (Shi et al., 2023). In particular, producing more powerful hardware could allow miners to solve more puzzles, encouraging them to buy more hardware. Reducing the difficulty of puzzles could also attract more miners and more computing power. Finally, Malfuzi et al. (2020) believe that miners will move to renewable energy but its production is subject to seasonality and climate factors. For these reasons, it may not be sufficient to meet the huge energy demand of mining. The remaining energy demand must be met from alternative sources, which in the worst case is an incentive to build new coal-fired power stations (de Vries, 2019).

Therefore, it is essential to consider other more energy-efficient consensus mechanisms. Eklund and Beck (2019) provide an overview of them. In particular, when we are talking about energy-efficient consensus mechanisms, we are referring to their carbon footprint. According to the Cryptowisser website[3], the carbon footprint of a cryptocurrency depends on its energy consumption, which arises from the process of validating nodes, and on the cleanliness of the energy used. The energy consumption is, indeed, contingent on the consensus mechanism. Furthermore, the cleanliness of the energy is determined by the sources used for its production.

To address the high energy consumption of PoW, an alternative proposed is Proof-of-Stake (PoS) (Irresberger et al., 2021). PoS eliminates traditional mining by selecting the next node to create a block based on its proportional "stake" in the blockchain, determined by the number of coins held and the duration of ownership (Bentov et al., 2014). This method does not require PoW's high computational power, and miners receive no reward except transaction fees. However, relying on nodes with the highest stake may lead to a more centralized blockchain (Bamakan et al., 2020).

The energy requirements of PoS blockchains are much lower than those of PoW blockchains. Recent measurements show that the most energy-intensive PoS blockchains consume less than 0.002% of the energy required to mine Bitcoin (Rieger et al., 2022). Meanwhile, PoS mechanisms offer a quite high level of security, albeit

---

[3] https://www.cryptowisser.com/crypto-carbon-footprint/

with some criticisms,[4] and faster transaction confirmation than PoW mechanisms (Nguyen et al., 2019). In terms of scalability, PoW requires a significant amount of resources, limiting its scalability. On the other hand, PoS is often considered more scalable because it does not require the same amount of energy.

The first PoS mechanism with a strict security system is "Ouroboros," presented by Kiayias et al. (2016), used by Cardano. It offers a high level of security and scalability, although its speed is less than other consensus mechanisms. Subsequently, other PoS consensus mechanisms were created, like the one used by Algorand and Bitgreen. Algorand uses Pure Proof-of-Stake, invented by Silvio Micali. PPoS randomly selects members independently of their stake, increasing the decentralization and security of the blockchain. Instead, Bitgreen employs a PoS protocol called "Green protocol." This protocol combines PoS and masternode mining,[5] ensuring energy efficiency and speed of transactions.

Delegate-Proof-of-Stake (DPoS), seen in Tron and EOS, is a variant of PoS. Unlike traditional PoS, DPoS lets the entire network, even those not meeting validator requirements, participate in transaction validation. Ordinary members vote for representatives to handle validation tasks, potentially centralizing the network (Bamakan et al., 2020). Representatives, determined by the delegate selection process where participants pool their tokens, validate blocks based on their stakes. Higher-staked representatives receive transaction fees as rewards, shared among supporting members. DPoS operates like a group decision-making process, expediting decisions and enhancing scalability.

Nano uses a variant of DPoS called Open Representative Voting (ORV). The peculiarity of Nano is that it uses a ledger structure based on parallel blockchains rather than a single blockchain, i.e., a block-lattice structure (Morais et al., 2020). Moreover, representatives do not earn transaction fees.

In short, all the PoS mechanisms and their improvement are more energy efficient than the PoW one, because they remove the competition that characterizes it with a selection of parties. There are also other alternatives to the PoS mechanism that are low energy consuming. These alternatives will be described below.

Bamakan et al. (2020) introduce the Proof-of-Burn mechanism, known for its energy efficiency. Users avoid wasting energy or time to earn rewards by "burning" some of their cryptocurrencies. They send these to an "eater address" to receive coins, tokens, or privileges. This consensus mechanism does not consume energy but leverages participants' willingness to accept short-term loss for long-term gain. Similar to PoS, the more coins burned, the higher the chance of creating the next block. Note that PoW cryptocurrencies are required for this process. Additionally,

---

[4]A detailed description of the security problems of the PoS mechanism is provided by Nguyen et al. (2019). The authors refer to: the nothing-at-stake attacks (i.e., when participant create different blocks at the same time and they continue two mine two different chains); the double-spending attacks (i.e., the actor spends the funds, and then attempts to remove the transaction from the blockchain in order to preserve the funds); and the long-range attack (in which an adversary wants to compromise a blockchain with a false chain).

[5]Masternode mining means that participants verify transactions without submitting new blocks.

Proof-of-Burn enhances decentralization, preventing centralization of investors and ensuring decisions are not monopolized by a few. It is considered more secure and relatively fast, exemplified by its use in Slimcoin.

Hashgraph is another consensus mechanism used by Hedera.[6] This consensus mechanism does not use computing power to verify transactions; instead, transactions are confirmed using a protocol based on node communication. Each member reaches a consensus using their knowledge of what the other participants may know. In addition, users do not need to use specialized hardware but can run hashgraph nodes on easily available hardware. In terms of security, the consensus algorithm is the Asynchronous Byzantine fault tolerant,[7] which offers the highest possible level of security for a distributed system. Additionally, by eliminating the traditional voting process, this consensus mechanism enhances the scalability and the rapidity of the process.

The IOTA Tangle, instead, uses another consensus mechanism called the Fast Probabilistic Consensus and uses the Directed Acyclic Graph algorithm (Popov & Buchanan, 2021). Consequently, the Tangle architecture does not assume the form of a chain but rather a network of nodes. This allows all transactions to occur simultaneously rather than sequentially, reducing the energy and time required to complete them. The IOTA Tangle is also designed to be highly scalable, and security is ensured by approving two previous transactions before making a new one.

Chase and MacBrough (2018) analyze the XRP Ledger Consensus Protocol (XRP is its native token). The underlying principle of the XRPL consensus protocol is trust. Each member of the network selects a set of validators who agree or disagree that a series of transactions should take place and that a given ledger is the result. If a high percentage of validators agree on a set of transactions, the server declares consensus. Conversely, the validators modify their proposal to come closer to the result of the other validators they trust. To validate an invalid transaction, more than 80% of the validators would have to agree. This entire process does not rely on mining, making the consensus mechanism energy efficient. At the same time, this consensus mechanism is also considered safer than the PoW.

Stellar Lumens (XLM) uses a different consensus protocol: the Stellar Consensus Protocol, proposed by Luu et al. (2015). This protocol is based on three concepts: the secure identification of network participants, their placement in different committees, and the execution of a consensus protocol within each committee (to propose blocks in parallel). Finally, a concluding committee is formed to combine the outcomes of the subcommittees to generate an ordered blockchain data structure. Processing more blocks simultaneously (and thus parallelizing the available computing power) increases scalability, rapidity, and sustainability, making energy consumption proportional to the result achieved.

[6] Source: Investopedia, https://www.investopedia.com/terms/h/hashgraph-consensus-mechanism.asp#toc-how-is-hashgraph-different-from-blockchain

[7] "This is a technical term meaning that no single node (or small group of nodes) can prevent the network from reaching consensus, nor can they change the consensus once it has been reached." Source: Hedera, https://hedera.com/learning/hedera-hashgraph/what-is-hashgraph-consensus

Holochain (Holo's cryptocurrency) diverges from traditional blockchains by utilizing a combination of cryptographic methods, including DHT (Distributed Hash Table), hash chains, and cryptographic signs. The DHT serves as the primary method for data sharing on Holochain, eliminating the need for mining or staking to verify transactions. Operating as a node on Holochain is energy efficient and viable on mobile devices, thanks to the DHT. This agent-based system verifies each transaction individually, with verified data distributed to other nodes for reverification using a validation scheme, known as "peer validation." Not relying on a consensus mechanism, this process enhances the network's resilience against consensus-based attacks.

In addition, there are some storage-based consensus mechanisms like Proof-of-Space and Proof-of-Space-Time. The former requires the miner to allocate a certain amount of memory for the proof (Ateniese et al., 2014), while the second requires both a certain amount of memory and time (Moran & Orlov, 2016). Filecoin is an example of a cryptocurrency using Proof-of-Space, while the Chia blockchain is the first to adopt the PoST consensus mechanism.

Proof-of-Space is sometimes referred to as Proof-of-Capacity (PoC) or Proof-of-Storage (Andoni et al., 2019). Proof-of-Commitment (PoC+), instead, represents the evolution of the PoC mechanism. This new consensus mechanism is used by Signum (Signa is its cryptocurrency) and combines two algorithms (PoC and PoS). Specifically, miners can enhance their effective storage capacity by holding a certain number of cryptocurrencies (the more cryptocurrencies held, the greater the storage capacity). The likelihood of successfully contributing to block creation increases in proportion to the quantity of dedicated storage space, discouraging malicious actors.

Unlike Bitcoin mining, which requires specialized hardware (e.g., ASICs), the consensus mechanisms described above use any standard hard drive, reducing energy consumption and e-waste, and improving decentralization.

In summary, we have described the main innovations in the consensus mechanisms of green cryptocurrencies. In particular, we have focused on the consensus mechanisms used by the cryptocurrencies selected as case studies, as reported below.

3 Methodology

This study uses a multiple case study approach to explore the potential of green cryptocurrencies to mitigate climate change. The Table 3 highlights the green cryptocurrencies selected as cases:

These cryptocurrencies were selected because they are considered to be the most environmentally friendly during 2022 (according to CoinMarketCap[8]). Indeed, these green cryptocurrencies use a different consensus mechanism from PoW,

[8] Source: CoinMarketCap, https://coinmarketcap.com/community/articles/38211/

Table 3 Green cryptocurrencies and their characteristics

Crypto	Foundation	Consensus mechanism	Characteristics	Decentralized	Open source
Cardano (ADA)	Founded in 2015 by Charles Hoskinson and launched in 2017.	Ouroboros protocol.	Considered the first truly secure PoS protocol.	V	V
Ripple (XRP)	Founded in 2011 by Jed McCaleb and Chris Larsen and launched in 2012.	XRP Ledger Consensus Protocol.	Based on trust.	V	V
Holo (HOT)	Founded by Arthur Brock and Eric Harris-Braun in 2017 and launched in 2018.	DHT protocol.	Agent-based protocol.	V	V
Hedera (HBAR)	Founded by Dr. Leemon Baird and Mance Harmon in 2018 and launched in the same year.	Hashgraph.	Transaction verification relies on node communication.	V	V
Chia (XCH)	Founded in 2017 by Bram Cohen and launched in 2021.	Proof-of-Space-and-Time.	It requires the miner to allocate a certain amount of memory for a certain amount of time.	V	V
Tron (TRX)	Founded in 2017 by a Singapore-based nonprofit organization, the Tron Foundation, and launched in the same year.	Delegated Proof-of-Stake.	Ordinary members vote on their representatives for the performance of validation tasks.	V	V
IOTA (MIOTA)	Founded by Sergey Ivancheglo, Serguei Popov, David Sønstebø, and Dominik Schiener and launched in 2015.	Fast Probabilistic Consensus.	It uses DAG structure and all the transactions occur simultaneously.	X	V
Nano (XNO)	Founded by Collin LeMahieu and launched in 2015.	Open Representative Voting (ORV).	Register technology called DAG. The structure is guaranteed by a network of delegates and is a block-lattice structure.	V	V

(continued)

Table 3 (continued)

Crypto	Foundation	Consensus mechanism	Characteristics	Decentralized	Open source
Algorand (ALGO)	Founded by Silvio Micali in 2017 and launched in 2019.	Pure proof-of-stake.	A block producer and a committee of validators are randomly selected from all ALGO holders.	V	V
Bitgreen (BITG)	Launched in 2017.	Green Protocol.	To ensure a fair distribution of coins, the Green Protocol uses PoS and masternode-only mining combined with fixed block rewards.	V	X
Signum (SIGNA)	Founded by Signum Network Association, a Swiss Not-for-Profit Organization and launched in 2014 as Burstcoin.	Proof of Commitment.	Combined PoC and PoS consensus mechanisms.	V	V
Stellar Lumens (XLM)	Founded by Jed McCaleb and Joyce Kim, and launched in 2014.	Stellar Consensus Protocol.	The existence of several committees, each running consensus protocols and proposing blocks in parallel.	V	V
EOS (EOS)	Founded by the company Block. one and launched in 2018.	Delegated Proof-of-Stake.	Ordinary members vote on their representatives for the performance of validation tasks.	V	V

reducing their energy consumption. Additionally, they are listed in the order reported by CoinMarketCap, ranked from the most to the least sustainable. It is essential to note that these green cryptocurrencies have garnered attention from both investors and the media.

We have collected data for the case studies from various sources, including the cryptocurrencies' websites, whitepapers, technical documentation, academic literature, and media reports. Furthermore, we have employed a qualitative approach to analyze the cases, aiming to identify common themes and patterns. Notably, the consensus mechanisms of these cryptocurrencies, discussed in the previous section, differ for almost all the cryptocurrencies.

Regarding the information collected for the qualitative analysis, we have focused on: the foundation, the characteristics of the consensus mechanism, the popularity, the market capitalization, the kWh and the CO_2 emissions for transaction, two different ESG ratings, and the pros and cons. However, not all information is available for all cryptocurrencies.

In addition, we conducted a preliminary statistical analysis to observe whether the variation in the number of green cryptocurrencies is related to the electricity costs and if there is a difference between green and non-green cryptocurrencies. For this reason, we collected information not only on green cryptocurrencies but also on non-green cryptocurrencies. Since there are 13 green cryptocurrencies that were taken as case studies for the qualitative analysis, we decided to use the same number of non-green cryptocurrencies. To select them, we followed the Cryptowisser website, which lists "beige" cryptocurrencies (for simplicity's sake, referred to as brown). Specifically, we selected only those cryptocurrencies based on the PoW or PoW hybrid mechanism, as these are the most energy intensive. We then compared this information with that of CoinMarketCap, also to analyze data availability. In conclusion, the list of brown cryptocurrencies consists of 13 cryptocurrencies.[9]

To perform our analysis, we have collected data on the closing price of cryptocurrencies, the market capitalization, and the cost of electricity[10] (taking into account the available data between January 01, 2018, and September 01, 2023). The ratio of market capitalization to the closing price of cryptocurrencies is used to calculate the amount of cryptocurrency in circulation for each month. Then, we calculated the return of cryptocurrencies on a monthly basis and the variation of the supply in circulation, also on a monthly basis. The total number of monthly observations is 1584 (735 for the green cryptocurrencies and 849 for the brown ones).

---

[9] We have voluntarily excluded Bitcoins from the sample due to their large market capitalization and the large number of cryptocurrencies in circulation. The final sample of brown cryptocurrencies include: Bitcoin Cash, Bitcoin SV, Dash, DigiByte, Dogecoin, Ethereum Classic, Horizen, Litecoin, Monero, Ravencoin, Siacoin, Syscoin, and Zcash. Note that our list does not exactly match Cryptowisser's "beige" cryptocurrencies, as there is an inconsistency between the latter source and CoinMarketCap. We have therefore decided to rely on the second source, as it is more frequently cited in the literature. In fact, eCash has implemented the PoS mechanism (while Cryptowisser states that it uses PoW). In addition, NEM does not rely on the PoW mechanism (therefore not included in our selection criteria). For these reasons we have selected two other cryptocurrencies from CoinMarketCap's PoW list (taking into account data availability between January 01, 2018, and September 01, 2023). Furthermore, as Cryptowisser only covers 12 "beige" cryptocurrencies, we have added another one using the same selection criteria.

In summary, Cryptowisser provides a list of 12 "beige" cryptocurrencies, and due to discrepancies we have replaced two cryptocurrencies, eCash and NEM, with Dogecoin and DigiByte, and we have also added Syscoin (since we have decided to use the same number of green and brown cryptocurrencies). It should also be noted that Dogecoin is not listed as "beige" by Cryptowisser. However, it has a high power consumption (according to Digiconomist, https://digiconomist.net/dogecoins-growing-energy-problem/).

[10] We used the average monthly cost of electricity per kWh in the USA because, according to Statista, most Bitcoin mining takes place in the USA. For simplicity, we assume this is the case for the other cryptocurrencies. Source: https://www.statista.com/statistics/1200477/bitcoin-mining-by-country/

Regarding the source of information for the quantitative analysis, we referred to the website Investing[11] for prices, CoinMarketCap[12] for market capitalization, and Federal Reserve Economic Data[13] for electricity costs.

Next, we present Graph 1, which compares the market capitalization of green (green line) and brown (brown line) cryptocurrencies. In particular, we can see that the trend is very similar, regardless of the total market capitalization. Indeed, there are certainly some common elements that could influence the success of cryptocurrencies.

In addition, brown cryptocurrencies considered in our sample usually need between 1 and 100 kWh per transaction, while green cryptocurrencies require less than 1 kWh per transaction. It is also interesting to note that a single Bitcoin transaction requires more than 600 kWh (according to Digiconomist). These numbers can help one understand the importance of the development and adoption of green cryptocurrencies.

The following graph represents the Bitcoin energy consumption (in terms of TWh per year). This makes it clear how the energy consumed by Bitcoin mining has increased over time (as previously mentioned in Sect. 2.1).

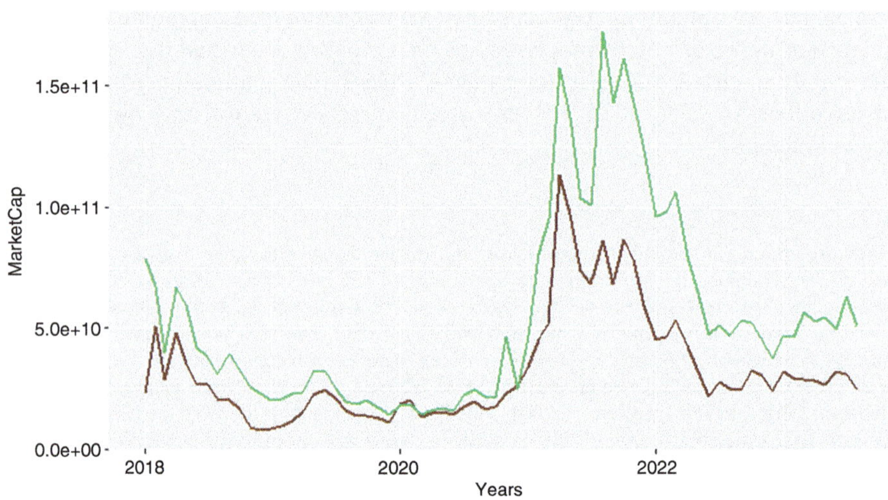

Graph 1 Market capitalization: Green vs Brown Cryptocurrencies

[11] https://it.investing.com/crypto/currencies

[12] https://coinmarketcap.com/

[13] https://fred.stlouisfed.org/tags/series?t=electricity

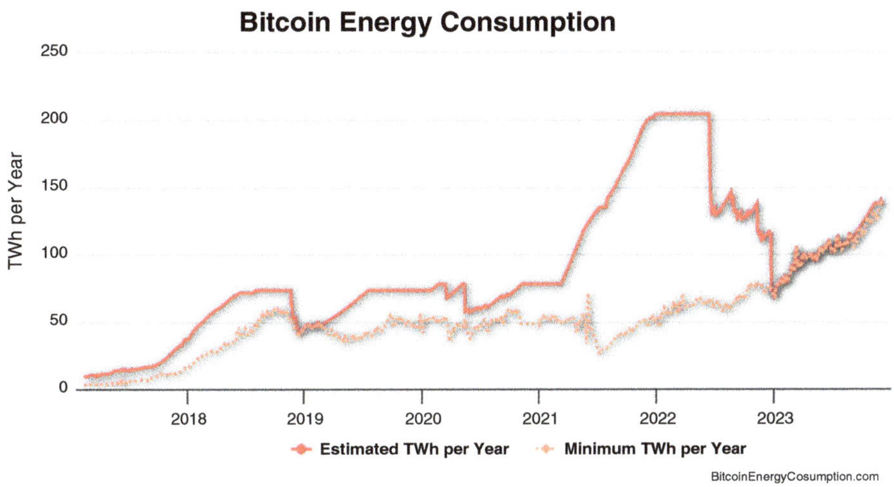

Source: Digiconomist

4 Results

The market capitalization and popularity of cryptocurrencies are collected from CoinMarketCap. In particular, the popularity is calculated based on how many users include the cryptocurrency in their watchlist. The rank includes 9.1K cryptocurrencies shown on CoinMarketCap (for example, Cardano is ranked 63rd among all cryptocurrencies).

Furthermore, we collect energy consumption and CO_2 emissions for transactions from various sources, particularly from Cryptowisser. Additionally, we obtain ESG ratings from CCData[14] and Green Crypto Research.[15]

Table 4 summarises the above aspects.

The European Parliament[16] defines carbon neutrality as maintaining an equilibrium between the release of carbon emissions and the absorption of carbon from the atmosphere through carbon sinks. In our case, as per Cryptowisser, a carbon-neutral cryptocurrency must emit emissions of ≤ 0 kg CO_2 per transaction, i.e., it should not produce any carbon emissions. Specifically, we can consider three of the above cryptocurrencies—Hedera, Algorand, and EOS—as carbon neutral.

Cryptocurrencies with ≤ 0.00475 kg CO_2 per transaction are considered "as, or more, energy efficient" than a VISA transaction, Google Search, and a regular email without attachment." In this case, we are referring to Ripple, Holo, IOTA, and Stellar Lumens. Instead, the other cryptocurrencies are defined as more energy

[14] https://ccdata.io/research/esg-rankings.

[15] https://www.greencryptoresearch.com/tokens.

[16] https://www.europarl.europa.eu/news/en/headlines/society

Table 4 Focus on green aspects

Crypto	Market Cap. (15-09-23)	Popularity	kWh/tr.	g CO_2e/tr.	ESG Rating: CCRI and CCData	ESG Rating: GCR
Cardano (ADA)	$8.74B	63rd	0.5479	≥ 4.75 ; < 475	A	A
Ripple (XRP)	$26.2B	70th	< 0.01	< 4,75	A	C
Holo (HOT)	$179.85M	195th	< 0.01	< 4,75	n.a.	n.a.
Hedera (HBAR)	$1.66B	156th	< 0.01	≤ 0	n.a.	B
Chia (XCH)	$231.77M	424th	0.023	≥ 4.75 ; < 475	n.a.	n.a.
Tron (TRX)	$7.47B	111th	≥ 0.01 ; <1	≥ 4.75 ; < 475	BB	B
IOTA (MIOTA)	$457.79M	196th	0.00011	< 4,75	n.a.	n.a.
Nano (XNO)	$81.66M	336th	<0.111	≥ 4.75 ; < 475	n.a.	n.a.
Algorand (ALGO)	$731.53M	132nd	< 0.01	≤ 0	BB	B
Bitgreen (BITG)	$125,623	3012th	0.4242	≥ 4.75 ; < 475	n.a.	n.a.
Signum (SIGNA)	$2.7M	3334th	0.5	≥ 4.75 ; < 475	n.a.	n.a.
Stellar Lumens (XLM)	$3.28B	110th	< 0.01	< 4,75	BB	B
EOS (EOS)	$621.12M	180th	< 0.01	≤ 0	BB	B

efficient than an email with a large attachment. In addition, Cryptowisser includes "beige" cryptocurrencies that generate emissions between 0.475 kg CO_2 and 47.5 kg CO_2 per transaction. The last category is referred to as "brown" and includes cryptocurrencies that produce more than 47.5 kg of CO_2 emissions per transaction. The only cryptocurrency in this category is Bitcoin.

The power consumption and carbon footprint per transaction of the analyzed cryptocurrencies are significantly lower than that of a single Bitcoin transaction. A Bitcoin transaction requires 620.17 kWh and produces 345.90 kgCO_2, equivalent to the energy used by the average US household for 21 days and the carbon footprint of 766,641 VISA transactions. In contrast, all the cryptocurrencies chosen as case studies are more sustainable than Bitcoin (including its variant, called Bitgreen).

Some sources have attempted to determine an ESG rating for cryptocurrencies to consider their sustainability. The main problem is the lack of regulation, and different sources use different methodologies (which are usually not fully explained as they may contain discretionary elements or may change over time). In this chapter, we have reported two ESG ratings: one from CCRI&CCData and one from GCR. The first gives a final score from AA to E, in which each category (the E, the S, and the G) represents 33.33% of the overall score. The second gives a final grade from A+ to D- and does not explain how the categories are taken into account. These elements make comparison difficult.

We will now conclude the discussion of the case studies by highlighting the main advantages and disadvantages of the different technologies employed (in Table 5).

Table 5 Advantages and disadvantages of the green cryptocurrencies

Crypto	Advantages	Disadvantages
Cardano (ADA)	● Protocol to prevent hacking ● Third-generation blockchain ● Credibility ● Fast transactions ● Minimum fees	● Existence of competing cryptographic tools
Ripple (XRP)	● Minimum fees ● Fast transactions ● Accepted by many banks ● Overtaking VISA	● Vertical management ● Many decentralized competitors (Ripple's consensus mechanism is actually not the most decentralized) ● Controversy with the SEC
Holo (HOT)	● Holochain is configurable and customizable ● Unique consensus algorithm ● Significantly lower fees than many other blockchain networks	● Huge supply of tokens (investors prefer to invest when the supply of tokens is low) ● Presence of many competitors ● Disagreement on the use of a technology other than the traditional one
Hedera (HBAR)	● Speed of transactions ● Very low and fixed costs ● Supports smart contracts ● Provides a high level of security	● Centralized network (managed by a consortium of 39 companies) ● It is not possible to set up your own validator node, as you must obtain permission from the board ● A maximum of 10 transactions per second can be processed
Chia (XCH)	● Innovative consensus protocol ● The farming process seems more intuitive	● Uncertain (huge success was expected but not achieved) ● Unsafe (PoST still has shortcomings) ● Increase in price of hard disks needed for farming ● Increase in e-waste
Tron (TRX)	● Processes 2000 transactions per second ● Low commissions ● Popular on social networks ● Easy DApp implementation	● The project is considered overrated ● Founder Justin Sun is surrounded by controversy
IOTA (MIOTA)	● No commissions ● Processes around 300 transactions per second ● Third largest distributed ledger project (behind Ripple and Stellar)	● Loss of enthusiasm for the project ● The IOTA network is seen as not secure enough
Nano (XNO)	● Fast transactions ● Free transactions ● Unique consensus protocol	● Coin adoption is slow and only a few exchanges have listed them ● DApps or smart contracts are not supported ● Nano is vulnerable to network attack by sending spam with more transactions than it can handle

(continued)

Table 5 (continued)

Crypto	Advantages	Disadvantages
Algorand (ALGO)	● Fast transactions ● Low fees ● Up to 6000 transactions per second ● Cryptocurrency with a long history of success ● Professional development team ● Solid blockchain structure	● Not fully decentralized ● Low level of adoption
Bitgreen (BITG)	● Fast transactions ● Low fees	[They are not highlighted]
Signum (SIGNA)	● Lower hardware requirements for mining	● Payment of higher fees to speed up transactions
Stellar Lumens (XLM)	● No disputes with regulators ● Relatively fast transactions ● Easy to access on exchange platforms ● Transactions typically cost less than a penny	● Aggressive competition with Ripple ● Not very popular ● No mining rewards ● Low trading volumes ● Low coin value
EOS (EOS)	● Low fees ● Strong governance model that enhances the security and transparency of the platform	● Some believe it is rather centralized ● When the network reaches maximum capacity it can limit transactions ● Validating a transaction can take 3 minutes ● Security issues in 2018

Green cryptocurrencies offer several advantages, including fast transaction speeds and low fees[17], with some not imposing fees at all. Moreover, they utilize distinctive and environmentally sustainable consensus protocols, ensuring minimal energy consumption for each transaction. Nevertheless, each method involves a trade-off between pros and cons (Andoni et al., 2019).

As we know, green cryptocurrencies share certain characteristics: low consumption per transaction; absence of PoW mechanisms; scalability; and, a tendency to be less decentralized than cryptocurrencies using PoW mechanisms. The latter two elements are critical points, as we will see shortly.

Scalability refers to a blockchain's ability to handle an increase in the amount of work as the number of participants or servers increases. Instead, decentralization is one of the most important attributes of blockchain because it promotes the participation of each individual in the network in the process, improving robustness against weak points and avoiding the costs of intermediate transactions (Wang et al., 2022).

With regard to these factors, the key challenge is to enable the scalability of the blockchain without compromising its decentralization and security (Sanka & Cheung, 2021). While PoS mechanisms are known for scalability and speed, PoW

[17] Transaction fees are paid by users, using the blockchain's cryptoasset, when they perform an action on the blockchain (such as adding data).

mechanisms are characterized by decentralization.[18] It is worth noting that scalability is inversely proportional to decentralization, because as the number of participants increases it takes more time to reach a consensus on a transaction.

The issue lies in alternative consensus mechanisms, like PoS, which can be more centralized, diminishing blockchain security. For instance, DPoS, as noted by Andoni et al. (2019), is prone to low voting participation, resulting in system centralization due to reduced energy consumption. This shift away from PoW's competitive nature leads to increased centralization, as highlighted by Sanka and Cheung (2021). They point out that PoS mechanisms, exemplified by EOS (using DPoS), are less secure, sparking debates over EOS's decentralization with 86% of tokens held by less than 1% of users and vulnerability issues resulting in substantial losses (Huang et al., 2020). Additionally, IOTA faced security problems, with its hash function being modified following a major hack in 2020 that resulted in the theft of millions of dollars.

Regulators need to consider the different forms of technology used and adopt different regulatory policies. While certain mechanisms may address climate change and enhance scalability, they might simultaneously give rise to safety or concentration issues, as seen with PoS mechanisms. Conversely, other mechanisms may be safer but pose environmental challenges, notably PoW mechanisms. For this reason, a better understanding by regulators can help them to identify risks and, consequently, to adapt to regulation (Bains et al., 2022).

Moreover, some previous studies have highlighted the necessity for regulatory intervention, particularly concerning the environmental aspect (Truby, 2018; Truby et al., 2022). Nevertheless, most blockchains prefer PoW because it is already well tested, while PoS mechanisms are relatively new. Truby et al. (2022) describe various interventions that regulators could implement to promote the spread of sustainable consensus mechanisms. They could incentivize the adoption of a more sustainable version of the PoW mechanism (e.g., Proof-of-Useful-Work) or promote a transition to a PoS mechanism. In addition, regulators may choose to impose regulations against the use of PoW or to use fiscal instruments (where the tax rate depends on the level of energy consumption).

Furthermore, the unique consensus mechanisms used by some of these green cryptocurrencies can lead to disagreements.

Finally, we report the results of our statistical analysis in Tables 6 and 7.

In Table n.4.3, we observe that the statistically significant result is represented by the relationship between the cost of electricity and the monthly variation in the amount of cryptocurrencies. Therefore, we decided to split the sample between green and brown cryptocurrencies to assess the impact of the same independent variable, differentiating between cryptocurrencies.

Summarizing, the increase in electricity cost produces a negative impact on the variation of brown cryptocurrencies but, at the same time, does not influence the variation of green ones. As expected, the change in the number of green

[18]According to Sanka and Cheung (2021), "Algorand's throughput was 125 times that of Bitcoin."

Table 6 Full sample

Dependent variables	Monthly returns of cryptocurrencies	Variation in circulating supply of cryptocurrencies
Independent variables	Monthly variation of electricity cost	Electricity cost
Results	We find no statistical significance between the return of cryptocurrencies and the variation in electricity cost. This means that returns do not seem to be affected by the variation in electricity cost. The results are still not statistically significant when we use a lagged independent variable.	We find a negative and statistically significant (5%) relationship between the monthly variation of cryptocurrencies and the cost of electricity. This means that when the cost of electricity increases, it has a negative impact on the variation of the amount of cryptocurrencies.

Table 7 Green vs Brown cryptocurrencies

Dependent variables	Variation in circulating supply of green cryptocurrencies	Variation in circulating supply of brown cryptocurrencies
Independent variables	Electricity cost	Electricity cost
Results	We find no statistical significance between the monthly variation of green cryptocurrencies and the electricity cost.	We find a negative and statistically significant (5%) relationship between the monthly variation of brown cryptocurrencies and the cost of electricity.

cryptocurrencies is not influenced by the cost of electricity. However, this issue requires further in-depth analysis, which we leave to future studies.

In conclusion, despite some disadvantages, the environmental benefits of green cryptocurrencies are evident, making their adoption crucial in addressing climate change.

5 Conclusions

According to the European Central Bank, the financial sector faces an increasing climate transition risk due to its growing exposure to cryptocurrencies, which have a significant carbon footprint. Green cryptocurrencies are designed to be more energy efficient and sustainable than traditional cryptocurrencies, employing various innovative consensus mechanisms and technologies to reduce their environmental impact. Analyses on electricity consumption and carbon dioxide emissions lead to the conclusion that further blockchain development should avoid PoW mechanisms to mitigate global temperature rise (Shi et al., 2023). Mora et al. (2018) suggest that any cryptocurrency development should prioritize reducing energy demand.

In this study, we investigated the potential of green cryptocurrencies to mitigate climate change. First of all, we provide a detailed literature review on the energy consumption and emissions of PoW cryptocurrencies. Thereafter, we compare the latter consensus mechanism with less energy-intensive ones, selecting some green cryptocurrencies as case studies. In particular, we have described their consensus mechanisms, their energy consumption and, finally, their advantages and disadvantages. In addition, we present a statistical analysis highlighting that the cost of electricity has a negative impact on the variation in the number of brown cryptocurrencies, but not on the green one. However, this is only a preliminary analysis and further research is needed. We leave this topic to future research. Furthermore, we recommend exploring alternative consensus mechanisms with the potential for being energy efficient.

However, we also identified several challenges that need to be addressed before green cryptocurrencies can achieve widespread adoption, including the need for greater regulation and improved security. The cryptocurrency industry is currently largely unregulated, posing risks for investors and consumers. Additionally, as green cryptocurrencies are still relatively new, there have been security breaches in the past, leading to significant losses for platforms such as EOS and MIOTA due to hacking attacks, amounting to millions of dollars. Therefore, it is crucial to develop and implement robust security measures to safeguard green cryptocurrencies and enhance the security of their consensus mechanisms.

For this reason, it is important to develop regulatory frameworks that support the growth of green cryptocurrencies while also protecting investors and consumers. Previous studies have suggested policy intervention as a solution. Specifically, Truby (2018) and Truby et al. (2022), focus on different proposals: regulators could use prohibitive regulations to "encourage" traders to avoid PoW cryptocurrencies or tax them according to the polluter-pays principle. Regulators can also consider how to motivate industry and miners to choose voluntary green cryptocurrencies. Of course, we have to ask ourselves if it is possible and how to promote a transition to a less energy-consuming blockchain without damaging this sector.

In any case, an environmentally friendly state can promote these tools, but the limitations underscore the necessity for international intervention. According to Truby et al. (2022), the switch to sustainable alternatives has already taken place due to energy costs, but at the same time, the increasing value of holding Bitcoin incentivizes mining activity. In particular, we must consider the possibility of actors moving to other countries to continue mining. For this reason, international cooperation in regulating digital currencies is particularly important. This is particularly crucial for policymakers who must establish global standards for cryptocurrency regulation, aiming to encourage the adoption and enhancement of green cryptocurrencies. Additionally, there is a need to raise awareness among investors. Indeed, cooperation is essential for reducing the environmental impact of cryptocurrencies.

In conclusion, mitigating the environmental impact of cryptocurrency mining is crucial, and green cryptocurrencies could provide a solution, albeit with areas for improvement related to decentralization, scalability, and security.

References

Alzoubi, Y. I., & Mishra, A. (2023). Green blockchain–A move towards sustainability. *Journal of Cleaner Production, 430*, 139541. https://doi.org/10.1016/j.jclepro.2023.139541

Andoni, M., Robu, V., Flynn, D., Abram, S., Geach, D., Jenkins, D., McCallum, P., & Peacock, A. (2019). Blockchain technology in the energy sector: A systematic review of challenges and opportunities. In *Renewable and Sustainable Energy Reviews* (Vol. 100, pp. 143–174). Elsevier Ltd. https://doi.org/10.1016/j.rser.2018.10.014

Ateniese, G., Bonacina, I., Faonio, A., & Galesi, N. (2014). Proofs of Space: When Space Is of the Essence. *Security and Cryptography for Networks*, 538–557. https://link.springer.com/chapter/10.1007/978-3-319-10879-7_31

Badea, L., & Mungiu-Pupăzan, M. C. (2021). The Economic and Environmental Impact of Bitcoin. *IEEE Access, 9*, 48091–48104. https://doi.org/10.1109/ACCESS.2021.3068636

Bains, P., Ismail, A., Melo, F., & Sugimoto, N. (2022). *Regulating the Crypto Ecosystem. International Monetary Fund.*

Bamakan, S. M. H., Motavali, A., & Babaei Bondarti, A. (2020). A survey of blockchain consensus algorithms performance evaluation criteria. *Expert Systems with Applications, 154*, 113385.

Bentov, I., Lee, C., Mizrahi, A., & Rosenfeld, M. (2014). Proof of activity: Extending Bitcoin's Proof of Work via Proof of Stake. *International Association for Cryptologic Research.* https://eprint.iacr.org/2014/452.pdf

Bondarev, M. (2020). Energy consumption of bitcoin mining. *International Journal of Energy Economics and Policy, 10*(4), 516–524. https://doi.org/10.32479/ijeep.9276

Calandra, D., Secinaro, S., Massaro, M., Dal Mas, F., & Bagnoli, C. (2023). The link between sustainable business models and Blockchain: A multiple case study approach. *Business Strategy and the Environment, 32*(4), 1403–1417. https://doi.org/10.1002/bse.3195

Campbell, R. (2020). *Former Mine in Norway is Providing a Sustainable alternative for bitcoin.* CoinJournal. Accessed Aug 22, 2022, from https://coinjournal.net/news/a-former-mine-is-norway-is-providing-a-sustainable-alternative-for-bitcoin/

Chase, B., & MacBrough, E. (2018). Analysis of the XRP ledger consensus protocol. *ArXiv.*. http://arxiv.org/abs/1802.07242

Chen, D. (2018). Utility of the Blockchain for Climate Mitigation. *The Journal of the British Blockchain Association, 1*(1), 1–9. https://doi.org/10.31585/jbba-1-1-(6)2018

de Vries, A. (2018). Bitcoin's Growing Energy Problem. In *Joule* (Vol. 2, pp. 801–805). Cell Press. https://doi.org/10.1016/j.joule.2018.04.016

de Vries, A. (2019). Renewable energy will not solve Bitcoin's sustainability problem. *Joule, 3*(4), 893–898. https://doi.org/10.1016/j.joule.2019.02.007

de Vries, A. (2020). Bitcoin's energy consumption is underestimated: A market dynamics approach. In *Energy Research and Social Science* (Vol. 70). Elsevier Ltd.. https://doi.org/10.1016/j.erss.2020.101721

de Vries, A. (2021). Bitcoin boom: What rising prices mean for the network's energy consumption. *Joule, 5*(3), 509–513. https://doi.org/10.1016/j.joule.2021.02.006

de Vries, A., & Stoll, C. (2021). Bitcoin's growing e-waste problem. *Resources, Conservation and Recycling, 175*, 105901. https://doi.org/10.1016/j.resconrec.2021.105901

Eklund, P. W., & Beck, R. (2019). *Factors that impact blockchain scalability.* 11th International Conference on Management of Digital EcoSystems, MEDES 2019, 126–133. https://doi.org/10.1145/3297662.3365818

Erdogan, S., Ahmed, M. Y., & Sarkodie, S. A. (2022). Analyzing asymmetric effects of cryptocurrency demand on environmental sustainability. *Environmental Science and Pollution Research, 29*(21), 31723–31733. https://doi.org/10.1007/s11356-021-17998-y

Gallersdörfer, U., Klaaßen, L., & Stoll, C. (2020). Energy Consumption of Cryptocurrencies Beyond Bitcoin. *Joule, 4*, 1839–1851. https://www.cell.com/joule/pdfExtended/S2542-4351(20)30331-7

Goodkind, A. L., Jones, B. A., & Berrens, R. P. (2020). Cryptodamages: Monetary value estimates of the air pollution and human health impacts of cryptocurrency mining. *Energy Research and Social Science, 59*. https://doi.org/10.1016/j.erss.2019.101281

Huang, Y., Wang, H., Wu, L., Tyson, G., Luo, X., Zhang, R., Liu, X., Huang, G., & Jiang, X. (2020). Understanding (Mis)Behavior on the EOSIO Blockchain. *ACM SIGMETRICS Performance Evaluation Review, 48*(1), 83–84. https://doi.org/10.1145/3410048.3410096

Huynh, A. N. Q., Duong, D., Burggraf, T., Luong, H. T. T., & Bui, N. H. (2022). Energy Consumption and Bitcoin Market. *Asia-Pacific Financial Markets, 29*(1), 79–93. https://doi.org/10.1007/s10690-021-09338-4

Irresberger, F., John, K., Mueller, P. C., & Saleh, F. (2021). *The public blockchain ecosystem: An empirical analysis*. Paris Webinar on Financial Technology and Crypto. https://ssrn.com/abstract=3592849

Jiang, S., Li, Y., Lu, Q., Hong, Y., Guan, D., Xiong, Y., & Wang, S. (2021). Policy assessments for the carbon emission flows and sustainability of Bitcoin blockchain operation in China. *Nature Communications, 12*(1). https://doi.org/10.1038/s41467-021-22256-3

Jones, B. A., Goodkind, A. L., & Berrens, R. P. (2022). Economic estimation of Bitcoin mining's climate damages demonstrates closer resemblance to digital crude than digital gold. *Scientific Reports, 12*(1). https://doi.org/10.1038/s41598-022-18686-8

Kiayias, A., Konstantinou, I., Russell, A., David, B., & Oliynykov, R. (2016). *Ouroboros: A provably secure proof-of-stake Blockchain protocol.* International Association for Cryptologic Research. https://eprint.iacr.org/2016/889.pdf

Köhler, S., & Pizzol, M. (2019). Life Cycle Assessment of Bitcoin Mining. *Environmental Science and Technology, 53*(23), 13598–13606. https://doi.org/10.1021/acs.est.9b05687

Krause, M. J., & Tolaymat, T. (2018). Quantification of energy and carbon costs for mining cryptocurrencies. *Nature Sustainability, 1*(11), 711–718. https://doi.org/10.1038/s41893-018-0152-7

Li, J., Li, N., Peng, J., Cui, H., & Wu, Z. (2019). Energy consumption of cryptocurrency mining: A study of electricity consumption in mining cryptocurrencies. *Energy, 168*, 160–168. https://doi.org/10.1016/j.energy.2018.11.046

Luu, L., Narayanan, V., Baweja, K., Zheng, C., Gilbert, S., & Saxena, P. (2015). *SCP: A Computationally-Scalable Byzantine Consensus Protocol For Blockchains*. Cryptology EPrint Archive. https://eprint.iacr.org/2015/1168

Malfuzi, A., Mehr, A. S., Rosen, M. A., Alharthi, M., & Kurilova, A. A. (2020). Economic viability of bitcoin mining using a renewable-based SOFC power system to supply the electrical power demand. *Energy, 203*. https://doi.org/10.1016/j.energy.2020.117843

McCook, H. (2018). *The cost & sustainability of Bitcoin*. Academia. https://www.academia.edu/37178295/The_Cost_and_Sustainability_of_Bitcoin_August_2018_

Mora, C., Rollins, R. L., Taladay, K., Kantar, M. B., Chock, M. K., Shimada, M., & Franklin, E. C. (2018). Bitcoin emissions alone could push global warming above 2°C. In *Nature Climate Change* (Vol. 8, pp. 931–933). Nature Publishing Group. https://doi.org/10.1038/s41558-018-0321-8

Morais, R., Crocker, P., & Melo de Sousa, S. (2020). A Tool for Implementing Privacy in Nano. *IEEE International Conference on Decentralized Applications and Infrastructures (DAPPS), 2020*, 159–163. https://doi.org/10.1109/DAPPS49028.2020.00021

Moran, T., & Orlov, I. (2016). *Simple proofs of space-time and rational proofs of storage*. International Association for Cryptologic Research. https://eprint.iacr.org/2016/035.pdf

Nguyen, C. T., Hoang, D. T., Nguyen, D. N., Niyato, D., Nguyen, H. T., & Dutkiewicz, E. (2019). Proof-of-Stake Consensus Mechanisms for Future Blockchain Networks: Fundamentals, Applications and Opportunities. *IEEE Access, 7*, 85727–85745. https://doi.org/10.1109/ACCESS.2019.2925010

Popov, S., & Buchanan, W. J. (2021). FPC-BI: Fast Probabilistic Consensus within Byzantine Infrastructures. *Journal of Parallel and Distributed Computing, 147*, 77–86. https://doi.org/10.1016/j.jpdc.2020.09.002

Rieger, A., Roth, T., Sedlmeir, J., & Fridgen, G. (2022). We need a broader debate on the sustainability of blockchain. *Joule, 6*(6), 1137–1141. https://doi.org/10.1016/j.joule.2022.04.013

Rogojanu, A., & Badea, L. (2015). *The issue of true money in front of Bitcoin's offensive.* Theoretical & Applied Economics. http://www.ectap.ro/the-issue-of-true-money-in-front-of-the-bitcoins-offensive-angela-rogojanu_liana-badea/a1084/

Saleh, F. (2020). Blockchain Without Waste: Proof-of-Stake. *The Review of Financial Studies, 34*(3), 1156–1190. https://doi.org/10.1093/rfs/hhaa075

Sanka, A. I., & Cheung, R. C. C. (2021). A systematic review of blockchain scalability: Issues, solutions, analysis and future research. *Journal of Network and Computer Applications, 195*, 103232. https://doi.org/10.1016/j.jnca.2021.103232
Schinckus, C., Nguyen, C. P., & Chong, F. H. L. (2022). Cryptocurrencies' hashrate and electricity consumption: evidence from mining activities. *Studies in Economics and Finance, 39*(3), 524–546. https://doi.org/10.1108/SEF-08-2021-0342
Schinckus, C., Nguyen, C. P., & Ling, F. C. H. (2020). Crypto-currencies trading and energy consumption. *International Journal of Energy Economics and Policy, 10*(3), 355–364. https://doi.org/10.32479/ijeep.9258
Sedlmeir, J., Buhl, H. U., Fridgen, G., & Keller, R. (2020). The Energy Consumption of Blockchain Technology: Beyond Myth. *Business and Information Systems Engineering, 62*(6), 599–608. https://doi.org/10.1007/s12599-020-00656-x
Shi, X., Xiao, H., Liu, W., Lackner, K. S., Buterin, V., & Stocker, T. F. (2023). Confronting the carbon-footprint challenge of Blockchain. *Environmental Science & Technology, 57*(3), 1403–1410. https://doi.org/10.1021/acs.est.2c05165
Stoll, C., Klaaßen, L., & Gallersdörfer, U. (2019). The carbon footprint of Bitcoin. *Joule, 3*(7), 1647–1661. https://doi.org/10.1016/j.joule.2019.05.012
Truby, J. (2018). Decarbonizing Bitcoin: Law and policy choices for reducing the energy consumption of Blockchain technologies and digital currencies. *Energy Research & Social Science, 44*, 399–410. https://doi.org/10.1016/j.erss.2018.06.009
Truby, J., Brown, R. D., Dahdal, A., & Ibrahim, I. (2022). Blockchain, climate damage, and death: Policy interventions to reduce the carbon emissions, mortality, and net-zero implications of non-fungible tokens and Bitcoin. *Energy Research & Social Science, 88*, 102499. https://doi.org/10.1016/j.erss.2022.102499
Wang, T., Hua, H., Wei, Z., & Cao, J. (2022). Challenges of blockchain in new generation energy systems and future outlooks. *International Journal of Electrical Power & Energy Systems, 135*, 107499. https://doi.org/10.1016/j.ijepes.2021.107499
Wendl, M., Doan, M. H., & Sassen, R. (2023). The environmental impact of cryptocurrencies using proof of work and proof of stake consensus algorithms: A systematic review. *Journal of Environmental Management, 326*, 116530. https://doi.org/10.1016/j.jenvman.2022.116530
Zade, M., Myklebost, J., Tzscheutschler, P., & Wagner, U. (2019). Is bitcoin the only problem? A scenario model for the power demand of blockchains. *Frontiers in Energy Research, 7*(MAR). https://doi.org/10.3389/fenrg.2019.00021

Irrigating Arid Lands: Sustainable Development Through Blue Sukuk

Muhammad Omer Rafique, Muhammad Asif Qureshi, Abdul Muhaimin, and Muzafar Hussain Shah

Abstract Amidst the escalating global water shortage crisis, Islamic financial instruments offer a viable and sustainable solution by utilizing blue sukuk. Grounded in the Islamic principle of Iḥya al-Mawāt, which advocates for the revitalization of barren lands, blue sukuk provides a framework for financing water infrastructure projects in arid regions. The proposed approach entails the issuance of blue sukuk to fund water infrastructure development, specifically irrigation systems, in parched lands. This strategy aligns with the Islamic principle of Iḥya al-Mawāt, enabling barren lands to be cultivated and transformed into productive agricultural areas. The authors used a content analysis of the Islamic law related to Iḥya and merged it with the existing Sukuk models to formulate a new solution for an old problem. Implementing blue sukuk projects holds immense potential to alleviate the water scarcity crisis, promising enhanced crop yields, poverty reduction, improved food security, and employment opportunities. However, realizing these benefits hinges on addressing the challenges like no regulatory framework, long-term financing, and the readiness of governments associated with blue sukuk projects, which include substantial investment requirements and regulatory adaptations. Despite the hurdles, the potential benefits of blue sukuk outweigh the challenges, positioning them as a promising tool to combat the water shortage crisis. By harnessing Islamic financial principles, Blue Sukuk offers a sustainable and practical approach to addressing this global challenge.

M. O. Rafique (✉)
Faculty of Business Administration, Muhammad Ali Jinnah University, Karachi, Pakistan

M. A. Qureshi
Faculty of Business Administration, Muhammad Ali Jinnah University Karachi, Karachi, Pakistan

Universiti Malaysia Perlis (UniMap), Arau, Malaysia

A. Muhaimin
Sindh Revenue Board Govt of Sindh, Karachi, Pakistan

M. H. Shah
Department of Business administration and Commerce, Newports Institute of communications and economics Karachi, Karachi, Pakistan

N. Naifar (ed.), *Climate Change and Finance*, Sustainable Finance,
https://doi.org/10.1007/978-3-031-56419-2_15

1 Introduction

Water shortage is a major challenge facing the world today. It is estimated that over 2 billion people live in water-stressed areas, and this number is expected to grow to 3.5 billion by 2050. Water shortage is a threat to human health, food security, and economic development. Irrigation is the largest user of water globally, accounting for about 70% of all freshwater withdrawals. However, irrigation practices are often inefficient, leading to water waste.

According to the United Nations Environment Programme (UNEP), barren lands cover about one-third of the Earth's land surface. They are home to a small fraction of the world's population, but they play an important role in the global environment. Barren lands help to regulate the Earth's climate, absorb carbon dioxide, and filter water. Barren lands are home to about 2% of the world's population. Barren lands are found in all parts of the world, but they are most common in arid and semi-arid regions. Major barren lands include the Sahara Desert, the Arabian Desert, the Gobi Desert, and the Australian Outback. Barren lands are important for the global environment. They help to regulate the Earth's climate, absorb carbon dioxide, and filter water. However, barren lands are also vulnerable to climate change and other environmental threats. Climate change is a major threat to barren lands. As the Earth's temperature rises, barren lands are becoming drier and less habitable. This is making it more difficult for plants and animals to survive in these areas.

Blue sukuk is a type of Islamic bond that can be used to finance water infrastructure projects. Blue sukuk are based on the Islamic principle of Iḥya al-Mawāt, which means "to bring dead land to life." This principle allows for the development of barren lands through irrigation. This research proposes a sustainable solution for water scarcity.

2 Problem Statement

Water shortage is a significant reason for destitution and struggle. At the point when individuals need more water, they are bound to become ill, lose their harvests, and be dislodged from their homes.

Some of the Water shortage facts show that 780 million individuals miss the mark on drinking water access. This implies that they do not approach clean water that is protected to drink. Moreover, 2.3 billion individuals live in water-focused nations. This implies that they need more water to meet their essential necessities. By 2025, 66% of the total populace might be confronting water deficiencies. This implies that they will not have sufficient water to meet their fundamental necessities, regardless of whether they approach it. Similarly, water shortage is a significant danger to biodiversity. At the point when streams, lakes, and wetlands evaporate, they obliterate the natural surroundings of plants and creatures. Nearly 700 million individuals could be dislodged by extraordinary water shortage by 2030. By 2040, approximately 1 of every 4 kids overall will be living in areas of very low water pressure.

Water dearth is a worldwide issue, however, it is especially intense in certain regions of the planet. The Center East and North Africa are the most water-focused locales on the planet, trailed by South Asia and Focal Asia. The Center East and North Africa: The Center East and North Africa are home to the driest puts on the planet. The locale is additionally confronting fast populace development and environmental change, which are fueling water shortage. South Asia is home to probably the most crowded nations on the planet, including India, Pakistan, and Bangladesh. The locale is likewise confronting a fast financial turn of events and environmental change, which are overwhelming water assets. Focal Asia is home to various enormous inland lakes, including the Aral Ocean and Lake Balkhash. In any case, these lakes are contracting because of overextraction of water for agribusiness (Fig. 1).

Water shortage is a squeezing worldwide issue that influences a huge number of individuals across the world. Notwithstanding the planet's bountiful water assets, over 2.2 billion individuals, or almost 30% of the worldwide populace, need admittance to safe drinking water. Besides, a stunning 4.2 billion individuals, or the greater part of the total populace, face serious water shortages for something like 1 month every year. This shortage is exacerbated by different elements, including populace development, environmental change, fumble of water assets, and contamination, which strain previously restricted freshwater holds.

The results of water shortage are broad and incorporate well-being emergencies, monetary precariousness, and social turmoil. It hampers rural efficiency, prompting

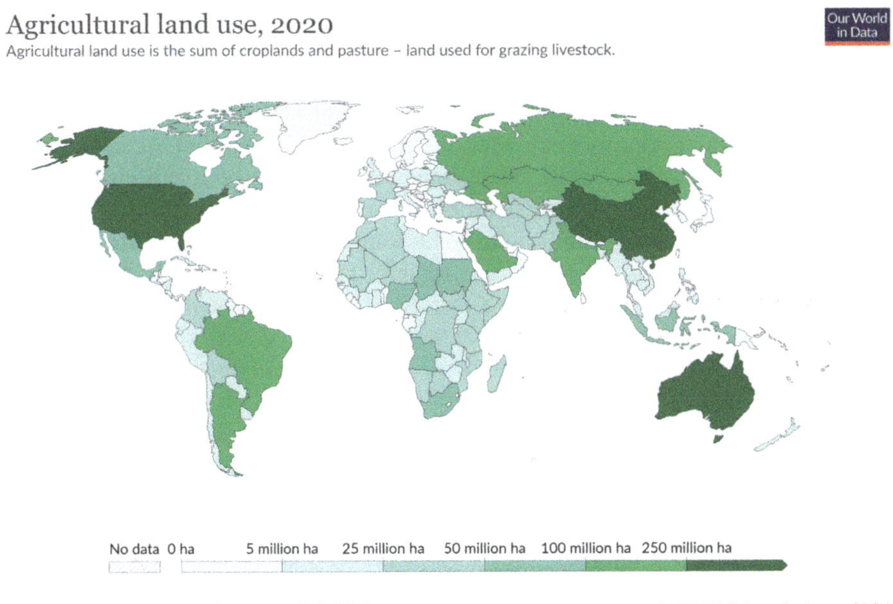

Fig. 1 Agri land use—worldwide

food deficiencies and cost climbs. Moreover, people groups without dependable admittance to clean water frequently experience the ill effects of waterborne sicknesses, making it a main source of disease and demise in numerous locales. Addressing water shortage requires purposeful worldwide work to focus on feasible water the board, put resources into the foundation, and elevate water preservation practices to guarantee that each individual approaches this crucial asset.

3 Methodology

This research uses a content analysis approach to probe into the Islamic law literature to find a solution to the pressing problem of water shortage and link the available infrastructure of the Islamic finance industry to solve this issue. The study outlines a probable framework for issuing Blue Sukuk to channel the investments to this venture and create a win-win situation for the investor as well as the barren land areas.

4 Literature Review

The relationship between water and finance is examined in the currency literature through the lens of historical and contemporary policies and practices, as well as new requirements arising from increasing water stress (i.e., droughts and floods) and climate change.

Kattel et al. (2019) concluded that China's south-to-north water diversion project has sparked a debate over the sustainability of water systems in the northern dry region, where water scarcity is a major obstacle to the functioning of ecosystems, farming, industry, and human life. They argue that big infrastructure projects have a positive impact on biodiversity protection and ecosystem services.

Bigger and Millington (2020) and Webber et al. (2022) examine the types of adaptive projects that are allocated funding through green borrowing. Drawing on research on the monetization of nature, as well as recent studies of racial capitalism and cost-cutting, they observe that, rather than undergoing transformative municipal transformation, NYC and Capetown cities are predominantly continuing with projects that reinforce existing inequalities within the city.

A modified MWS (modified water resource stress index) related to virtual water outflow is proposed in Lu et al. (2023) study. The results suggest that food energy output and virtual water flow are on the rise in southwest China. This puts pressure on water resources both in the location and in the output region. Green Financing for Water Security in China (Colven, 2023) showed how financial speculation can lead to flood risk and overuse of water, which can lead to different levels of risk in different parts of the city. These changes in Jakarta's water system can hurt the city's reputation as a good place to invest.

Guy J. Alaerts looked at the relationship between the water sector and the financial sector. We look at past and present policies and practices, as well as new needs driven by increasing water insecurity (e.g., drought and floods), and climate change. He focused on emerging markets and emerging economies. He concluded that the "conventional" water supply and sanitation agenda has been increasingly successful in providing safe drinking water. The more recent water resources agenda, which covers flood, drought, and irrigation, has to deal with rapid changes in the landscape caused by the destruction of catchments and rapid growth in water use (Alaerts, 2019).

As far as the Islamic finance and water shortage literature is concerned, very few and less focused literature can be found in the current studies (Rafique & Abdullah, 2023). More literature is targeted at social finance in general not focusing on the water shortage issue (Hureri et al., 2022). However, Musari (2022) concluded that by using Islamic blended finance, Nigeria could close the funding gap in the water, sanitation, and hygiene sector, which would help the country reach its sustainability goals, national ambitions, and the higher goals of Shariah. Ahmed (2021) looked at what the key tools are, what sustainable solutions they offer, and what the future holds for this traditional form of financing. We also look at the growing role of Islamic social finance instruments in humanitarian work.

The topic of green Sukuk is also related to our Topic. The findings of the research on green sukuk can be divided into three distinct sub-categories. The first sub-categories include articles that explore green sukuk from a Maqasid Sharia perspective, articles that focus their research on the potential and difficulties of issuing green sukuk, and articles that explore how green sukuk can be used as a tool for achieving Sustainable Development Goals (Liu & Lai, 2021). None of these have a primary focus on the water shortage issue.

5 Explaining Sukuk

A sukuk is a kind of Islamic bond or endorsement that follows Sharia regulation, the Islamic legitimate system. Sukuk are like traditional bonds in that they are used to raise money for projects; however, sukuk holders do not get paid interest. Instead, they share in the profits that are generated by the assets that are the foundation of the sukuk (Musari, 2022).

Sukuk can be organized in various ways, however, they all offer the accompanying key elements (Zuhaili, 2011):

- Asset-backed: Real assets like real estate, infrastructure, or commodities serve as the backing for sukuk. Sukuk holders are entitled to a portion of the underlying assets' profits as a result of their ownership stake in them.
- Shariah-compliant: The structure of sukuk is in accordance with Sharia law. This indicates that they do not involve maysir (gambling), gharar (uncertainty), or riba (usury).

- Pass-through: Sukuk holders get a portion of the benefits produced by the basic resources, however, they did not have responsibility for resources. As a result, holders of sukuk do not bear the risk of loss in the event that the underlying assets lose value.

Investors looking to diversify their portfolios and align their investments with Islamic values are increasingly turning to sukuk, a growing asset class. Sukuk can be utilized to fund a wide assortment of undertakings, including foundation improvement, land improvement, and environmentally friendly power projects.

6 The Potential of Blue Sukuk for Irrigating Barren Lands

Blue sukuk can help to bring water to arid lands and make them productive. This can lead to increased crop yields, which can help to reduce poverty and improve food security. By bringing water to arid lands, blue sukuk can help to reduce desertification and improve soil quality. This can help to protect the environment and ensure the long-term sustainability of agricultural production.

Water Sukuk can increase the GDP of a developing country. Irrigation allows farmers to grow crops in areas that would otherwise be too dry, and it can also help to increase crop yields in areas that already have some rainfall. This can lead to increased agricultural production, which can boost the economy and create jobs (Onu et al., 2022). In addition, irrigation can help to reduce poverty and improve food security. When farmers have access to irrigation, they are able to produce more food, which can lead to lower food prices and improved nutrition for the population (Zhao et al., 2022).

A study by the World Bank found that irrigation can increase agricultural GDP by up to 50%. The study also found that irrigation can create up to 10 jobs per hectare of irrigated land (Obaideen et al., 2022). Another study, by the International Food Policy Research Institute, found that irrigation can help to reduce poverty by up to 20%. The study also found that irrigation can help to improve food security by up to 10% (Alaerts, 2019).

Water Sukuk is particularly important for developing countries, where agriculture is often a major sector of the economy. In many developing countries, farmers do not have access to the credit they need to invest in irrigation systems. This is where Water Sukuk can play a role. Water Sukuk can help farmers invest in irrigation systems, which can lead to increased agricultural production, economic growth, and poverty reduction (Onu et al., 2022). There is a great potential for growth in GDP of developing countries, especially Muslim world, which is dependent on agri-exports and imports of food items. Countries like Malaysia and Saudia Arabia with heavy outlook in Sukuk and bulk land inventory to transform in a green way.

In India, the government's Pradhan Mantri Krishi Sinchai Yojana (PMKSY) has been credited with increasing agricultural GDP by over 5% per year. The PMKSY provides subsidies to farmers to invest in irrigation systems (Jana & Tamang, 2023). In Pakistan, the World Bank-funded Pakistan Irrigation Improvement Project has helped to increase agricultural GDP by over 10%. The project has provided

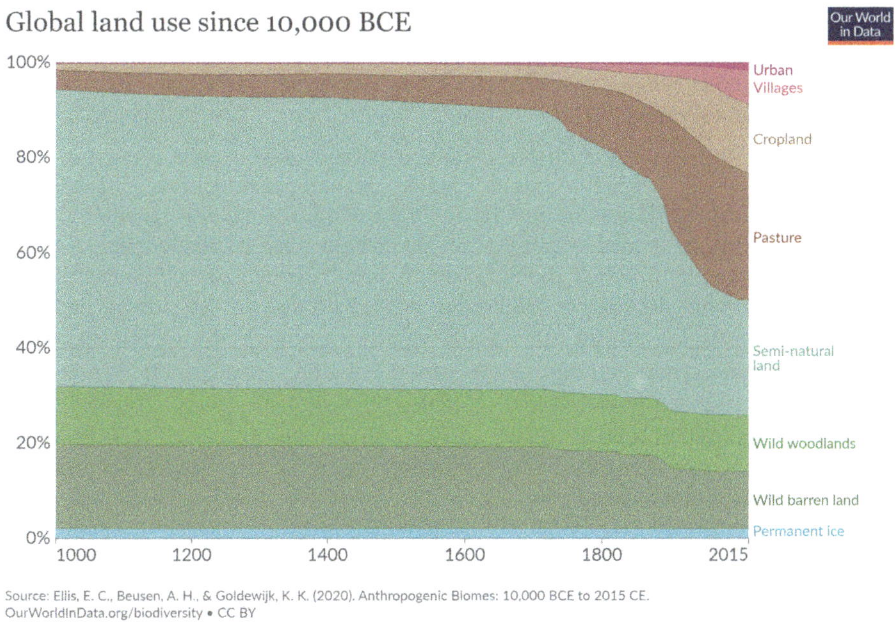

Fig. 2 Barren lands since last 1000 years

financing for the construction of new irrigation canals and the rehabilitation of existing irrigation infrastructure (Blakeslee et al., 2023). In Ethiopia, the government's Irrigation Development Project has helped to increase agricultural GDP by over 20%. The project has provided financing for the construction of new irrigation systems and the rehabilitation of existing irrigation infrastructure (Juturu et al., 2023).

These are just a few examples of how Water Sukuk can increase the GDP of developing countries. Water Sukuk is a critical tool for developing countries to achieve their economic development goals (Fig. 2).

This chart shows the potential of irrigating the barren lands to create opportunities in the rural and barren areas by moving investment into these areas.

7 Islamic Solution on Irrigating Facilities: Iḥya Mawāt

Iḥya means to make something alive, and Mawāt means the dead land. According to the Arabic Dictionary:

الإحياء لغة: جعل الشيء حياً، أي ذا قوة حساسة أو نامية

Translation: Bringing something to life/Giving something the power of growth and feeling. Explaining the literal meaning of the word Mawāt He says:

والموات: ما لا روح فيه، أو الأرض التي لا مالك لها، أو الأرض الخراب الدارسة غير العامرة

Translation: That thing which has no soul/or the land which has no owner and is uninhabited, barren, and desolate. The scholars defined the Mawāt land as those uninhabited forests and useless mountain lands that are not included in any particular city and are not related to any settlement and are not actually arable and usable, such lands are called "Mawāt land" in terms of Shariah.

The dead land is that land upon which there are no signs of ownership such as fencing, planting, habitation, or the like. Cultivation of land includes making it suitable for farming at once. Every piece of dead land once cultivated by a person becomes his ownership.[1] Thus, the Shari'ah gives it to the one who cultivates it. This is according to what Bukhari related from 'Aisha (ra) that the Messenger of Allah (pbuh) said: "Whosoever cultivated a land that nobody owns, then He deserved it more." Abu Dawud narrated that the Messenger of Allah (pbuh) said: "Whosoever fenced a (dead) land it becomes his" and Bukhari narrated from Umar (ra) that He (pbuh) said: "Anyone who cultivated a dead land it becomes his" (Al-Bukhari, 1978). Muslims and non-Muslim Dhimmi are equal in this matter because the Hadith is general in its words.

But it is only for unowned land.[2] Thus, whether the land is barren or inhabited if its owner/heir is known, it is illegal to dispose of it without his permission.[3] Similarly, non-owned land which is useful for the common and common needs of the inhabitants of the settlement near the population, such as streets and playing grounds within the settlement or every cemetery, Eid Gah, pasture, etc. are not included in it. The Shariah ruling is that the possession and disposal of this kind of land is not permissible, nor can a Muslim sultan himself become their owner, nor can he make someone else the owner, nor can he give it to someone else as a fief. The citizens' properties will always be protected for the common needs of the country like waqf (Zuhaili, 2011).

---

[1] الكلام في موضعين في بيان أنواع الأراضي وفي بيان حكم كل نوع منها.(أما) الأول: فالأراضي في الأصل نوعان: أرض مملوكة، وأرض مباحة غير مملوكة، والمملوكة نوعان: عامرة وخراب، والمباحة نوعان أيضا: نوع هو من مرافق البلدة محتطبا لهم ومرعى لمواشيهم ونوع ليس من مرافقها وهو المسمى بالموات بدائع الصنائع في ترتيب الشرائع (6/192).

The lands are originally of two types: owned land, and permissible land that is not owned, and owned are of two types: full and ruined, and permissible are also of two types: a type that is one of the town's facilities as a wood for them and pasture for their livestock and a type that is not one of its facilities, which is called Al-Mawat.

[2] الفقه الإسلامي وأدلته للزحيلي (6/4625)ولا يشترط عند الجمهور (الحنفية والمالكية والحنابلة) (1) كون المحيي مسلماً، فلا فرق بين المسلم والذمي في الإحياء، لعموم قول النبي صلّى الله عليه وسلم: «من أحيا أرضاً ميتة، فهي له»، ولأن الإحياء أحد أسباب التمليك، فاشترك فيه المسلم والذمي، كسائر أسباب الملكية.

It is not required according to the scholars (Hanafi, Maliki and Hanbali) that the reviver is a Muslim, so there is no difference between a Muslim and a dhimmi in revival, because of the general saying of the Prophet (peace and blessings of Allaah be upon him): "Whoever revives a dead land, it is his", and because revival is one of the reasons for ownership, so the Muslim and the dhimmi participated in it, like all other reasons for ownership.

[3] الأراضي المملوكة العامرة: فليس لأحد أن يتصرف فيها من غير إذن صاحبها؛ لأن عصمة الملك تمنع من ذلك، وكذلك الأرض الخراب الذي انقطع ماؤها ومضى على ذلك سنون لأن الملك فيها قائم وإن طال الزمان حتى يجوز بيعها وهبتها وإجارتها وتصير ميراثا إذا مات صاحبها. (أما)

(As for the land owned by the owner: no one can dispose of it without the permission of the owner, because the infallibility of the king prevents this from happening, as well as the ruined land whose water has been cut off and years have passed since that, because the ownership in it exists, even if it is long until it is permissible to sell, gift and rent it, and it becomes an inheritance if its owner dies. spiepr par404).

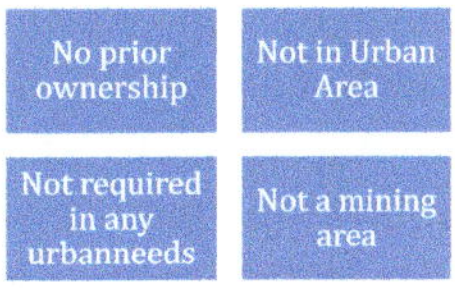

Fig. 3 Conditions for a Mawāt land

Similarly, the land of salt, tar, oil, and the like, is not to be spared by Muslims; it is not a land of Mawāt. It is not permissible for the Imam to cut it off for anyone or consider it as Mawāt; Because it is the right of the Muslim community, and the abrogation of their rights is not permissible.[4]

In other words, not every land is rehabilitated, only that land is rehabilitated in which the following conditions are found:

1. It should not be owned by anyone.
2. That land is outside the city.
3. There should not be any right of the people of a particular settlement with this land, i.e., it should not be a place for them to settle their garbage, nor a way for them to come and go, nor a cemetery for them, nor a pasture or a stable, etc. (Kasani, 1989).
4. Similarly, mines of salt, and springs of kerosene oil or petrol, etc., which are included in the necessities of life of the common people, whether they are near a settlement or not, cannot become a specific land or jagir of a person. Neither the Sultan can dispose of the ownership of the Muslim houses nor can they give them to anyone else, but they will be managed under the supervision of the government for the common good (Fig. 3).

8 The Techniques of Iḥya

Cultivation is related to the dead land upon which there is no apparent ownership. There are no signs of fencing, planting, building, or the like. Cultivation of such land means to populate it with anything that indicates inhabitation. Fencing the land is equal to its cultivation. This is due to the saying of the Messenger of Allah (pbuh), "Whosoever fenced a land with a wall then it becomes his" and his (pbuh) saying, "Whosoever fenced

[4] بدائع الصنائع في ترتيب الشرائع (6/ 194)فالأرض الموات هي أرض خارج البلد لم تكن ملكا لأحد ولا حقا له خاصا فلا يكون داخل البلد موات أصلا، وكذا ما كان خارج البلدة من مرافقها محتطبا بها لأهلها أو مرعى لهم لا يكون مواتا حتى لا يملك الإمام إقطاعها؛ لأن ما كان من مرافق أهل البلدة فهو حق أهل البلدة كفناء دارهم وفي الإقطاع إبطال حقهم.وكذلك أرض الملح والقار والنفط ونحوها مما لا يستغني عنها المسلمون لا تكون أرض موات حتى لا يجوز للإمام أن يقطعها لأحد؛ لأنها حق لعامة المسلمين وفي الإقطاع إبطال حقهم وهذا لا يجوز

The Mawat land is a land outside the country that did not belong to anyone and has no special right, so it is not inside the city favorable in the first place, as well as what was outside the town of its facilities that were wooded for its people or pasture for them is not favorable so that the imam does not own its fiefdom, because what was one of the facilities of the people of the town is the right of the people of the town as the courtyard of their house and it is feudalism to invalidate their right. The imam cannot cut it off for anyone, because it is the right of the common Muslims, and in feudalism they invalidate their right, and this is not permissible spiepr par425.

anything with a wall, it becomes his thereby." Also, his (pbuh) saying, "Whosoever reached a thing first that no other Muslim reached before, He deserves it more."

According to Islamic scholars, "Iḥya" will be established by adopting one of the following methods (whereas the conditions of proof are coming forward):

1. Seeds may be cultivated in it.
2. Trees and plants may be planted.
3. A house etc. may be constructed.
4. It may be driven by shoe/plough, etc.
5. Drains, etc. may be made in it (for irrigation of the land).
6. A shelter or a bridge may be made (for protection from (extra) water).
7. A wall may be built around it.
8. A well or spring may be dug in it and water should be extracted.[5]

9 Fencing/Taḥjir

Taḥjir means that stones are placed like a wall around the land only for the purpose of symbolic demarcation, not actual construction of a building, etc. is done. This is called "Taḥjir." The person who does this will be more entitled to the "revival (IḤYA)" of this land than others.[6]

Thus, by fencing, the fencer gains the right of disposal of the land. The fencer also has the right to prevent anyone from cultivating the land which He has fenced. If another person overpowered him and managed to inhabit the land that He had fenced, He would not own that land and it would be returned back to the original fencer (Al-Marghinani, 2006). Fencing is also like cultivation with regard to the disposal of the land and possession of it. If the person who fenced a land later sold it He owns its price, because it is a right that can be recompensed with property, so

[5] درر الحكام في شرح مجلة الأحكام (3/ 284)
وَالْخُلَاصَةُ أَنَّ الْأُمُورَ الَّتِي تُعَدُّ إحْيَاءً كَمَا ذُكِرَ فِي الْمَادَّةِ الْآنِفَةِ وَفِي مَتْنِ وَشَرْحِ هَذِهِ الْمَادَّةِ هِيَ عِبَارَةٌ عَنْ تِسْعَةِ أُمُورٍ:
(1) زَرْعُ الْبَذْرِ (2) غَرْسُ الشَّتْلِ (3) بِنَاءُ الْأَبْنِيَةِ (4) كِرَابُ الْأَرْضِ (5) سَقْيُ الْأَرْضِ (6) شَقُّ جَدْوَلٍ وَمَجْرًى لِلسَّقْيِ (7) إحَاطَةُ الْأَرْضِ بِجِدَارٍ (8) بِنَاءُ مُسَنَّاةٍ (9) حَفْرُ بِئْرٍ حَتَّى خُرُوجِ الْمَاءِ فَلِذَلِكَ
لَوْ حَفَرَ أَحَدٌ فِي أَسْفَلِ الْجَبَلِ الْمَوَاتِ بِئْرًا فَيَمْلِكُ ذَلِكَ الْجَبَلَ حَتَّى أَعْلَاهُ (الْهِنْدِيَّةُ فِي الْبَابِ الْأَوَّلِ مِنْ الْمَوَاتِ).

(3/ 284) The conclusion is that the matters that are considered revival, are: planting the seed (2) planting seedlings (3) building any building (4) tilling the land (5) watering the land (6) Digging a stream and channel for irrigation (7) Surrounding the land with a wall (8) Building a stake (9) Digging a well until the water comes out, so that if someone digs at the bottom of the The Mountain, He possesses that mountain even to its highest point (Hindiyya).

[6] الفقه الإسلامي وأدلته للزحيلي (6/ 4626)
هل يحصل الإحياء بالتحجير؟ التحجير أو التحويط: هو الإعلام بوضع الأحجار حول الأرض، أي وضع سور من الأحجار والأشواك ونحوها على جوانب الأرض، وقد اتفق الفقهاء على عدم صلاحيته للإحياء، لكن المتحجر يكون أحق بها من غيره.

Does revival take place by quarrying? Quarrying or hedging: It is the notification of the placement of stones around the ground, that is, the placement of a fence of stones, thorns and the like on the sides of the earth, and the jurists have agreed that it is not suitable for revival, but the fossilized is more entitled to it than others.

it can be exchanged. If this person dies, the ownership of this fenced land is transferred to his inheritors like any other of his properties and they gain the right of disposal over it and it is divided among them according to the divine rules like other inherited properties (Zuhaili, 2011). However, fencing a land does not mean just putting stones around it, it is rather building anything around it which indicates holding a hand over it, which indicates ownership. Fencing may be by placing stones around the land, putting dry branches, clearing it, burning the thistles, cutting the spikes and grass, or placing other such items around it to prevent people from entering it. It could also be by preparing the canals even if one did not irrigate them, or any other similar thing.

From the Hadith, it is clear that fencing like cultivation must only be with regard to the dead land, and it would not be for other than that. The saying of Umar (ra) "a fencer has no right after three years" (Hanbal, 1995) refers to the fencer having no right in the dead land. The non-dead land cannot be possessed by fencing nor by revival, it is rather possessed by a grant from the Government. This is because revival and fencing came connected with the dead land. The Messenger of Allah (pbuh) said: "Whoever revived a dead land…(Al-Bukhari, 1978)." The word "dead" is an adjective, so it has a concept that is usable as a restriction on the word land (This means that the land that is other than dead land cannot be owned by walling or revival). Al-Baihaqi also narrated from Amir ibn Shuaib "that Umar made fencing for three years (Al-Bayhaqi, 1994)." If He left it (the land) for 3 years and another person cultivated it then He becomes more deserving of it. This means that the non-dead land is not owned by fencing or cultivation.

This differentiation between the dead and used land indicates that the Messenger of Allah (pbuh) allowed the people to own the dead land by habitation and fencing. So it became one of the Mubah things. Therefore, it does not need permission from the Imam for habitation or fencing, because the Mubah things do not need permission from the Imam.

Is It Established by Rehabilitating the Right of Usufruct, or the Ownership of the Land That Includes the Right to Dispose of, use, and Exploit?

The jurist Abu al-Qasim Ahmad al-Balkhi said: What is established by the revival(Iḥya) of the inheritance is the right of usufruct, not the right of ownership, as compared to a man sitting in a permissible place, if he has the usufruct, so if he rises from it, and offers it, it invalidates his right (Al-Marghinani, 2000). Where the general jurists said: Iḥya is the right to absolute ownership, based on the hadith: "From the dead earth, it is for him." For he added the right to the dead, not for him, in his saying, "It is for him." Thus, the right that is obtained due to revitalization (Iḥya) is not only the right of usufruct but also the right of ownership. Therefore, the settler can dispose of this land in any way by sale, gift, lease, etc.

AAOIFI (Accounting and Auditing Organization for Islamic Financial Institutions) also stated a standard statement regarding this fencing in standard number 42.

"Right of Taḥjir 9/1 Taḥjir means taking possession of a piece of land and marking it with certain recognised markers with the permission of the government.

9/2 Taḥjir results in exclusivity and priority over others, but it does not own its own confer ownership.

9/3 A person who has carried out Taḥjir on a piece of land is permitted to waive his entitlement in favor of another person for consideration, waive his entitlement in favor of another person for consideration, but he is not allowed to sell the land as he does not yet own it.

9/4 The right of Taḥjir expires if the land is not utilized for three years, or as provided by applicable regulations."

The basis for the right of Taḥjir and the rights established by first use is the Sunnah. Ibn Qudamah states: "Anyone who carries out Taḥjir of previously unappropriated land and begins to make it productive but did not complete it, he is more entitled to it than others, based upon the statement of the Prophet (peace be upon him): "Anyone who first uses what no other Muslim has previously used is more entitled to it". If he transfers it to someone else, then the second person is more entitled to it than others to someone else, because the first owner of the right has granted it to him. If the owner of the right dies, it transfers to his heirs, based on the statement of the Prophet (peace be upon him): The heirs of a deceased person inherit his estate."

However, it is not permitted for the owner of a right of Taḥjir to sell it because his entitlement is confined to his personal use like the right of Shuf 'ah (preemption). [An alternative view is that] its sale could be considered valid because he is more entitled to it. Al-Mardawi states: "Anyone carrying out Taḥjir over previously unappropriated land will not own it…but he is more entitled to it than others, as well as his heirs after him and whomever he transfers it to, without any dispute. However, he cannot sell it, and this is the position of our School (مجدي محمد فتحي الهلالي and et al. 2023).

10 Taxes on the Owned Barren Lands

Whoever cultivates a dead land of the 'Ushri land, He owns it and its benefit, whether He is Muslim or non-Muslim. For such land, the Muslim landlord is obliged to pay the Zakat ('Ushr) of the plants and fruits which are entitled to Zakat once the amount of the harvest reaches the Nisab. As for the non-Muslim landlord of such land, He is not obliged to pay either Zakat or Kharaj, because He is not one of those who are subject to pay Zakat and because there is no Kharaj on 'Ushri land (Al-Marghinani, 2000).

Whoever cultivates a dead land in the Kharaji area where no Kharaj has been put over it before, He owns its neck and its benefit if He is Muslim. If He is non-Muslim He owns its benefit only. The Muslim landlord of such land is obliged to pay the 'Ushr with no Kharaj on him. While the non-Muslim landlord has to pay the Kharaj, similar to that put on its kuffar inhabitants at the time of its conquest.

Whoever cultivates a dead land in the Kharaji area where Kharaj has been levied before it became dead, He owns its benefit only without owning its property, whether the landlord is Muslim or non-Muslim. Such a landlord is obliged to pay the Kharaj because it is a conquered land. Therefore, the Kharaj remains on it at all times, whether owned by a Muslim or non-Muslim (Zuhaili, 2011).

This is the case if the land was cultivated for farming. If, however, the land is cultivated or fenced for the purpose of housing, industry, stores or sheds, then no 'Ushr or Kharaj is due, whether it was originally 'Ushri or Kharaji land. This is the situation because when the Sahabah conquered Iraq and Egypt they built Kufa, Basra, and Al-Fustat and they lived there at the time of Umar ibn Al-Khattab. Other people (Muslims and non-Muslims) joined them in these cities. Yet Kharaj was not levied on them, nor did they pay Zakat, because Zakat is not due on homes and buildings (Zuhaili, 2011).

Abu Yusuf said: If the land is a Muslim, then if that land is one of the lands of the tenth, then the tenth is obligatory on it, and if it is one of the lands of the Kharaj, then the Kharaj is obligatory on it. Muhammad said: If I live with the water of the last month like the rain or the big river, it is ushery, and if I live with the water of the Kharaj, the words taken from the river of the holes of non-Muslims, then it is tributary. This opinion is what was followed by Sahib Al-Hidayah.[7]

11 Hareem (Sanctuary)

Siphoning from a well in a water table spring brings down the water table close to the well. This region is known as a cone of despondency. The land region over a cone of sadness is called the area of impact. Groundwater streams toward the well into the cone of sorrow which can adjust the normal flow of groundwater stream inside the area of impact around the well.

For the deceased, additional ownership is determined: it is the sanctuary of the land,[8] so that he can revive what he needs from the adjuncts such as the space in front of the al-Dar and extending from its sides (Fig. 4).[9]

---

[7] الفقه الإسلامي وأدلته للزحيلي (6/ 4629)

(.قال أبو يوسف: إن أحياها مسلم، فإن كانت تلك الأرض من الأراضي العشرية فالواجب فيها العشر، وإن كانت من حيّز الأراضي الخراجية فالواجب فيها الخراج

وقال محمد: إن أحياها بماء العشر كماء المطر أو الأنهار الكبيرة فهي عشرية، وإن أحياها بماء الخراج، كالماء المأخوذ من نهر حفره غير المسلمين فهي خراجية. وهذا الرأي هو ما مشى عليه صاحب الهداية.

Abu Yusuf said: If it is revived by a Muslim, if that land is one of the ten lands, then it is obligatory to do the tithe, and if it is from the area of the abscessive lands, then it is obligatory to go out.

[8] وَالأَصل فِي مَشْرُوعِيَّةِ الْحَرِيمِ أَنَّ النَّبِيَّ صَلَّى اللَّهُ عَلَيْهِ وَسَلَّمَ جَعَل لِلْبِئْرِ وَالْعَيْنِ وَكُل أَرْضٍ حَرِيمًا، بِقَوْلِهِ صَلَّى اللَّهُ عَلَيْهِ وَسَلَّمَ: مَنْ حَفَرَ بِئْرًا فَلَهُ أَرْبَعُونَ ذِرَاعًا عَطَنًا لِمَاشِيَتِهِ الموسوعة الفقهية الكويتية (17/ 213) .

The basic principle regarding the legality of harems is that the Prophet, may God's prayers and peace be upon him, made the well, the spring, and every land a harem, by saying, may God's prayers and peace be upon him: He who digs a well, and he has forty cubits of fodder for his livestock.

[9] لاَ خِلاَفَ بَيْنَ الْفُقَهَاءِ فِي أَنَّهُ لاَ يَجُوزُ إِحْيَاءُ حَرِيمِ الْبِئْرِ وَالنَّهْرِ، وَالْعَيْنِ، وَكُل مَمْلُوكٍ لاَ يَجُوزُ إِحْيَاءُ مَا تَعَلَّقَ بِمَصَالِحِهِ، لِقَوْلِهِ عَلَيْهِ الصَّلاَةُ وَالسَّلاَمُ: مَنْ أَحْيَا أَرْضًا مَيِّتَةً فِي غَيْرِ حَقِّ مُسْلِمٍ فَهِيَ لَهُ - الموسوعة الفقهية الكويتية (17/ 213)

There is no disagreement among the jurists that it is not permissible to revive a well, a river, a spring, or any property whose interests it is not permissible to revive. And to him, may the peace and blessings of God be upon him: Whoever revives dead land without the right of a Muslim, it belongs to him.

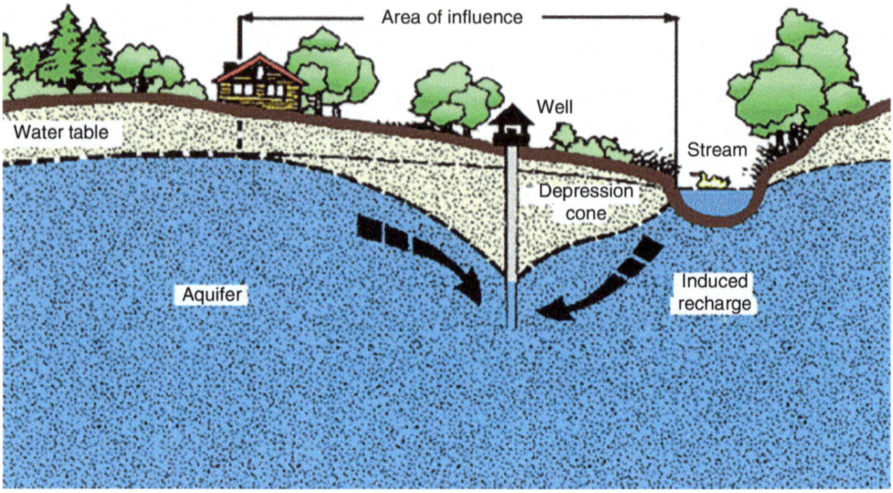

Fig. 4 Illustration for Hareem concept. Source: https://wellwater.oregonstate.edu/groundwater/understanding-groundwater/groundwater-and-wells

According to Islamic scholars, the Sanctuary of the groundwater is 500 yards from each side.[10] The harem of the well of Al-'Atn (the one from which the water is drained by hand): is forty cubits on each side according to Scholars's agreement, as some narrations indicate, but they are strange, meaning they are not proven, as Al-Zayla'i said.[11] The harem of the well of al-Nadh (which is the one from which water is drained by a camel and the like) is 40 cubits, like the well of al-'Atan at Abu Hanifa, in accordance with the needs of the people. According to the two companions: 60 cubits, according to what was narrated: "And the harem of Bir al-Nadh is sixty cubits." What is correct is that its harem is as much as needed from all sides.

---

[10] حريم العين الجارية: الأصح أن حريمها خمس مئة ذراع من كل جانب والذراع ست قبضات، كل قبضة أربع أصابع، وهو من المرفق إلى الأنامل. لقول الزهري: «وحريم العين خمس مئة ذراع، من كل ناحية» وبناء عليه يمنع غير صاحب الحريم من الحفر ونحوه في مسافة الحريم، لأنه ملك له، فله تضمين المعتدي أو ردم الحفرة. الفقه الإسلامي وأدلته للزحيلي (6/ 4632)

Harem of the running stream: It is more correct that its harem is five hundred cubits on each side and the arm is six fists, each fist is four fingers, which is from the elbow to the fingers. Al-Zuhri says: "The harem of the stream is five hundred cubits on each side" and accordingly it is forbidden for the non-owner of the harem to dig and so on in the distance of the harem, because it belongs to him, so he may include the aggressor or fill the hole.

[11] وحريم بئر الناضح وحريم بئر العطن(هي التي ينزح منها الماء باليد): أربعون ذراعا من كل جانب باتفاق الحنفية، كما دلت بعض الروايات لكنها غريبة أي لم تثبت، كما قال الزيلعي - (وهي التي ينزح منها الماء بالبعير ونحوه) أربعون ذراعا كبئر العطن عند أبي حنيفة، عملا بحاجة الناس. وعند الصاحبين: ستون ذراعا لما روي: «وحريم بئر الناضح ستون ذراعا» والصحيح: أن حريمها على قدر الحاجة من كل الجوانب. بشرط أن يحفرها في موات بإذن الإمام أو في ملكه، فلو حفر في ملك الغير لا يستحق الحريم.

1The harem of Bir al-Nadooh and the harem of Bir al-'Atan (from which water drains by hand): forty cubits from each side according to the agreement of the tap, as indicated by some narrations, but they are strange, i.e. they have not been proven, as al-Zayla'i said. (from which water drains with camels and so on) forty cubits like the well of al-'Atan according to Abu Hanifa, in accordance with the needs of the people. And when the two companions: sixty cubits for what was narrated: «And the harem of the well of the exudation sixty cubits» and the correct: that her harem is as much as needed from all sides. Provided that he digs it in a favorable place with the permission of the imam or in his property, if he digs in the property of others, he does not deserve the harem.

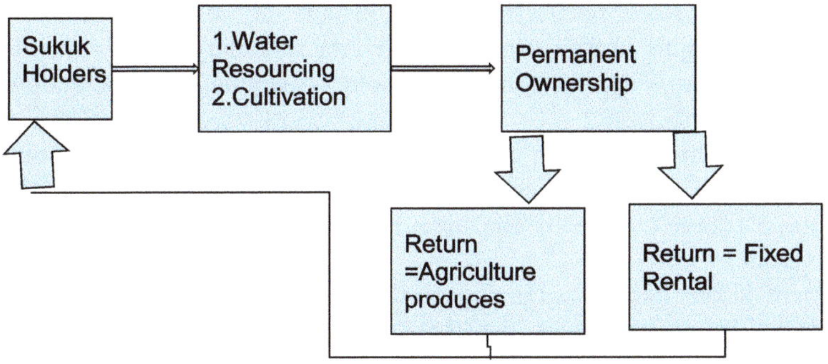

Fig. 5 The above-drawn framework includes four integral steps

Provided that he digs it in a dead place with the permission of the imam or in his possession.

Harem of the Canal(Karez): It is an underground water channel. Canal harem is limited to what was suitable for throwing mud and the likely needs.[12]

Harem of River: The river has a sanctuary as much as it needs to dump mud and the likely needs. Abu Yusuf and Muhammad said the river has a harem on both sides, because the harem's entitlement is due to need, and the owner of a river needs him like the owner of a well, because he needs to walk on the banks of the river, and he needs a place to dump mud on when the river dribbles. Half of the total river breadth is the amount of the Hareem of Canal.[13]

12 Linking Hareem and Iḥya Mawāt to Sukuk Finance (Fig. 5)

Component 1: The investors shall buy the Sukuk and invest the amount to the SPV. It is a special-purpose vehicle. This SPV is a legal entity that on behalf of holders conducts all the infrastructure building and irrigation process and returns the amount to the Sukuk holders after the investment in the SPV. It shall work on the water resourcing into barren land and the cultivation procedure itself or through any expert organization.

Component 2: After infrastructure setting the permanent ownership of the Barren land shall be titled to the SPV as per the law of Shariah which mentions that

---

[12] .الفقه الإسلامي وأدلته للزحيلي (6/ 4633)حريم القناة: وهي مجرى الماء تحت الأرض. ولم يقدر حريمها بشيء يمكن ضبطه، فحريمها بقدر ما يصلحها لإلقاء الطين ونحوه

Harem of the canal: It is the underground stream of water. Her harem could not have anything to control, as her harem was as much as he could fit her to throw mud and the like. River Harem:

[13] .حَرِيمُ النَّهْرِ: الموسوعة الفقهية الكويتية (17/ 217) الأَصَحُّ عِنْدَ الْحَنَفِيَّةِ أَنَّ لِلنَّهْرِ حَرِيمًا بِقَدْرِ مَا يَحْتَاجُ إِلَيْهِ لإِلْقَاءِ الطِّينِ وَنَحْوِهِ، فِيمَا لَوْ أَحْيَاهُ فِي أَرْضٍ مَوَاتٍ، وَقِيل: لاَ حَرِيمَ لَهُ عِنْدَ أَبِي حَنِيفَةَ

It is more correct according to the tap that the river has a harem as much as it needs to throw mud and the like, if it lives in a favorable land, and it is said: There is no harem for it according to Abu Hanifa.

"Whoever cultivates the Barren land or installs the infrastructure of irrigation such as the digging of the river the land and the sanctuary of that cultivation area is entitled to him. Therefore, the Sukuk holders eventually get the entitlement of ownership of this land.

Component 3: This cultivation and irrigation procedure shell generate income in the form of agricultural production as appropriate to the region or it will generate fixed rental income if the area is given on rental to the experts and cultivators.

Component 4: This fixed rental income or the varying agri products income periodically shall be returned to the investors.

13 Challenges

It is advised that the returns should be given yearly as the installation of the infrastructure for irrigation needs a lot of effort as well as time to yield some income. Moreover, the initial heavy investments shall be needed because the infrastructure costs a lot of money, maybe billions of dollars. However, the return on infrastructure shall be on the very high side that can in some cases reach 50 or 60% of the investment annually, which can be divided into 6% monthly income after 3 years.

This kind of Sukuk is more appropriate for governments and regulatory authorities to initiate as it requires thousands of investors and billions of dollars for the irrigating processes. However, a small project of irrigating and getting the ownership of that land is applicable for a medium-sized commercial bank as well.

As discussed in the introduction of the Iḥya Mawāt rules and regulations of Islamic law the already owned land by anyone cannot be recultivated for ownership. Regulatory authorities and property rights regulations are involved in this matter to allow investors to work in the barren land and create job opportunities as well as financing activities for their own return.

14 Execution Steps for Sukuk Issuance

A unique method of issuing is required due to the unique nature of the financial instrument. Typically, the following steps can be observed during the issuing procedure (Fig. 6):

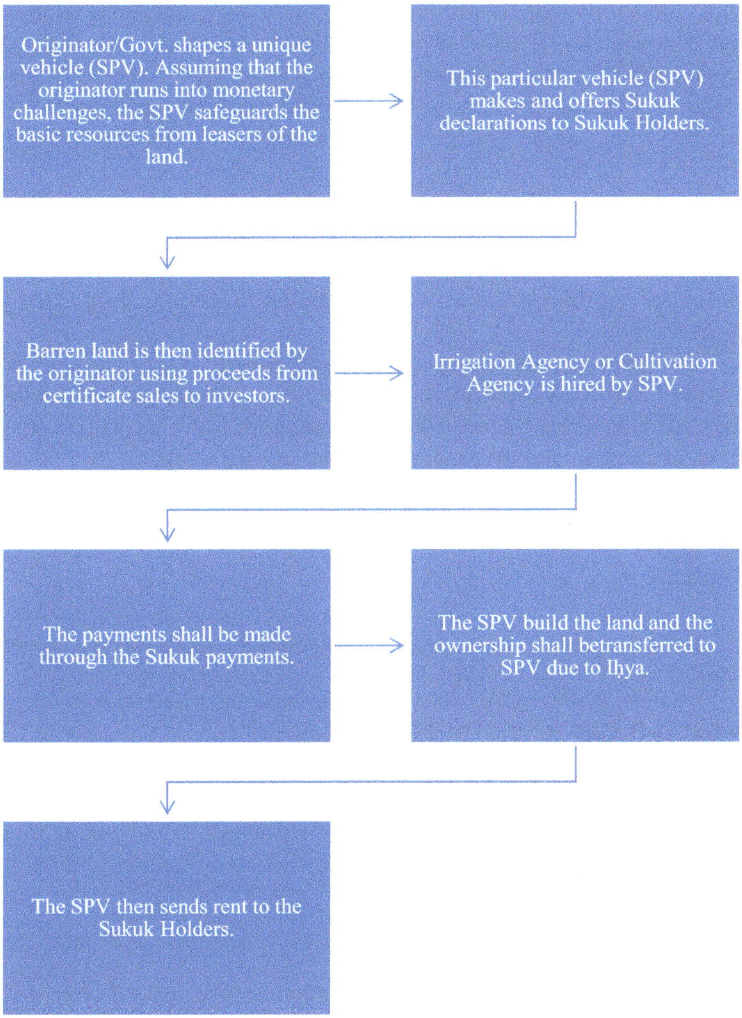

Fig. 6 Executing blue sukuk

15 Conclusion

Blue sukuk is a type of Islamic bond that can be used to finance water infrastructure projects. Blue sukuk are based on the Islamic principle of Iḥya al-Mawāt, which means "to bring dead land to life." This principle allows for the development of barren lands through irrigation.

Blue sukuk has the potential to play a major role in addressing the water shortage crisis and achieving the Sustainable Development Goals. By financing irrigation projects in arid regions, blue sukuk can help bring water to barren lands, increase crop yields, reduce poverty, improve food security, and create jobs. Blue sukuk is also an environmentally friendly way to develop water infrastructure. They can help to reduce desertification and improve soil quality.

The Islamic principle of Iḥya al-Mawāt provides a strong foundation for the development of blue sukuk. This principle is based on the idea that it is a meritorious act to bring dead land to life. Iḥya al-Mawāt has been used to develop irrigation systems in many parts of the world, and it can continue to be used to address the water shortage crisis in the future. In addition to the benefits mentioned above, blue sukuk can also help to promote sustainable water management practices, reduce water pollution, improve water access and sanitation, and create a more resilient water infrastructure.

Thus, Blue Sukuk is a promising tool for addressing the water shortage crisis and achieving the Sustainable Development Goals. By mobilizing capital from investors, Blue Sukuk can help finance the development of water infrastructure projects that will benefit people and the environment around the world.

References

Ahmed, Maram. 2021. "Islamic social finance." In *Innovative humanitarian financing: Case studies of funding models*, Maram Ahmed, 101–134. : Springer International Publishing.

Alaerts, G. J. (2019). Financing for water—water for financing: a global review of policy and practice. *Sustainability: Science Practice and Policy, 11*(3), 821.

Al-Bayhaqi, A. H. (1994). *Sunan Al-Bayhaqi Al-Kubra*. Maktabat Dār Al-Bāz.

Al-Bukhari, M. (1978). *Sahih Al-Bukhari*. academia.edu. 1978. https://www.academia.edu/download/46412378/shahih-al-bukhari.pdf

Al-Marghinani, B. (2000). *Al-Hidayah: Sharh Bidayat Al-Mubtadi*. MaktabahRashidiyyah.

Bigger, P., & Millington, N. (2020). Getting Soaked? climate crisis, adaptation finance, and racialized austerity. *Environment and Planning E: Nature and Space, 3*(3), 601–623.

Blakeslee, D., Dar, A., Fishman, R., Malik, S., Pellegrina, H. S., & Bagavathinathan, K. S. (2023). Irrigation and the spatial pattern of local economic development in India. *Journal of Development Economics, 161*(March), 102997.

Colven, E. (2023). A political ecology of speculative urbanism: The role of financial and environmental speculation in Jakarta's water crisis. *Environment & Planning A, 55*(2), 490–510.

Hanbal, I. (1995). Al-Musnad. *Qohiroh. Dār Al-Hadīs*.

Hureri, M. A., Rafique, M. O., Alia, K., Ullah, M. Z., Ahmad, M., & Majid, A. (2022). Holistic Islamic Model for Securing Wealth in Light of Maqasid-E-Shariah. *AQ, 8*(2) http://alqantara-journal.com/index.php/Journal/article/view/113

Jana, S. K., & Tamang, P. (2023). Prospects of rehabilitation of ancient irrigation systems in India–A case study from Coastal Saline Zone of West Bengal. *Agricultural Systems, 207*(103638), 103638.

Juturu, P., Conlon, T., Zhang, Y., Avraam, C., Siddiqui, S., Simane, B., & Zaitchik, B. (2023). Optimal grid expansion under future electricity demand for groundwater irrigation in Ethiopia. *Energy for Sustainable Development, 72*(February), 351–377.

Kasani, A. D. (1989). *Badae Al-Sanae*.

Kattel, G. R., Shang, W., Wang, Z., & Langford, J. (2019). China's South-to-North water diversion project empowers sustainable water resources system in the North. *Sustainability: Science Practice and Policy, 11*(13), 3735.

Liu, F. H., & Lai, K. P. (2021). Ecologies of green finance: Green sukuk and development of green Islamic finance in Malaysia. *Environment and Planning A: Economy and Space, 53*(8), 1896–1914.

Lu, S., Wenjing, L., Meng, X., Taghizadeh-Hesary, F., & Tang, Y. (2023). Water-Energy-Food Security under Green Finance Constraints in Southwest China. *Energy Economics, 118*(February), 106478.

Al-Marghinani, B. D. (2006). Al-Hidayah: The Guidance. *Trans. Imran Ahsan Khan.*

Musari, K. (2022). Taxonomy of Islamic Green Financing to Address the Climate Change, from Circular Economy to Sustainable Energy: A Conceptual Paper. In *A Paper Was Presented at the 5th International Congress on Economics, Finance & Energy (EFE).* The International University of Sarajevo (IUS), University of New York Tirana (UNYT), and Ibn Haldun University. researchgate.net. https://www.researchgate.net/profile/Khairunnisa-Musari/publication/363737586_Taxonomy_of_Islamic_Green_Financing_to_Address_the_Climate_Change_From_Circular_Economy_to_Sustainable_Energy_A_Conceptual_Paper/links/632be93970cc936cd327b7ad/Taxonomy-of-Islamic-Green-Financing-to-Address-the-Climate-Change-From-Circular-Economy-to-Sustainable-Energy-A-Conceptual-Paper.pdf

Obaideen, K., Yousef, B. A. A., AlMallahi, M. N., Tan, Y. C., Mahmoud, M., Jaber, H., & Ramadan, M. (2022). An Overview of Smart Irrigation Systems Using IoT. *Energy Nexus, 7*(September), 100124.

Onu, U. G., Silva, G. S., Zambroni, A. C., de Souza, B., Bonatto, D., Ferreira, V. B., & da Costa. (2022). Integrated Design of Photovoltaic Power Generation Plant with Pumped Hydro Storage System and Irrigation Facility at the Uhuelem-Amoncha African Community. *Renewable Energy, 198*(October), 1021–1031.

Rafique, M. O., & A. S. C. Abdullah, A. S. C. (2023). Hiyal (Legal Stratagems) in Islamic finance: Systematic literature review. *Aquatic Microbial Ecology: International Journal.* https://www.openaccessojs.com/JBReview/article/view/1752

Webber, S., Nelson, S., Millington, N., Bryant, G., & Bigger, P. (2022). Financing Reparative Climate Infrastructures: Capital Switching, Repair, and Decommodification. *Antipode, 54*(3), 934–958.

Zhao, P., Ma, M., Yaqi, H., Wenyong, W., & Xiao, J. (2022). Comparison of International Standards for Irrigation with Reclaimed Water. *Agricultural Water Management, 274*(December), 107974.

Zuhaili, W. (2011). Fiqih Islam Waadillatuhu Jilid 5. *Jakarta: Gema Insani.*

Green Finance Initiative: Accelerating Sustainable Development for a Green Future

Sumi K V

Abstract This chapter explores the pivotal role of Green Finance Initiatives (GFIs) in driving sustainable economic development. It delves into the multifaceted components of GFIs, including renewable energy investments, sustainable infrastructure projects, and socially responsible financial practices. The chapter highlights the significance of regulatory support, such as tax incentives and environmental, social, and governance (ESG) reporting, in fostering a conducive environment for green finance. It examines the evolving landscape of green bonds and financial instruments, emphasizing their impact on shaping environmentally conscious investment strategies. The chapter also discusses the importance of risk management through climate risk assessments and integrating GFIs into global cooperation frameworks. Furthermore, it explores the educational and capacity-building aspects of GFIs, underscoring their role in cultivating a knowledgeable and environmentally conscious financial sector. Through a comprehensive analysis, this chapter aims to provide insights into how GFIs catalyze sustainable economic development, contributing to a harmonious balance between financial prosperity and ecological responsibility.

1 Introduction

In recent years, global attention has increasingly shifted toward sustainable development, driven by the urgent need to address environmental challenges. As a response to this imperative, the Green Finance Initiative (GFI) emerges as a critical avenue, leveraging financial mechanisms to propel the world toward a sustainable and green future. This chapter explores the intricate interplay between finance and sustainability, focusing on the GFI's role in accelerating sustainable development.

S. K V (✉)
Institute of Management in Kerala, University of Kerala, Karyavattom, Thiruvananthapuram, Kerala, India

University of Kerala, Thiruvananthapuram, Kerala, India

N. Naifar (ed.), *Climate Change and Finance*, Sustainable Finance,
https://doi.org/10.1007/978-3-031-56419-2_16

The global community has recognized the urgent need to address environmental challenges and accelerate the transition to a sustainable and green future. Climate change, resource depletion, and biodiversity loss are just a few of the pressing issues that demand immediate attention. One of the key drivers for achieving these sustainability goals is the Green Finance Initiative (GFI). In this chapter, we will explore the significance of GFI in advancing sustainable development and its potential to shape a greener future. In the face of mounting environmental challenges, the world finds itself at a critical juncture. Climate change threatens ecosystems and economies alike, resource depletion looms on the horizon, and social inequalities persist (Dietz & Stern 2015). The urgency to address these global issues has never been greater, and the need for transformative solutions is undeniable. In a world facing ever-increasing environmental challenges and the urgent need for sustainable solutions, the Green Finance Initiative (GFI) stands as a beacon of hope and innovation. The GFI represents a dynamic and transformative force in the realm of finance, offering a pathway to harmonize economic growth with environmental responsibility. This chapter serves as an introduction to the overarching theme of the GFI—accelerating sustainable development for a green future.

The urgency of the environmental crisis, characterized by climate change, habitat loss, and resource depletion, has brought to the forefront the need for a fundamental shift in our approach to finance. The GFI embodies this shift, redefining the way we invest, allocate resources, and conduct business to create a more sustainable and resilient global economy. This chapter is an exploration of the GFI, highlighting the critical challenges we face, the compelling reasons for embracing green finance, and the overarching vision of a future where financial activities are intrinsically tied to the well-being of the planet and its inhabitants.

2 Research Motivation

The escalating threats posed by climate change and environmental degradation underscore the need for innovative and effective solutions. Against this backdrop, the research is motivated by a desire to comprehensively understand the transformative potential of the Green Finance Initiative. By examining its key components and strategies, we aim to shed light on how financial mechanisms can be strategically employed to address pressing environmental issues and contribute to sustainable development goals.

3 Problem Statement

Despite growing acknowledgment of the importance of green finance, a comprehensive understanding of its practical implications and contributions to sustainable development is still nascent. This research seeks to address this gap by delving into the nuances of the GFI. Specifically, we aim to identify challenges, gaps, and opportunities within the current landscape of green finance, providing a nuanced analysis that informs both theory and practice.

4 Main Contributions

This chapter makes significant contributions to the existing body of knowledge on the GFI. Firstly, it synthesizes and critically evaluates the current state of green finance, offering a comprehensive overview of its key components. Secondly, by identifying practical challenges and opportunities, this research provides valuable insights for policymakers, financial institutions, and practitioners seeking to enhance the effectiveness of green finance initiatives. Lastly, it contributes to the academic discourse by advancing our understanding of the role finance plays in driving sustainable development, thereby bridging the gap between environmental studies and financial literature.

In summary, this chapter positions itself as a crucial exploration of the Green Finance Initiative's impact on accelerating sustainable development. By combining theoretical insights with practical considerations, it aims to provide a robust foundation for future research and policy interventions in the realm of sustainable finance. In this context, the Green Finance Initiative (GFI) emerges as a beacon of hope and a harbinger of change. GFI represents a comprehensive and innovative approach to redefining the relationship between finance and sustainability. It is a powerful response to the pressing need to align our financial systems with environmental and social goals. This chapter embarks on a journey to explore the multifaceted and pivotal role of GFI in reshaping our world toward a sustainable and green future.

1. Understanding the Green Finance Initiative

The Green Finance Initiative is a comprehensive strategy that encompasses various financial instruments, policies, and practices designed to support environmentally friendly projects, investments, and economic activities. At its core, GFI aims to align financial flows with environmentally sustainable, low-carbon, and socially responsible objectives. It encourages the integration of environmental, social, and governance (ESG) factors into decision-making processes across the financial sector. In an era marked by growing environmental concerns and the urgent need to address climate change, the Green Finance Initiative (GFI) has emerged as a critical mechanism for channeling financial resources toward sustainable and environmentally responsible projects. This initiative represents a paradigm shift in the world of finance, with the aim of aligning investments with global sustainability goals. To truly grasp the significance of the GFI, it is essential to explore its core principles, key components, and the transformative impact it is having on the global economy.

2. Defining the Green Finance Initiative

The Green Finance Initiative, often referred to as GFI, is a multifaceted, global effort that seeks to integrate environmental and sustainability considerations into financial decision-making. It serves as a bridge between the financial sector and environmental objectives, facilitating the redirection of capital toward projects and activities that promote sustainability while mitigating harm to the environment.

3. Key Principles of the GFI

At its core, the GFI aims to incorporate sustainability criteria into all aspects of financial decision-making. This means that when evaluating investments, lenders and investors consider not only the financial returns but also the environmental and social impacts of their choices. One of the primary goals of the GFI is to fund and support projects that have a positive environmental impact Eccles et al. (2014). This includes investments in renewable energy, clean technologies, sustainable agriculture, green infrastructure, and other activities that reduce greenhouse gas emissions and protect natural ecosystems. Transparency is a cornerstone of the GFI. It encourages financial institutions to disclose the environmental and social impact of their investments. This transparency enables investors to make informed decisions and holds institutions accountable for their commitments to sustainability. The GFI recognizes the financial risks associated with climate change and environmental degradation. It encourages financial institutions to assess and manage these risks, ensuring the long-term viability of their investments.

4. Key Components of the GFI

The Green Finance Initiative (GFI) comprises several key components that work in harmony to promote sustainable finance and drive the transition toward a more environmentally responsible and socially equitable global economy. These components encompass a diverse range of strategies and mechanisms, including the development and promotion of green financial products like green bonds and sustainability-linked loans that mobilize capital for sustainable projects Esty (2013). The GFI also engages in policy advocacy and regulatory support to create an enabling environment for sustainable finance, advocating for tax incentives and regulatory frameworks that encourage green investments. Capacity building and knowledge sharing are vital components, as the GFI provides education and training programs to equip financial professionals, governments, and investors with the necessary skills to navigate sustainable finance effectively. Furthermore, the initiative emphasizes global collaboration by fostering multi-stakeholder engagement, ensuring that sustainable finance is a collective effort involving governments, financial institutions, NGOs, and other stakeholders. Inclusivity is another key aspect, with the GFI encouraging investments that benefit underserved communities, addressing social equity and environmental justice issues. Additionally, the initiative plays a crucial role in raising public awareness by educating and engaging the public, empowering individuals to make sustainable choices and advocating for sustainable development. These components collectively form the backbone of the GFI's mission to accelerate sustainable development for a greener and more prosperous future.

Green Financial Products: The GFI has led to the development of innovative financial products designed to support sustainability. Green bonds, for instance, are a notable example. These bonds are issued to fund environmentally friendly projects and provide investors with the opportunity to support sustainability while earning returns.

Policy Advocacy: The GFI often works closely with governments and regulatory bodies to advocate for policies that incentivize green finance and discourage investments in carbon-intensive industries. This includes the development of tax incentives, subsidies, and regulatory frameworks that promote sustainability.

Capacity Building: Recognizing that the transition to green finance can be complex, the GFI offers capacity-building programs and educational resources. These initiatives help financial institutions, investors, and governments navigate the intricacies of sustainable finance.

5. Impact of the GFI

The Green Finance Initiative has already made significant contributions to the global economy and the environment. Through its innovative financial products, advocacy for policy changes, and commitment to transparency, the GFI has mobilized substantial capital toward sustainable projects. These investments have not only created jobs and driven economic growth but have also accelerated the transition to a low-carbon and environmentally responsible global economy Kolk (2010). Moreover, the GFI's emphasis on risk assessment and management has helped financial institutions navigate the challenges posed by climate change and environmental degradation, making the financial sector more resilient. In tandem with its efforts to reduce environmental and social risks, the GFI is fostering a culture of sustainability in finance, shaping a future where financial practices are intrinsically tied to the well-being of the planet and its inhabitants.

Mobilizing Capital: The GFI has played a pivotal role in mobilizing trillions of dollars of capital for green projects. This funding has accelerated the deployment of renewable energy, energy-efficient technologies, and sustainable infrastructure worldwide.

Mainstreaming Sustainability: Sustainability considerations have become mainstream in the financial sector. Investors increasingly factor in environmental and social criteria when making investment decisions, leading to a shift away from carbon-intensive industries.

Encouraging Innovation: The GFI has fostered innovation in financial products and services, spurring the development of green bonds, sustainability-linked loans, and other mechanisms that attract investors seeking both financial returns and environmental impact.

Global Collaboration: The GFI's collaborative approach has brought together governments, financial institutions, and civil society organizations from around the world. This global network facilitates knowledge sharing, best practices, and the scaling up of green finance efforts.

6. Objectives of GFI

The Green Finance Initiative (GFI) has several key objectives aimed at promoting sustainability, combating climate change, and aligning financial practices with environmental and social goals. The following are the primary objectives of the GFI mission and guide its activities and initiatives. The following are the primary objectives of the GFI:

Mobilizing Capital for Green Projects: One of the central objectives of the GFI is to mobilize financial resources, both public and private, to support green and sustainable projects. This involves redirecting investments away from activities that harm the environment and toward projects that promote sustainability. By increasing the flow of capital to green initiatives, the GFI accelerates the transition to a low-carbon and environmentally responsible global economy.

Enhancing Market Transparency: The GFI advocates for greater transparency in financial markets, especially concerning the environmental and social impact of investments. This objective includes encouraging financial institutions and businesses to disclose relevant information about their investments and activities. Transparent reporting enables investors to make informed decisions, fosters accountability, and ensures that sustainability criteria are considered in financial decision-making.

Developing Green Financial Products: To facilitate investments in green projects, the GFI encourages the development of innovative financial products and instruments. These products may include green bonds, sustainability-linked loans, green mortgages, and other financial mechanisms that are specifically designed to attract investors interested in both financial returns and positive environmental outcomes. By creating these products, the GFI helps bridge the gap between capital providers and sustainability initiatives.

Capacity Building and Knowledge Sharing: Recognizing that the transition to green finance can be complex, the GFI is committed to building capacity within financial institutions, governments, and other stakeholders. It provides training, educational resources, and knowledge-sharing platforms to help these entities navigate the evolving landscape of sustainable finance. Capacity building ensures that financial professionals have the necessary expertise to incorporate sustainability into their practices effectively.

Advocating for Policy and Regulatory Support: The GFI collaborates with governments and regulatory bodies to advocate for policies and regulations that promote green finance. This includes the development of tax incentives, subsidies, and regulatory frameworks that incentivize sustainable investments and discourage environmentally harmful activities. Advocacy efforts help create a conducive environment for the growth of green finance.

Promoting Risk Assessment and Management: Environmental and climate-related risks are becoming increasingly important considerations in financial decision-making. The GFI encourages financial institutions to assess and manage these risks effectively. By integrating risk assessment processes, financial institutions can make more informed and resilient investment choices, reducing exposure to climate-related financial risks.

Fostering Inclusivity: The GFI aims to ensure that green finance benefits all segments of society. This includes promoting investments in projects that address social equity and environmental justice issues. By fostering inclusivity, the GFI seeks to ensure that sustainable development benefits not only the environment but also vulnerable and underserved communities.

Increasing Public Awareness and Engagement: Public awareness and engagement are critical to the success of the GFI. The initiative works to raise awareness about the importance of green finance among the general public, investors, businesses, and policymakers. It encourages individuals and organizations to actively engage with sustainable finance practices and investments.

Mitigating Climate Change: GFI seeks to channel funds toward projects and technologies that reduce greenhouse gas emissions, such as renewable energy, energy efficiency, and carbon capture and storage.

Preserving Biodiversity: GFI supports investments in activities that protect and restore ecosystems, promote sustainable land use, and prevent biodiversity loss.

Enhancing Social Equity: In addition to environmental sustainability, GFI emphasizes social responsibility. It aims to promote fair and inclusive economic growth, creating opportunities for vulnerable populations and reducing inequality.

Strengthening Resilience: GFI encourages investments in infrastructure and initiatives that build resilience against the impacts of climate change and other environmental challenges.

7. Key Components of GFI

GFI comprises several key components that work in harmony to promote sustainable finance:

The Green Finance Initiative (GFI) comprises several key components that work in harmony to promote sustainable finance, ushering in an era where economic growth is intricately linked with environmental responsibility. At its core, the GFI seeks to transform the traditional financial landscape by fostering a paradigm shift toward sustainability. One of its fundamental components involves the development and promotion of innovative green financial products, such as green bonds and sustainability-linked loans. These financial instruments redirect capital toward environmentally responsible projects, thereby driving investment in renewable energy, sustainable infrastructure, and conservation efforts.

In tandem with product innovation, the GFI engages in policy advocacy and regulatory support to create an enabling environment for sustainable finance. This involves collaborating with governments and regulatory bodies to establish frameworks that incentivize green investments through tax incentives, favorable policies, and stringent environmental and social standards. By aligning financial regulations with sustainability objectives, the GFI ensures that the financial sector becomes a proactive agent of positive change.

Capacity building and knowledge sharing represent another integral facet of the GFI's approach. The initiative recognizes the need to equip financial professionals, governments, and investors with the necessary skills and knowledge to navigate the complexities of sustainable finance effectively. Through educational programs, workshops, and knowledge-sharing initiatives, the GFI empowers stakeholders to make informed decisions and integrate sustainability principles into their financial strategies.

Furthermore, the GFI emphasizes the importance of global collaboration as a driving force behind sustainable finance. By fostering multi-stakeholder engagement that includes governments, financial institutions, NGOs, and other stakeholders, the initiative creates a unified front to address global environmental and social challenges. This collaborative approach not only amplifies the impact of sustainable finance efforts but also fosters international cooperation in the pursuit of shared sustainability goals.

Inclusivity is a central tenet of the GFI's mission, underscoring the importance of ensuring that sustainable finance benefits all segments of society. The initiative encourages investments that prioritize underserved communities, addressing social equity and environmental justice issues. In doing so, it seeks to bridge the gap between economic development and social well-being, fostering a more equitable and sustainable future.

The Green Finance Initiative's multifaceted approach, encompassing innovative financial products, policy advocacy, capacity building, global collaboration, and inclusivity, constitutes a comprehensive strategy to promote sustainable finance. By harmonizing these key components, the GFI paves the way for a financial landscape where profitability converges with environmental and social responsibility, accelerating progress toward a more sustainable and prosperous world for present and future generations.

The Green Finance Initiative (GFI) is a multifaceted effort that comprises several key components working in harmony to promote sustainable finance and accelerate the transition toward a more environmentally responsible global economy. These components play complementary roles and reinforce each other in achieving the GFI's objectives. Here are the primary components of the GFI:

Green Financial Products: Green financial products are investment instruments specifically designed to fund environmentally sustainable projects. These products include:

Green Bonds: Green bonds are debt securities where the proceeds are earmarked for green projects such as renewable energy, energy efficiency, or clean transportation. They enable investors to support sustainability initiatives while earning a return on their investment.

Sustainability-Linked Loans: These loans have interest rates that are linked to the borrower's performance on sustainability goals. If the borrower meets predetermined sustainability targets, they may benefit from lower interest rates, incentivizing sustainability improvements.

Green Mortgages: Green mortgages provide incentives to homebuyers or property owners who invest in energy-efficient or environmentally friendly features for their properties. These mortgages may offer lower interest rates or other financial benefits.

Policy Advocacy: The GFI collaborates with governments and regulatory bodies to advocate for policies and regulations that support sustainable finance. This component involves:

Tax Incentives: Lobbying for tax incentives that encourage investments in green projects or sustainable financial products.

Subsidies: Promoting government subsidies for renewable energy projects or other sustainable initiatives.

Regulatory Frameworks: Encouraging the development of regulatory frameworks that require financial institutions to consider environmental and social factors in their decision-making processes.

Capacity Building and Education: Recognizing that the transition to green finance can be complex, the GFI provides capacity-building programs, educational resources, and knowledge-sharing platforms to various stakeholders. These initiatives help:

Financial Institutions: Equip financial professionals with the necessary expertise to incorporate sustainability criteria into their practices effectively.

Governments: Build capacity within governmental bodies to develop and implement green finance policies and regulations.

Investors: Educate investors about the benefits and risks associated with sustainable investments, empowering them to make informed decisions.

Transparency and Reporting: Transparency is fundamental to green finance. The GFI encourages financial institutions and businesses to disclose relevant information about the environmental and social impact of their investments. Transparent reporting enables investors to make informed decisions and fosters accountability. This component includes:

Environmental, Social, and Governance (ESG) Reporting: Encouraging companies to report on their ESG performance, which includes metrics related to environmental impact, social responsibility, and corporate governance.

Impact Reporting: Promoting the measurement and reporting of the tangible environmental and social impacts of investments.

Risk Assessment and Management: The GFI acknowledges the financial risks associated with climate change and environmental degradation. It encourages financial institutions to assess and manage these risks effectively. Integrating risk assessment processes ensures the resilience of investments and reduces exposure to climate-related financial risks.

Global Collaboration: The GFI fosters global collaboration by bringing together governments, financial institutions, non-governmental organizations (NGOs), and other stakeholders. This collaborative network facilitates knowledge sharing, best practices, and the scaling up of green finance efforts.

Innovation: Encouraging innovation in financial products and services is a crucial component of the GFI. This includes the development of new financial instruments that attract investors seeking both financial returns and positive environmental impact.

Green Bonds: Green bonds are a fundamental tool of GFI. These are financial instruments specifically earmarked for environmentally friendly projects, such as renewable energy installations, sustainable agriculture, and clean transportation. Green bonds provide investors with a transparent way to support sustainable initiatives.

Sustainable Investment Criteria: GFI encourages financial institutions to incorporate sustainability criteria into their investment decisions. This includes assessing ESG factors when evaluating the risk and return of investments.

Regulatory Support: Governments and regulatory bodies play a crucial role in GFI. They can introduce policies, regulations, and incentives that favor green investments and discourage environmentally harmful practices.

Capacity Building: GFI supports capacity-building efforts within the financial sector to enhance the knowledge and skills necessary to identify, assess, and manage environmental and social risks.

8. The Role of GFI in Sustainable Development

The United Nations' Sustainable Development Goals provide a global blueprint for addressing pressing challenges, and GFI aligns closely with many of these goals. By directing financial resources toward sustainable and socially responsible projects, GFI contributes significantly to achieving targets related to climate action, clean energy, life below water and on land, and responsible consumption and production, among others Krueger (2012), Lo (2018). The Green Finance Initiative (GFI) plays a pivotal role in advancing sustainable development by aligning financial practices with environmental and social goals. Its actions and initiatives contribute significantly to addressing pressing global challenges and fostering a more sustainable and equitable future. Here is an exploration of the key roles the GFI plays in sustainable development:

Mobilizing Capital for Sustainable Projects: The GFI facilitates the flow of capital into renewable energy projects such as solar, wind, and hydroelectric power. By doing so, it accelerates the transition from fossil fuels to clean and sustainable energy sources, reducing greenhouse gas emissions and mitigating climate change.

Energy Efficiency: Investments in energy-efficient technologies and practices are crucial for sustainable development. The GFI channels capital into initiatives aimed at improving energy efficiency in industries, buildings, and transportation, thereby conserving resources and reducing environmental impact.

Sustainable Infrastructure: Infrastructure projects that promote sustainability, such as public transportation systems and green urban planning, receive financial support through the GFI. These investments enhance the quality of life for communities while minimizing environmental harm.

Green Bonds: The GFI has been instrumental in popularizing green bonds, which raise funds specifically for environmentally friendly projects. These bonds provide investors with an avenue to support sustainability while earning returns, thus encouraging sustainable development.

Sustainability-Linked Loans: These loans incentivize borrowers to meet sustainability targets, offering reduced interest rates for achieving specific environmental and social performance goals. Such loans encourage businesses to embrace sustainable practices.

Tax Incentives: The GFI collaborates with governments to advocate for tax incentives that promote green investments. These incentives stimulate private sector involvement in sustainable projects.

Regulatory Frameworks: The GFI encourages the development of regulatory frameworks that require financial institutions to consider environmental and social factors in their decision-making processes. This ensures that sustainability remains a central focus of financial activities.

Environmental, Social, and Governance (ESG) Reporting: The GFI promotes ESG reporting, encouraging companies to disclose their sustainability performance. This transparency provides investors with valuable information to make informed decisions aligned with sustainable development goals.

Impact Reporting: By advocating for the measurement and reporting of the tangible environmental and social impacts of investments, the GFI ensures that financial decisions lead to meaningful outcomes in terms of sustainable development.

Climate and Environmental Risk Management: The GFI helps financial institutions assess and manage climate-related and environmental risks. This proactive approach safeguards investments and contributes to the resilience of the financial sector.

Capacity Building and Knowledge Sharing:

Education and Training: Through capacity-building programs, the GFI equips financial professionals, governments, and investors with the knowledge and skills necessary to navigate sustainable finance effectively.

Knowledge Sharing: The GFI fosters knowledge sharing by facilitating collaboration between stakeholders globally. This sharing of best practices and experiences enhances the implementation of sustainable finance initiatives.

Multi-Stakeholder Engagement: By bringing together governments, financial institutions, NGOs, and other stakeholders, the GFI creates a collaborative environment that drives sustainable development efforts on a global scale. This diverse network ensures that sustainable finance is a collective effort.

Promoting Inclusivity:

Supporting Underserved Communities: The GFI emphasizes the importance of sustainable development being inclusive. It encourages investments that benefit underserved communities, addressing social equity and environmental justice issues.

Raising Public Awareness:

Educating and Engaging the Public: The GFI plays a role in raising public awareness about the importance of sustainable finance and investments. Engaging the public empowers individuals to make sustainable choices and advocate for sustainable development.

Green finance not only supports environmental goals but also fosters innovation and economic growth. Investments in renewable energy, clean technology, and sustainable agriculture drive job creation, technological advancement, and economic resilience Ruggie (2014). GFI acts as a catalyst for private sector investments in these areas, helping societies transition to greener and more sustainable economies. In an era defined by environmental challenges and the imperative to transition to a sustainable global economy, green finance emerges as a powerful tool to drive innovation and foster economic growth. This chapter explores how the integration of green finance principles can unlock innovation across sectors, spur economic development, and set the stage for a more sustainable and prosperous future.

9. Green Finance: A Catalyst for Innovation

Renewable Energy Revolution: Green finance has played a pivotal role in the rapid growth of the renewable energy sector. Investments in solar, wind, and hydroelectric power have surged, driven by innovative financing mechanisms such as green bonds and sustainable loans. This influx of capital has propelled technological advancements, making renewable energy sources more efficient, affordable, and accessible.

Energy Efficiency Solutions: Sustainable finance has incentivized businesses to invest in energy-efficient technologies and practices. This drive for efficiency has led to breakthroughs in energy-saving technologies, from smart buildings to electric vehicles, revolutionizing industries and reducing environmental impact.

Circular Economy and Sustainable Supply Chains: Green finance encourages businesses to adopt circular economy models and sustainable supply chain practices. These innovations prioritize resource efficiency, waste reduction, and recycling, leading to more sustainable production processes and reduced environmental footprint.

Green Infrastructure Development: Investments in green infrastructure, such as public transportation systems, clean water projects, and sustainable urban planning, have sparked innovation in urban development and infrastructure design. These innovations enhance the quality of life for communities while reducing pollution and resource consumption.

Technological Advancements: Green finance supports research and development in clean technologies, such as carbon capture and storage, sustainable agriculture, and water management solutions. These innovations not only address environmental challenges but also create new markets and job opportunities.

10. Economic Growth through Green Finance

Job Creation: Investments in green sectors generate employment opportunities across a wide range of industries, from renewable energy installation to green infrastructure development. The green economy has become a driver of job growth, supporting livelihoods and economic stability.

Market Expansion: Green finance has led to the creation of new markets and industries, such as the electric vehicle market, green building sector, and sustainable agriculture. These markets offer growth prospects for businesses and entrepreneurs, driving economic expansion.

Increased Productivity: Sustainable practices often lead to increased productivity and cost savings for businesses. Energy-efficient technologies, for example, reduce operational expenses, contributing to improved profitability.

Resilience to Climate Risks: Green investments enhance economic resilience by reducing vulnerability to climate-related risks. Sustainable infrastructure is better equipped to withstand extreme weather events, reducing the economic impact of disasters.

Attracting Investment: Countries and regions that prioritize green finance often attract foreign investment seeking sustainable opportunities. This influx of capital stimulates economic development and fosters international cooperation.

Long-Term Value: Sustainable finance principles encourage a long-term perspective in decision-making. Businesses that integrate sustainability into their strategies are better positioned for long-term success, creating enduring economic value.

11. Reducing Environmental and Social Risks

One of the key advantages of GFI is its ability to mitigate environmental and social risks in the financial sector. By encouraging institutions to assess and address ESG risks in their portfolios, GFI helps prevent potential financial crises stemming from unsustainable practices and investments in industries that harm the environment or exploit vulnerable communities. The integration of green finance principles into financial systems and investment decisions represents a significant step in mitigating environmental and social risks. This chapter delves into the pivotal role that green finance plays in identifying, addressing, and reducing these risks, ultimately fostering a more sustainable and socially responsible global economy. In the world of finance, environmental and social risks have emerged as critical factors that can significantly impact the performance and sustainability of investments and financial institutions. Understanding these risks is essential for responsible decision-making and ensuring that investments align with broader environmental and social goals. This chapter delves into the concept of environmental and social risks, their significance, and the tools and approaches used to assess and manage them.

Environmental Risks: Environmental risks pertain to the potential adverse impacts on investments or financial institutions resulting from environmental factors. These risks can encompass a wide range of issues, including:

Climate Change Risk: This is perhaps the most prominent environmental risk. It includes both physical risks (e.g., damage from extreme weather events) and transition risks (e.g., policy changes and market shifts in response to climate change).

Natural Resource Scarcity: Risks related to the availability and price volatility of critical resources such as water, minerals, and energy sources.

Biodiversity and Ecosystem Risks: Risks associated with the degradation of ecosystems, which can affect supply chains, land use, and regulatory compliance.

Pollution and Contamination Risks: Risks related to pollution, waste management, and contamination of air, water, and soil, which can lead to legal and reputational issues.

Social Risks: Social risks, on the other hand, encompass factors related to the social and human aspects of investments or financial activities. These risks include:

Labor Practices and Human Rights: Risks related to unethical labor practices, human rights violations, and workplace safety issues within a company's supply chain or operations.

Community Relations and Social License: Risks arising from conflicts with local communities, land rights disputes, and the potential loss of a “social license” to operate in a particular region.

Health and Safety Risks: Risks associated with the health and safety of employees, customers, and communities affected by a company's operations or products.

Income Inequality and Social Equity: Risks associated with income disparities, discrimination, and social inequalities within a company, industry, or region.

12. Significance of Environmental and Social Risks

Environmental and social risks are no longer peripheral concerns; they have moved to the center of financial decision-making. These risks can have material impacts on financial performance, reputation, and legal liabilities, making them increasingly important considerations for investors and financial institutions. Governments and regulatory bodies worldwide are introducing stricter environmental and social regulations, which can directly affect businesses and financial markets. Investors, customers, and the public are placing greater emphasis on corporate responsibility and sustainability, influencing the perception and value of financial products and companies. Climate-related events, social unrest, and reputational crises can lead to market volatility and impact the financial sector.

13. Assessing and Managing Environmental and Social Risks

Understanding and managing environmental and social risks are integral aspects of responsible finance. As the world grapples with environmental challenges and seeks to address social inequalities, integrating these considerations into financial decision-making is not just a moral imperative; it is a strategic necessity for investors and financial institutions to thrive in a rapidly changing landscape. By effectively assessing and managing these risks, finance can become a positive force for both profit and the betterment of society and the environment. Assessing and managing environmental and social risks in finance involve several key steps and approaches:

Environmental and Social Due Diligence: Conduct thorough due diligence to identify potential risks associated with investments or financing activities.

Risk Assessment Models: Develop and utilize risk assessment models and tools to quantify and measure environmental and social risks.

Environmental and Social Impact Assessments: Conduct impact assessments to understand how investments may affect the environment and communities.

Scenario Analysis: Use scenario analysis to evaluate how different environmental and social scenarios could impact investments over time.

Risk Mitigation Strategies: Develop strategies to mitigate environmental and social risks, which may include diversification, sustainability-linked loans, or responsible supply chain management.

Transparency and Reporting: Enhance transparency by disclosing information related to environmental and social risks, allowing stakeholders to make informed decisions.

Engagement and Stakeholder Dialogue: Engage with stakeholders, including local communities, NGOs, and industry peers, to build trust and address environmental and social concerns collaboratively.

14. Green Finance: Mitigating Environmental and Social Risks

In an era defined by escalating environmental challenges, such as climate change, habitat loss, and resource depletion, as well as growing concerns about social inequalities, the concept of green finance emerges as a powerful instrument to mitigate environmental and social risks Schaltegger et al. (2012). This essay explores the pivotal role that green finance plays in identifying, addressing, and reducing these risks, ultimately fostering a more sustainable and socially responsible global economy Sullivan et al. (2012). Environmental and social risks are multifaceted challenges that have far-reaching consequences for both the financial sector and society at large. To better grasp the significance of these risks, it is essential to define them:

Climate Stress Testing: Green finance institutions are increasingly conducting climate stress tests to evaluate the resilience of investments to climate-related risks. This proactive approach helps identify and address vulnerabilities.

Environmental and Social Due Diligence: Financial institutions that adhere to green finance principles perform rigorous environmental and social due diligence before funding projects. This process helps identify and mitigate potential risks.

Environmental, Social, and Governance (ESG) Reporting: The integration of ESG factors into investment analysis promotes transparency. Companies reporting on their ESG performance are better equipped to manage and reduce risks associated with environmental and social issues.

Incentivizing Responsible Practices:

Green Bonds: Green bonds raise funds exclusively for environmentally responsible projects. These bonds incentivize companies and governments to prioritize sustainability in their operations.

Sustainability-Linked Loans: These loans tie interest rates to a borrower's sustainability performance. Borrowers are rewarded with lower interest rates for achieving predefined sustainability targets.

Responsible Supply Chain Financing: Financial institutions offer financing solutions that promote responsible sourcing and supply chain practices. This helps reduce the risk of labor exploitation and environmental degradation in supply chains.

Encouraging Circular Economy Practices:

Circular Economy Investments: Funding circular economy initiatives, such as recycling programs and product lifecycle management, reduces the risk of resource depletion and waste generation.

Social Impact Bonds: These financial instruments tie returns to the achievement of social outcomes, such as reduced recidivism rates or improved educational attainment. They encourage investments that address social risks and inequalities.

Capacity Building and Education:

Training and Knowledge Sharing: Green finance initiatives offer training and educational resources to financial institutions, governments, and investors. This capacity-building helps stakeholders better understand and address environmental and social risks.

Green finance is a powerful tool for reducing environmental and social risks. By emphasizing responsible investment, transparent reporting, and sustainable practices, green finance helps identify, address, and mitigate risks associated with climate change, natural resource depletion, labor exploitation, and other pressing challenges. Through the integration of green finance principles, financial institutions, businesses, and governments can contribute to a more sustainable, socially responsible, and resilient global economy, where environmental and social risks are effectively managed for the benefit of present and future generations.

15. Challenges and Future Outlook

While green finance has made significant strides in promoting sustainability and mitigating environmental and social risks, it faces various challenges and opportunities on the path to a more sustainable future. In this chapter, we explore the key challenges and provide insights into the future outlook for green finance.

Policy and Regulatory Uncertainty: The lack of consistent and robust policy frameworks and regulations can hinder the growth of green finance. Investors and businesses require clear guidelines and incentives to make sustainable investments.

Data and Reporting Quality: Reliable and standardized data on environmental and social factors is essential for effective decision-making in green finance. Ensuring the quality, comparability, and transparency of this data remains a challenge.

Risk Assessment: Assessing and pricing environmental and social risks can be complex. Developing comprehensive methodologies and models for risk assessment is an ongoing challenge in green finance.

Access to Green Finance: Many small and medium-sized enterprises (SMEs) and developing countries face challenges in accessing green finance. Ensuring inclusivity and broadening access to sustainable funding sources is vital.

Financial Product Innovation: While green bonds and sustainability-linked loans have gained prominence, there is a need for continued innovation in green financial products to cater to evolving sustainability needs and emerging technologies.

Market Volatility and Perception: Green investments can be subject to market volatility and may be perceived as riskier by some investors. Encouraging long-term commitment to green finance, despite short-term fluctuations, is a challenge.

Greenwashing: The misrepresentation of investments or products as "green" when they do not meet sustainability criteria remains a concern. Robust verification mechanisms and clear definitions are necessary to combat greenwashing.

Socioeconomic Equity: Ensuring that green finance benefits all segments of society, including marginalized communities, is a challenge. Strategies to address social and environmental justice issues need to be developed.

5 Future Outlook

The future of green finance is likely to see increased regulatory and policy support, creating a more consistent and transparent framework for sustainable investments. Regulatory bodies may implement stricter disclosure requirements and standardized taxonomies to facilitate green finance. Improved data quality, availability, and harmonized reporting standards will be crucial. Technological advancements, such as blockchain and big data analytics, may further enhance the accuracy and transparency of environmental and social data. Green finance will continue to innovate, offering a broader range of financial products and instruments. Sustainable derivatives, green insurance, and nature-based solutions are areas with growth potential. Sustainability considerations will become even more integrated into mainstream finance. Environmental, social, and governance (ESG) factors will be central to investment decisions, and sustainable finance will no longer be seen as a niche market. The future will witness increased investments in green technologies, such as clean energy, sustainable agriculture, and circular economy solutions. Technological innovation will drive down costs and enhance efficiency. Collaborative efforts between governments, financial institutions, businesses, and civil society will grow stronger. International cooperation will be essential to address global environmental challenges, such as climate change and biodiversity loss. Efforts to improve access to green finance for SMEs and developing countries will expand. Innovative financing mechanisms, such as green microloans and blended finance, will play a role in increasing inclusivity. Impact investing, which prioritizes both financial returns and positive societal or environmental impact, will gain prominence. Investors will increasingly seek investments that align with their values and impact objectives.

6 Conclusion

The exploration of the Green Finance Initiative reveals a pivotal strategy in our collective pursuit of sustainable development and a green future. This chapter has underscored key findings that position the initiative as a transformative force in reshaping financial landscapes toward environmental and social sustainability.

6.1 Key Findings

The Green Finance Initiative, through its core principles and strategies, facilitates the integration of sustainable practices into financial systems. It harnesses the power of finance to drive positive change, channeling investments toward eco-friendly projects and fostering innovation in the pursuit of environmental and social goals. The initiative signifies a paradigm shift, highlighting the interconnectedness of economic prosperity and environmental responsibility.

Implications

The implications of the Green Finance Initiative extend beyond the financial sector. Its emphasis on responsible investment and environmentally conscious practices has far-reaching consequences for businesses, governments, and individuals alike. By aligning financial decisions with sustainable principles, the initiative not only mitigates environmental risks but also contributes to long-term economic resilience and social well-being.

Potential Areas for Future Research

As we look ahead, there are promising avenues for future research in the realm of green finance. Exploring the nuanced impact of green investments on economic indicators, assessing the effectiveness of policy frameworks supporting sustainable finance, and investigating the role of technology in advancing green financial solutions are areas ripe for exploration. Additionally, understanding the behavioral aspects influencing sustainable investment choices and evaluating the scalability of green finance across diverse economies present intriguing avenues for future inquiry.

The dynamic nature of both finance and sustainability ensures that the Green Finance Initiative will continue to evolve. Future research endeavors should focus on refining strategies, enhancing measurement frameworks, and addressing emerging challenges to ensure the initiative's sustained impact.

In essence, the Green Finance Initiative represents not just a chapter in our journey toward sustainability but a cornerstone in the foundation of a green future. As we conclude this exploration, it is evident that the initiative has set in motion a positive trajectory, and future research holds the key to unlocking its full potential, driving us closer to a world where financial systems are not just instruments of economic growth but catalysts for a thriving, resilient, and environmentally conscious global community.

References

Dietz, S., & Stern, N. (2015). Endogenous growth, convexity of damages and climate risk: How Nordhaus' framework supports deep cuts in carbon emissions. *The Economic Journal, 125*(583), 547–620.

Eccles, R. G., Ioannou, I., & Serafeim, G. (2014). The impact of corporate sustainability on organizational processes and performance. *Management Science, 60*(11), 2835–2857.

Esty, D. C. (2013). *Modernizing the World Trade Organization for the twenty-first century: Building support for the global marketplace*. Yale Center for Environmental Law and Policy.

Kolk, A. (2010). Trajectories of sustainability reporting by MNCs. *Journal of World Business, 45*(4), 367–374.

Krueger, A. B. (2012). The rise of green entrepreneurship. In Innovation policy and the economy, Volume 12 (pp. 47–68). University of Chicago Press.

Lo, S. F. (2018). How does corporate social responsibility contribute to firm financial performance? The mediating role of competitive advantage, reputation, and customer satisfaction. *Corporate Social Responsibility and Environmental Management, 25*(5), 744–757.

Ruggie, J. G. (2014). The social contract for a green economy. *International Environmental Agreements: Politics, Law and Economics, 14*(3), 265–276.

Schaltegger, S., Lüdeke-Freund, F., & Hansen, E. G. (2012). Business cases for sustainability: The role of business model innovation for corporate sustainability. *International Journal of Innovation and Sustainable Development, 6*(2), 95–119.

Sullivan, R., Feitelson, E., & Gal, G. (2012). *Environmental finance and investments*. Routledge.

www.ingramcontent.com/pod-product-compliance
Ingram Content Group UK Ltd.
Pitfield, Milton Keynes, MK11 3LW, UK
UKHW022215051026
15624UKWH00001B/172
* 9 7 8 3 0 3 1 5 6 4 1 8 5 *